Geography

Revise
AS & A2

Contents

Specification lists

Unit	Specification topic	Chapter reference
Unit 1: Physical and Human Geography	Core: Rivers floods and management Population change	4 4, 9
	Options: Cold environments Coastal environments Hot deserts Food supply energy issues Health Issues	7, 3 5 6, 3 12, 3 13
Unit 2: Geographical Skills	Based on the Unit 1 content. Includes investigative, cartographic, graphical, ICT and statistical skills.	15

Examination analysis

Geography Unit 1: Physical and Human Geography
Structured short and extended questions.
2 hr exam   70% of AS

Geography Unit 2: Geographical Skills
Structured skills and generic fieldwork questions.
1 hr exam   30% of AS

Unit	Specification topic	Chapter reference
Unit 3: Contemporary Geographical Issues	Three topics selected. At least one from either the physical or human sections: Physical:	
	● Plate tectonics and associated hazards	1
	● Weather and climate and associated hazards	2
	● Ecosystems: change and challenge	3
	Human:	
	● World cities	8
	● Development and globalisation	10/11
	● Contemporary conflicts and challenges	13
Unit 4A: Geography Fieldwork Investigation	Candidates analyse and evaluate their fieldwork in response to set questions OR	15
Unit 4B: Geographical Issue Evaluation	A pre-release document is issued that extends specification content. In the exam skills of analysis, synthesis and evaluation are tested	15

Examination analysis

Geography Unit 3: Contemporary Geographical Issues
Two high value structured/extended questions (one on human geography and one on physical geography) and one essay (from a different area).
2 hr 30 min exam   60% of A2

Geography Unit 4A: Geography Fieldwork Investigation
Section A extends the study undertaken and Section B assesses fieldwork skills with stimulus material. Short structured and extended questions
1 hr 30 min exam   40% of A2
OR
Geography Unit 4B: Geographical Issue Evaluation
Short structured and extended questions based on advanced information.
1 hr 30 min exam   40% of A2

OCR AS

Unit	Specification topic	Chapter reference
Unit 1: F761 – Managing Physical Environments	River environments Coastal environments Cold environments Hot and semi-arid environments	4, 3 5, 3 7, 3 6, 3
Unit 2: F762 – Managing Human Environments	Managing urban change Managing rural change Energy issues The growth of tourism	8 8 13 14

Examination analysis

Unit 1: F761 – *Managing Physical Environments*

Divided into two sections:
- *Section A: Two questions from four structured data-response questions. Choose one from River environments or Coastal environments, and one from Cold environments or Hot and semi-arid environments.*
- *Section B: Answer one question from four extended writing questions. The question chosen must be different to Section A.*

1 hr 30 min exam   50% of AS

Unit 2: F762 – *Managing Human Environments*
Divided into two sections:
- *Section A: Two questions from four structured data-response questions. Choose one from Managing urban change or Managing rural change, and one from Energy issues or The growth of tourism.*
- *Section B: Answer one question from four extended writing questions. The question chosen must be different to Section A.*

1 hr 30 min exam   50% of AS

OCR A2

Unit	Specification topic	Chapter reference
Unit 3: F763 – Global Issues	Environmental issues: • Earth hazards (Option A1) • Ecosystems and environments under threat (Option A2) • Climatic hazards (Option A3) Economic issues: • Population and resources (Option B1) • Globalisation (Option B2) • Development and inequalities (Option B3)	1, 2, 4 3 2 8, 9, 12, 13 10, 11 10
Unit 4: F764 – Geographical Skills	Geographical skills: • Identifying a suitable geographical question or hypothesis for investigation • Developing a plan and strategy for conducting the investigation • Collecting and recording appropriate data • Presenting the data collected in appropriate forms • Analysing and interpreting the data	15

Examination analysis

Unit 3: F763 – *Global Issues*
Divided into two sections:
- *Section A: Answer three questions, including at least one from Environmental issues and one from Economic issues. Issues are identified from the data set, and strategies for their management have to be suggested.*
- *Section B: Two essays to be written, one from Environmental issues and one from Economic issues. This is a synoptic unit.*

2 hr 30 min exam   60% of A2

Unit 4: F764 – *Geographical Skills*
Divided into two sections:
- *Section A: One structured data response question using stimulus material that may have been developed during the research phase of preparations.*
- *Section B: Two extended writing questions. Focus on skills and techniques used during geographical research. This is a synoptic paper.*

1 hr 30 min exam   40% of A2

Edexcel AS

Unit	Specification topic	Chapter reference
Unit 1: Global Challenges	World at risk	1, 5, 7
	Going global	8, 9, 10, 11, 13
Unit 2: Geographical Investigations	Physical topics – A: ● Extreme weather ● Crowded coasts	1, 2 5
	Human topics – B: ● Rural and urban disparity: Regenerating rural and urban places	8

Edexcel A2

Unit	Specification topic	Chapter reference
Unit 3: Contested Planet	The use and management of resources in six compulsory areas: ● Energy ● Water ● Biodiversity ● Superpowers ● Development ● Technological fix	3, 10, 11, 12, 13
Unit 4: Geography Research	One from: ● Tectonic activity ● Cold Environments ● Food supply problems ● Leisure and Tourism	1 7 6, 12 14

Examination analysis

Unit 1: Global Challenges
Divided into two sections:
- Section A: Objective items, data-response and short answer questions.
- Section B: Choice of World at risk or Going global longer/guided essay questions
1 hr 30 min exam   60% of AS

Unit 2: Geographical Investigations
Divided into two sections; four options are offered. Answer one Physical topics question from A and one Human topics question from B. Long responses are expected and the questions are designed to test data-response, investigative and evaluative skills. Impact and management issues will also be tested.
1 hr 15 min exam   40% of AS

Examination analysis

Unit 3: Contested Planet
Divided into two sections:
- Section A: two essays from six choices.
- Section B: All questions to be answered (that is, one question in three parts); it is in the format of a synoptic investigation.
2 hr 30 min exam   60% of A2

Unit 4: Geography Research
One question related to the student's choice of topic.
1 hr 30 min exam   40% of A2

WJEC AS

Unit	Specification topic	Chapter reference
Unit G1: Changing Physical Environments	Climate change Tectonic and hydrological change	2 1
Unit G2: Changing Human Environments	Population change Settlement change	9 8

Examination analysis

Unit G1: Changing Physical Environments
Three structured questions with stimulus material, one of which tests research/fieldwork.
1 hr 30 min exam   50% of AS

Unit G2: Changing Human Environments
Three structured questions with stimulus material, one of which tests research/fieldwork.
1 hr 30 min exam   50% of AS

WJEC A2

Unit	Specification topic	Chapter reference
Unit G3: Contemporary Themes and Research in Geography	One from: ● Deserts and tundra ● Glacial or coastal ● Climatic hazards And one from: ● Development ● Globalisation ● India or China	3, 6, 7 1, 5, 7 2 10 10, 11
Unit G4: Sustainability	Food supply Water supply Energy Cities	12 4, 13 13 8

Examination analysis

Unit G3: Contemporary Themes and Research in Geography
Divided into two sections:
– Section A: Answer one question from three physical geography options, and one from three human geography options.
– Section B: A two-part question based on enquiry research/pre-release information (one from ten topics).
2 hr 15 min exam   60% of A2

Unit G4: Sustainability
Divided into two sections:
– Section A: Tests the four areas of Food supply, Water supply, Energy and Cities, and uses pre-release information (also in Section B).
– Section B: Draws on the topics Food supply, Water supply, Energy and Cities and how they link with other topics studied in the course.   *1 hr 45 min exam   40% of A2*

CCEA AS

Unit	Specification topic	Chapter reference
Unit 1: Physical Geography (including Fieldwork Skills)	• Fluvial processes/features • Ecosystems • Atmospheric systems Human Interaction with the above systems	4 3 2
Unit 2: Human Geography (Including Fieldwork Skills)	Population data, structure and resources Challenges for rural environments Challenges for urban environments Development and issues in development.	9 8 8 10

Examination analysis

Unit AS 1: Physical Geography
Six questions from three sections:
– Section A: Multi-part question, which assesses fieldwork skills.
– Section B: Three compulsory short structured questions.
– Section C: Answer two out of three extended questions.
                    1 hr 30 min exam   50% of AS

Unit AS 2: Human Geography
Six questions from three sections:
– Section A: Multi-part question, in which secondary sources are investigated.
– Section B: Three compulsory short structured questions.
– Section C: Two out of three extended questions.
                    1 hr 30 min exam   50% of AS

CCEA A2

Unit	Specification topic	Chapter reference
Unit 1: Human Geography and Global Issues	Natural population change Migration causes streams and impacts Population policies Sustainable development Urban land use and planning Ethnic diversity Ethnic conflict Air pollution Nuclear energy Agricultural change and impact Tourism	9 9 9 Various chapters 8 9 9 13 13 12 14
Unit 2: Physical Geography and Decision-making	Demands on fluvial and coastal environments River basin management Coastal processes Tropical biomes Processes in tropical forest environments Management of tropical ecosystems Plate tectonics Volcanic activity Earthquake activity	4, 5 4 5 3 3 3, 6 1 1 1

Examination analysis

Unit A2 1: Human Geography and Global Issues
Divided into two sections:
– Section A: Answer two questions from three structured questions with an extended element.
– Section B: Answer one question from four structured questions with an extended element.
                    1 hr 30 min exam   50% of A2

Unit A2 2: Physical Geography and Decision-making
Divided into two sections:
– Section A: Answer two from three questions, all structured.
– Section B: Decision-making synoptic exercise.
                    2 hr 30 min exam   50% of A2

AS/A2 Level Geography courses

AS and A2

All A Level qualifications comprise two units of AS assessment and two units of A2 assessment. This offers Geography students the opportunity to complete a freestanding AS course or to complete their geographical education, and to develop ideas, themes and concepts further into a full A Level course via the much more demanding and challenging A2 course.

How will you be tested?

Assessment units

For AS Geography, you will be tested by two assessment units. For the full A Level in Geography, you will take a further two A2 units.

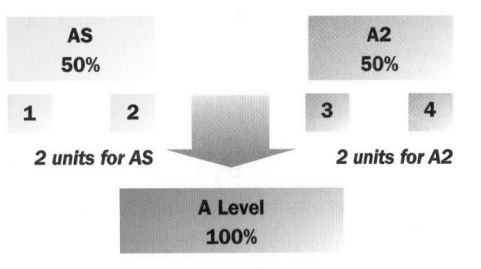

Each unit can normally be taken in either January or June. Alternatively, you can study the whole course before taking any of the unit tests. There is a lot of flexibility about when exams can be taken and the diagram below shows just some of the ways that the assessment units may be taken for AS and A2 Level Geography.

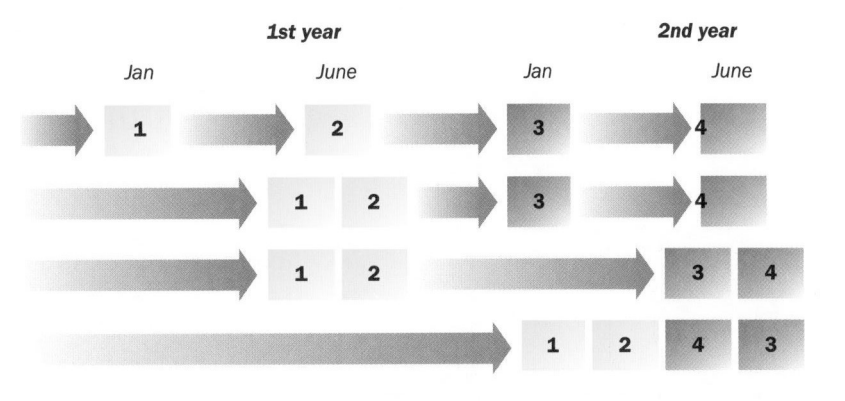

Remember that the combinations shown might not be the route chosen by your school to go through the examination. There are many other possible combinations.

If you are disappointed with a module result, you can re-sit each module taken and most exam boards will offer papers in both January and June. The higher mark always counts.

A2 and synoptic assessment

Having studied AS Geography you will then hopefully take on the more challenging A2 Geography course. For this assessment you will need to take two further units of Geography. Similar assessment arrangements to AS apply, except some units draw together different parts of the course in extended writing or essays and much of this work will be assessed synoptically.

Investigative work

Fieldwork / coursework used to form an integral part of all the old geography courses. However, for the most part it has now been replaced by **investigative work** that is tested in examination conditions.

Some exam boards use pre-released materials; some exam boards use data response exercises; others use research exercises and decision-making exercises.

See the specification information on pages 4–8 to help you determine what is expected of you by the specification that you are following.

Key skills

AS and A2 Geography specifications identify opportunities for developing and assessing key skills, where these are appropriate to the subject. The range of key skills is broad, incorporating communication, ICT, application of number, improving own learning and performance, working with others and problem solving. However, the only key skill component that has to be assessed through Geography is communication.

Different types of questions

Structured questions and data response questions

Structured questions:

- Are sub-divided and tend to increase in difficulty as the question progresses.
- Based on a support resource such as a diagram, photograph or map.
- There is more of a focus on description rather than explanation.

You need to be concise – look for trends and anomalies. You must try to fit your answers to the lines offered and to the mark weighting; practising such questions is an absolute must. You will probably have an opportunity to use exemplars and to include ideas from your fieldwork.

Data response questions:

- Test your geographical skills.
- Are different to structured questions in that the standard expected is probably a little higher.

Precision in description is expected, and some elementary analysis might be called for. You might have to summarise a table, label (annotate) a diagram or photo or describe in a more technical way.

Here are some important points to remember when answering structured or data response questions:

- Structured questions are a quick test of knowledge and understanding. Ensure you have practised structured questions before the exam.
- Avoid writing out the question.
- Use the detail in any resource (map, photo, table or graph) that is given to you.
- Use case studies in your answers.
- Structured questions (especially the higher scored questions) are level-marked; here the correct use of terminology is vital.

The use of Ordnance Survey

Most exam boards use OS maps, though to a variable degree. Most seek to use them for analytical and interpretative purposes. Those boards using OS skills at A2 utilise analysis and interpretation of the OS map to both extend and develop geographical intellect. On the whole, maps are usually chosen to represent problem areas for which an interpretation can be offered on the basis of printed evidence from both the physical landscape and/or the rural and urban landscape. Typically, questions at A2 ask for a description, analysis or synthesis of evidence presented by the OS map, or any combination of these three.

- **Descriptive** accounts usually involve looking at physical or cultural elements on the map. The features are described in terms of their relief, vegetative cover, or settlement and communications might also be described.
- **Analysis** in which map evidence is systematically analysed at an elementary level and in relation to other mapped distributions.
- **Synthesis** in which analyses are collated to form a statement or interpretation of the physical and cultural features observed.

For example, an excerpt of an OS Boscastle map highlights in purely physical terms why it is so prone to the effects of flash flooding – with its narrow / steep valleys, the proximity of high, hard geology and the fact the Boscastle sits at the confluence of three rivers… amongst other factors.

Extended prose

This type of question requires you to write several sentences or up to a page in your answer. Such questions attempt to assess your ability to communicate ideas clearly and logically. Your answer must be packed with facts and you must know your examples/case studies well. Above all you must respond to command words (see below). You can use diagrams sparingly, but make sure they contribute to the answer! Some planning pays dividends in these questions. Go for depth – don't be superficial. Use mnemonics to remember facts / number trends. These questions test knowledge and understanding.

Essay questions

At A2, because candidates are expected to have a deeper understanding, greater knowledge and to be able to make links between different parts of their specification it is expected these candidates can handle the more complex technique of essay writing. A2 essays are a step up from extended prose, therefore one might rightly expect there to be an incline in the complexity and type of language and command words used. The lists below attempt to show how the step up is made between AS and A2; there are of course many other command words used.

AS and A2:
- **Describe:** details of appearance and characteristics needed.
- **How:** process or mechanism recognition.
- **Why:** explain.
- **State:** briefly, perhaps one word.
- **Illustrate:** with examples and case studies.
- **Comment:** a balanced view or judgement.

A2:
- **Examine:** investigate in detail.
- **Criticise:** perhaps explore weaknesses in arguments.
- **Discuss:** consider both sides of an argument.
- **Justify:** argue in support of a particular view.

Essays are a mainstay of the exam system, and are prominent in many of the new specifications.

Invariably examiners are looking for those who can 'effectively argue'. Those who 'think outside the box' and write in a synoptic fashion usually score highly.

Here are some important points to remember when answering essay questions:
- There is a correct and appropriate response to command words.
- Remember there is rarely a definitive answer in an essay! Your viewpoint is often required such as rich v poor, old v young, MEDC v LEDC.
- Work should be paragraphed, with one relevant and strong theme running through the essay.
- Structure the essay, working from an appropriate plan (written on the paper that is handed in – remember it might be looked at if you run out of time!).
- Case studies should be used but not overused, and should draw on a variety of scales (local, regional and global, if required).
- Balance must be achieved between fact and the more discursive aspects.
- Diagrams and sketches may be used that move the essay on.
- A basic introduction is important; it must show your understanding of the question. An expansion of ideas in the middle section of the essay (the 'meat' of the essay) should appear next, followed by a conclusion.
- Try to give yourself time to read through your essay and correct any mistakes.

Stretch and Challenge

'Stretch and Challenge' refers to the new provision by the exam boards to add extension material for the top candidates at A Level. Stretch and Challenge will be introduced within compulsory questions: more extended writing questions, fewer structured questions, and bringing the old extension award questions into A Level. In geography it considers:
- Candidates' use of command words.

- An ability to make synoptic connections.
- The skilful use of case studies by candidates.
- Responses to open-ended questions and extended writing tasks.
- How candidates use resource materials.

Questions set to stretch and challenge candidates may well be the ones that decide the A* award.

Exam technique

AS and on to A2

In order to study Geography you will probably have studied the subject to GCSE level, although many schools don't insist on this. It is of course necessary to have studied Geography to AS Level to move on to A2! At A2, many study areas will be new and other areas will be developing ideas in a new or different way. A2 will extend and develop your knowledge and understanding still further. It is likely that your teachers will be looking at A2 for students who want

to develop and refine their knowledge and understanding of the subject matter covered at AS, and to further their studies with the more advanced ideas, concepts and processes offered at A2. Students who are willing to explore ideas in an enquiring and lively way, and who can communicate findings and ideas effectively, are likely to get the most out of Geography at A2.

What are examiners looking for?

All exam questions seek to assess your appreciation and attitude to the content – be it physical, human or regional in nature – of the specification you are studying. There are certain common qualities that all exam boards look for in candidates:

Common qualities

- A knowledge of facts, basic vocabulary, geographical concepts, processes and theories.
- An ability to use information in an organised way, supported by appropriate case studies and examples.
- The appreciation that all geography content is dynamic.
- A range of skills understood and used in a range of geographical contexts.
- An excellent ability to comment on and evaluate world issues and problems.

Practical qualities

- In A2 essay work especially, utilise every word that you write, display knowledge and understanding, use case studies, respond to command words, be organised, balanced and evaluative and give a conclusion. Most A2 essays will look for a 40–45 minute answer.
- Examiners are looking for answers that are clear, legible and concise.
- The mark schemes examiners use enable candidates to offer a wide range of interpretations to questions. They are not looking for a specific or pre-determined answer.
- Diagrams must make a real contribution. They have to be both neat and accurate.

Quality of English is also now integrated into the marking of papers. You must be clear and accurate in your use of English punctuation, grammar and spelling.

Some dos and don'ts

Examiners try to assess the degree to which you can demonstrate as many of the qualities listed above as possible. But there are many common problems encountered by examiners when they are marking examination work. Make sure you follow this advice:

- **Do** plan answers, whether in response to synoptic questions or structured questions. Quick simple plans are a must, rewriting the essay title on your answer sheet can also help.
- **Do** respond to command words (see page 12).
- **Do** respond to the vocabulary in the question, i.e.

channels are different to valleys, glaciation is different to peri-glaciation, environmental hazards could mean pollution and / or biological and / or geomorphologic hazards, etc.

- **Do** have a range of case studies to draw upon. Make sure your case studies illustrate points rather than obscure them.
- **Do** make yourself aware of the ploys to reach the highest level or tier in a question.
- **Do** read the rubric of the exam, and organise your time effectively. Those who fail to do so invariably

show all the signs of panicking, i.e. questions unfinished, poor quality of language and major and avoidable omissions. It is important to read through your answers!

- **Do** understand your specification and how knowledge, understanding and skills fit each module/paper you sit. Failure to practise past papers is also very obvious.
- **Do** seek out and outline inter-relationships.
- **Do** give requested information, i.e. a named area within a city.
- **Do** offer supporting diagrams and sketches.
- **Don't** spend too much time on one question or a part of a question.

- **Don't** let words like coast or river trigger an 'all I know' type of response, avoiding the focus of the question.
- **Don't** over-learn favourite topics in the hope they will appear in the exam. At A2, especially when synoptic issues are covered, you must have an extensive knowledge of a range of topics. Narrowing your choice is dangerous.
- **Don't** attempt questions that lead to the snake pit! If you don't understand flocculation or frontogenesis, avoid these questions!

What grade do you want?

Your final A Level grade depends on the extent to which you meet the assessment objectives of the exam board you are studying with. To gain the best possible mark you will have worked hard throughout the course and will be highly determined and motivated. If you have identified weaknesses in your knowledge and understanding, you will balance this by improving and building on your performance in other areas of the subject.

A* or A grade

To achieve a grade A* or a grade A you will:

- show in-depth knowledge and critical understanding of places, themes and environments and understand how physical and human processes affect these areas
- understand and be able to use a wide range of Geographical terminology
- understand how all of the above connect and be able to convey your understanding.

You have to be a very good all-rounder to achieve a grade A* or A. The exams test all areas of the syllabus, and any weaknesses in your understanding of Geography will be found out.

C grade

To achieve a grade C you will:

- produce sound answers, which make use of Geographical terminology
- synthesise and communicate your ideas and views well, but less effectively than an A* / A grade candidate.

Make sure you identify and improve on your weaknesses. Prepare fully by reading around the subject and keeping up with current affairs. Make sure you practise answering past exam questions to give yourself valuable exam practice and familiarise yourself with the different types of questions.

Practice questions

This book is designed to help you get better results.

- Look at the **sample questions and answers**; see if you understand where the candidates have done well or where they have slipped up.
- Try the **exam practice questions** and then look at the answers. Make sure you understand why the answers given are correct.

If you perform well on the questions in this book you should do well in the exam. Remember that success in exams is all about hard work, not luck.

1 Earth challenges

The following topics are covered in this chapter:

- **The lithosphere and tectonic processes**
- **Rock types and weathering processes**
- **Killer tsunamis**
- **Assessing hazard hotspots**

1.1 The lithosphere and tectonic processes

LEARNING SUMMARY

After studying this section, you should be able to understand:

- the contribution continental drift and plate tectonics theory offers to our understanding of volcanic and earthquake activity
- that volcanoes and earthquakes are an ever present danger
- that tectonic activity creates, as well as destroys
- that man can predict and prepare for volcanoes and earthquakes
- that differences exist in human responses to tectonic hazards based on levels of economic development

> Isostacy: is the principle of flotation or buoyancy. Differences in density between continental and oceanic areas explain why continents (low density) stand high above oceanic basins (high density).

The Earth has been in existence for approximately 4.6 billion years. During that time the character, position and distribution of the land and sea has changed many times. However, we have a good understanding of the internal structure of the Earth from earthquakes, volcanoes and plate tectonic activity.

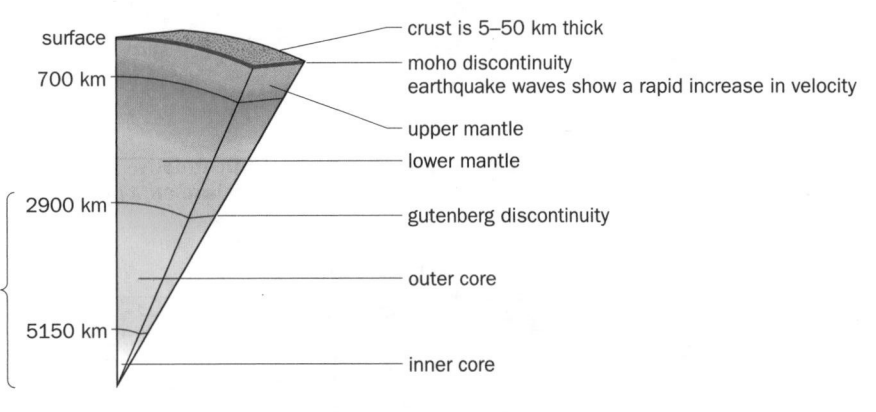

> NIFE – Alloys of Nickel (Ni) and Iron (Fe) at 4000°C. The outer core is liquid and spins with the Earth. Thought to be the source of Earth's magnetism.

The internal structure of the Earth

Continental drift and plate tectonics

AQA	**A2 U3**
OCR	**A2 U3**
Edexcel	**A2 U4**
WJEC	**AS UG1**
CCEA	**A2 U2**

> Many theories are comparatively recent!

Alfred Wegener was the first to produce the theory of continental drift in his book *The origin of continents and oceans*. He noted the jigsaw fit of the continents, the similarity of the rock structures and fossils on either side of the Atlantic and the fact that there is abundant coal in Antarctica. Arthur Holmes and Alexander Du Toit kept the hypothesis alive. They envisaged rafts of lighter granitic continental rocks (SIAL – Si+Al) floating on denser basaltic oceanic rocks (SIMA – Si + Ma).

It was only when good maps of the ocean floor were produced showing underwater mountains, ridges, island arcs, trenches and plains, that scientists became aware of plate structures. Plate tectonics has proved a unifying theory and was widely accepted by the late 1960s.

The Earth's outer layer is made up of irregular-shaped plates. There are seven major ones and a dozen or so smaller ones. The plates are about 5–70 km in depth, rigid and are part crust and part upper mantle (**lithosphere** = the solid outer crust). They sit above a less rigid layer, the **asthenosphere**, where the rock is almost at its melting point.

> There is a strong link between volcanoes, earthquakes and plate boundaries.

Most of the world's great mountain ranges, most destructive earthquakes and volcanoes occur at the plate boundaries (e.g. the Pacific Ring of Fire). There are three types of boundary: converging, diverging and conservative. While most of the present day ocean floor is less than 200 million years old, some continental rocks have an estimated age of 4000 million years.

Converging boundaries

> Alternative terminology = destructive/collision and subduction zones.

Plates move together on destructive boundaries and material is destroyed. One plate rides over the top of another pushing it back deep into the Earth to be remelted. The continents are not dense enough to be pushed under. When continents carried on two separate plates collide, the ground buckles and folds and a new mountain chain is born. Two distinctive types of converging boundary occur: subduction zones, with ocean trenches and island arcs and collision zones with mountain ranges.

> You must know the different outcomes on different boundaries.

> Labelling a simple diagram is common at AS Level.

Ocean trenches are typically 1600 km or more in length, 96 km wide, 2–3 km below the surrounding ocean and are V-shaped. The deepest soundings have been made in the Marianas Trench at 11 040 m. The descending plate dips at an angle of about 45°. At 720 km the plate loses loses its rigid characteristics and has heated up sufficiently to blend in with the surrounding mantle.

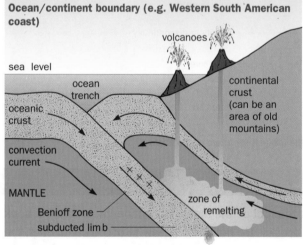

Ocean/continent boundary (e.g. Western South American coast)

Collision zone

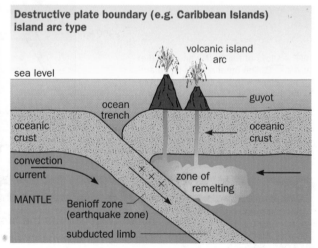

Destructive plate boundary (e.g. Caribbean Islands) island arc type

Subduction zone

> Most volcanoes and earthquakes are associated with converging boundaries.

Volcanic island arcs are situated on the over-riding plate 100 to 200 km behind the trench. Two-thirds of known volcanoes are associated with island arcs. Evidence suggests that island arcs played an important part in shaping the face of the Earth by contributing much of the material that makes up the continents.

Diverging boundaries

Alternative terminology = constructive/spreading ridges.

The plates move apart, magma seeps to the surface from the mantle to cool and form a new lithosphere. Almost all these boundaries lie deep beneath the oceans where they form mid-ocean ridges. There is a continuous system twisting and turning for 75 000 km around the world.

Constructive boundary (Mid-Atlantic Ridge)

symmetrical magnetic profile

ocean trench

sea level

younger rock

ocean ridge

older rock

earthquakes

new crust created

oceanic crust

convection currents

convection currents

Mantle

Constructive boundary

The geologist Sir Edward Bullard measured the heat coming to the surface along the Mid-Atlantic ridge and found that it was significantly higher than elsewhere along the sea bottom. The ridge head has many deep cracks, and a rift valley with an average width of 32 km and is up to 1525 m beneath the crest. Most tremors take place in or near it.

Palaeomagnetic studies

Frequent questions ask for a knowledge of the evidence for seafloor spreading, continental drift and plate tectonics.

The discovery of sea-floor spreading is attributed to the work of Vine and Matthew.

The Earth has a weak magnetic field that behaves strangely. The magnetic poles wander slowly about and from time to time the **magnetic field reverses** itself (171 times in the last 76 million years). There is a record of this, frozen in rocks containing iron oxides. Lava shows no magnetism, but as the rock cools it becomes magnetised in the direction of the north magnetic pole and then retains this record throughout time. Magnetic readings from the ocean floor form a zebra-striped pattern proving sea-floor spreading. Spreading rates vary from 2 to 20 cm per year. The distance between Europe and North America has increased by about 15 m in nearly 500 years.

Conservative boundaries

Alternative terminology = transform boundary.

No crustal material is created or destroyed. Two plates grind and slip past each other in a series of jerky movements. Faulting occurs widely in the fracture zones associated with spreading ridges; pressure, rotation and contraction cause.

The mechanism

An understanding of the mechanisms that drive the plates ensures a complete knowledge and understanding of plate tectonics.

The Hawaiian Islands have been formed by the hotspot process.

The force that drives the plates is to a certain extent still a mystery, but their movement certainly suggests convection currents in the mantle. In 1971, J.W. Morgan developed the **Hotspot (or Plume) Theory**. He suggested that giant pipes that carry up hot material from deep inside the Earth to the surface powered the plates. These plumes are up to 100 km in diameter and 99% of the material spreads out beneath the plate in a circular pattern like a thunderhead cloud. Only 1% reaches the surface. Where several hot spots punch the surface, the lithosphere may break. This would form a crack and lava could seep to the surface to form a diverging boundary. Morgan located 20 possible hot spots including seven main ones beneath the Mid-Atlantic ridge and four along the East Pacific Rise. Most rock found in the spreading ridges is basalt from the upper asthenosphere.

The pattern of change

The continents were once joined to form one landmass; **Pangaea** (all lands), made up of **Gondwanaland** and **Laurasia**. It began splitting about 200 million years ago. It is predicted that a new super-continent will form again 200 million years from now with the centre of the old Pangaea becoming the coastlines of the new landmass.

Volcanoes

AQA	**A2 U3**
OCR	**A2 U3**
Edexcel	**A2 U4**
WJEC	**AS UG1**
CCEA	**A2 U2**

Volcanoes play a key role in shaping the planet. They probably created our atmosphere and they certainly released water vapour onto the surface of the planet, creating the oceans. Every year about 50 or so of the Earth's 700 active volcanoes erupt. Simply, volcanoes are openings in the Earth's crust through which magma, ash and gas erupt onto the Earth's surface.

> Few questions are asked about features, though it's wise to know about calderas, fumaroles, geysers and the different types of volcano. Most questions want information on causes of volcanic activity, use by man and effects on man.

> Simple and effective diagrams should appear in your essays.

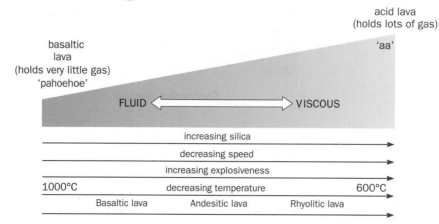

The nature of lava

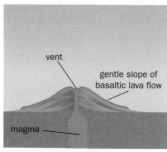

Shield volcano

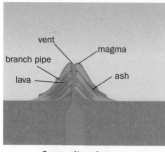

Composite volcano

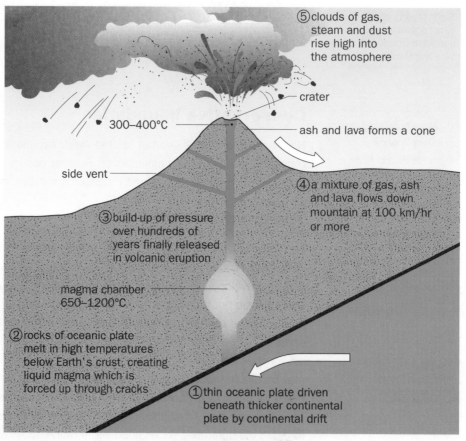

The structure of a volcano

⑤ clouds of gas, steam and dust rise high into the atmosphere

crater

ash and lava forms a cone

300–400°C

side vent

④ a mixture of gas, ash and lava flows down mountain at 100 km/hr or more

③ build-up of pressure over hundreds of years finally released in volcanic eruption

magma chamber 650–1200°C

② rocks of oceanic plate melt in high temperatures below Earth's crust, creating liquid magma which is forced up through cracks

① thin oceanic plate driven beneath thicker continental plate by continental drift

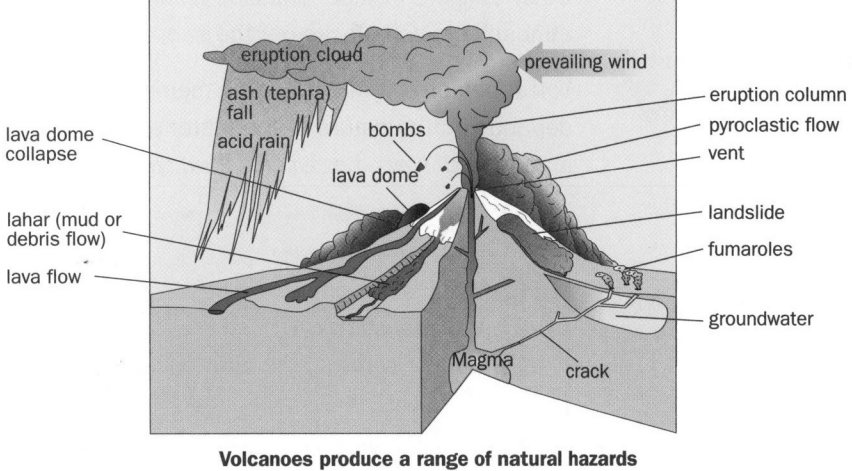

Volcanoes produce a range of natural hazards

A knowledge of these examples of volcanic hazard is important at AS and A2 Level along with appropriate case studies.

The hazards resulting from volcanic activity are divided thus:

Hazards	Details	Examples
Lava flows	Two main types are *aa* and *pahoehoe*. When hot enough lava flows like water.	• Mount Etna, Sicily is a frequent eruptor; basalt lava often runs down its flanks.
Lahars	These are torrents of mud, moving at 90 km/hr. There are two main reasons for their development. • Increased rain because of volcanic activity washes soot and ash downhill. • Heat from the eruption melts snow and ice, which picks up debris.	• Armero, at the head of the Langunilla Valley in Colombia, was buried under a mudflow 20+m thick in November 1985. • Mount Mayon, lahars buried Padang in November 2006.
Ballistic and tephra clouds	This is the solid material such as dust, ash and cinders that is thrown from the volcano.	• During the eruptions of Mount Pinatubo in the Philippines in 1991 rocks the size of tennis balls were hurled 50 km away!
Gases and acid rain	Gases released encompass a wide range of chemicals, from carbon dioxide and sulphur dioxide	• In August 1986 a huge volume of carbon dioxide escaped from Lake Nyos in Cameroon, West Africa. • Eyjafjallajökull, Iceland 2010.
Nuée ardentee	A mixture of superheated rock and gas.	• Mount St. Helens, USA, May 1980. • More dramatic was the eruption of Mount Pelee, Martinique, May 1902.
Jökulhlaups	Sudden outbursts of water due to a build-up of sub-glacial meltwater because of volcanic activity.	• Caused by the meltdown of ice by the volcano Loki, Iceland under the Vatnajökull ice-sheet. • Huge jökulhlaups realeased as Eyjafjallajökull erupted in 2010.
Tsunamis	Form when water is vertically displaced. Caused by volcanic eruptions (and earthquakes). Tsunamis move at 800 km/hr with a height near land of up to 30 m.	• Noteworthy here is the eruption of Krakatoa in 1883. • In 2009 16 tsunamis were recorded. • One of the biggest historically was the Boxing Day tsunami of 2004.

At present 20% of the world's volcanoes are watched 24hr/day. Most are in MEDCs.

Other dangers include landslides, fire, disease, and volcanic winters (where atmospheric ash blocks out the sun, and crop yields drop).

Volcanoes on the whole display themselves typically as a conical shape, but this depends on the nature of the material and type of eruption. They are classified in a number of ways, i.e. type of flow, type of eruption and level of activity.

PROGRESS CHECK

1. What is the difference between continental and oceanic crust?
2. What is isostacy?
3. Outline some of the differences between basic and acid lava.

Answers: 1. Oceanic crust is denser. 2 Relates to the buoyancy of the crustal materials. 3. Basic lava has less silica, moves quickly and is hotter.

The management of natural hazards

AQA	A2 U3
OCR	A2 U3
Edexcel	A2 U4
WJEC	AS UG1
CCEA	A2 U2

The control, planning and management of natural hazards – the 'four Rs'

In numerical terms, a disaster may be defined as that which causes more than 10 deaths, injures over 100 people and causes damage worth $16 000 000.

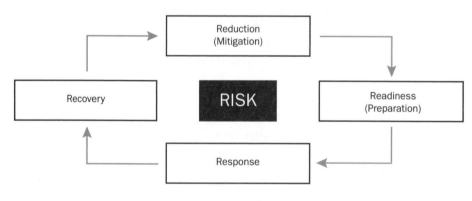

The four R's

Examiners will be excited to read about how people prepared for and responded to, specific events. Ensure you provide actual case studies with all the relevant facts and figures.

Reduction

Reduction refers to all the activities that result in a risk or hazard being reduced. Such activities might include:

- hazard planning and risk management analysis
- community and civil contingency plans by government agencies
- schools and families undertaking awareness sessions led by government agencies and the emergency services
- strict building regulations.

Readiness

This refers to the prediction of, and prevention of, hazards and preparing the community/population to deal with hazard events. This might involve:

- training officials and volunteers
- practising and testing family and school defence plans
- public education
- practising and maintaining communications systems, such as the emergency services, and checking and testing warning systems.

Response

This refers to dealing with the situation. This involves:

- coordination and control
- evacuations
- dealing with injuries
- maintaining public health
- restoring infrastructure
- communications.

Recovery

This refers to re-establishing the community to its pre-event condition. This involves:

- restoration and rebuilding
- making emergency finance available and providing insurance support
- emergency shelters
- planning and emergency reviews.

The **impact** of any tectonic hazard depends on:

- the levels of economic/socio-economic development (MEDC or LEDC)
- technology deployment
- the density of the population and whether it is prepared for hazards
- training and education about all risks involved.

> Do offer examiners examples from more economically developed countries (MEDCs) and less economically developed countries (LEDCs).

Predicting volcanic activity

AQA	A2 U3
Edexcel	A2 U4
WJEC	AS UG1
CCEA	A2 U2

Significant amounts of money have been invested in predicting volcanic activity by MEDCs where volcanic activity is common. Funding is high for such agencies in the USA, Japan and Italy. Prediction is based on a number of well-founded principles.

- That several monitored features must coincide before an eruption.
- It is assumed that active volcanoes will eventually erupt!
- All volcanoes behave in different ways and eruptions are unexpected.
- Their volcanic history has to be relied upon.

Despite all of this prediction techniques have not been totally refined.

Some approaches to the monitoring of volcanoes are outlined below:

Approaches	Detail	Usefulness
Seismological	As a volcano prepares to erupt there is increased quake activity. Short-period quakes as magma rises and long-period quakes as the magma increases in the volcano. Harmonic quakes occur as the magma pushes through the crust.	Difficult to decipher, but is a good indicator of increased volcanic activity. Successfully used to evacuate Mexico City in 2000 as Popocatépetl erupted.
Changes in gas emissions	As magma rises gas pressure usually drops and gas escapes. Most commonly emitted gas is sulphur dioxide.	Extensively and successfully monitored around the world. Pinatubo's (Philippines) eruption was accurately predicted using gas analysis.
Hydrological changes	Water levels can rise or fall in the ground. Water flow in rivers can increase causing unexpected flooding.	Both of these hydrological changes are frequently seen.
Remote sensing	Involves charting tree growth or stunting, thermal changes, gas sensing and assessment of ground deformation.	All these techniques are used to successfully monitor volcanoes around the world.
Mass movement	Movements can be predicted using conductivity measurements in rocks.	Measurement successfully used to predict the Mount St. Helens, USA pre-eruption slip.

People continue to live in volcanically active areas.

- Lava weathers to give fertile soils, e.g. in Indonesia.
- Precious stones and minerals form in such areas, e.g. in South Africa.
- Superheated water can be used for geothermal power, e.g. in Iceland.
- These areas can be tourist attractors, e.g. Tenerife and Lanzarote.
- Products of vulcanicity can be used for building, e.g. ignimbrite blocks.

PROGRESS CHECK

1. What is a natural hazard?
2. Why are gas levels monitored in volcanoes?
3. How can the products of volcanic eruptions be successfully used?

Answers: 1. It has 'potential to cause loss of life or property'. 2. Increase in gas levels suggest an imminent eruption. 3. Building blocks and fertile soils power generation.

Case study: Responding to the volcano hazard – Sakurajima, Japan

Sakurajima is a monitored active volcano situated on the southern island of Kyushu near Kogoshima in Japan. The volcano has 7000 people living on its flanks and half a million residents living in Kogoshima. Sakurajima is a three peaked acidic stratovolcano and is 70 km wide and over 2000 metres high. The volcano erupts frequently (there were 400 eruptions in 2009), but constant monitoring means it has not killed anyone since 1991.

The Sakurajima Volcanic Observatory monitors the volcano 24 hrs/day. Information comes into the research centre from equipment such as extensometers, tiltmeters, seismometers, magnetometers and tide gauges. Gas emissions are constantly monitored and water levels in the ground assessed. Residents are updated daily.

As well as being monitored Sakurajima is managed too. Land use planning is used and concrete storm channels are used to direct landslides and ash slides away from Kogoshima. Evacuation drills are practised and children wear hard hats to and from school. Farmers protect crops by using systems of poly tunnels.

Case study: The impact of volcanic activity – three examples from across the world

Prediction, response and outcome is a frequently visited area at A-Level. You must know a range of case studies and examples.

Mount Mayon, Luzon, Philippines, 2006–10

Mount Mayon is an active stratovolcano that has erupted more than 50 times in the last 400 years. It is part of the Pacific Ring of Fire. It is about 10 km from major cities and towns and is the Philippines' most monitored volcano (by the Philippine Institute of Volcanology and Seismology – PHIVOLCS.

Chronology of the eruption

In July 2006 sulphur dioxide emissions increased and the size and number of rock falls from the lava dome began to increase. By August 2006 low frequency earthquakes were detected and the government ordered the evacuation of 20 000 people living near the volcano and a further 20 000 who lived within an 8 km danger zone. In late August Mayon went 'quiet', but ground surveys showed it to be highly 'swollen' (the ground swells before an eruption as magma is forced through the Earth's crustal layers). By September 2006 activity slowed completely and in November 2006 typhoon Durian caused mudslides from the slopes of Mayon that killed 1300 and filled the village of Padang to roof height.

There was a further eruption in 2008 when ash was ejected to 200 m and there was increased seismic activity. Earthquake activity continued through 2009 and at the beginning of December increases in sulphur dioxide and earthquakes led to an evacuation of approximately 34 000 people (72% of the total number of people that needed to be evacuated). Lava reached down 1500 m below the summit and ash was thrown into the atmosphere up to 2000 m. By the end of December increased lava flow and sulphur dioxide emissions (up from 700 tonnes to 9000 tonnes per day) led to another wave of evacuations of 45 000 people. By January 2010 the eruptions began to subside and evacuation centres were emptied. The ejected volcanic material erupted by Mayon was estimated to be 23 million cubic metres of rocks and volcanic debris.

Consequences of the eruption
- Tourism decreased by 20%.
- Forced evictions had to be carried out.
- USA committed $100 000s in aid.
- Philippines' military deployed to aid evacuations.
- The United Nations World Food Programme (UN-WFP) was deployed.
- The United States Agency for International Development (USAID) and the Red Cross were the principal non-governmental agencies (NGOs) deployed.
- The total cost to the Philippines was estimated to be $1.3 billion.

Chaitén, Chile, 2008–09
Chaitén is a small volcanic caldera (a former volcano that has collapsed around the vent) located in southern Chile. Before its last eruption in 2008 the volcano was last active 9400 years ago.

Chronology of the eruption
The Chaitén authorities began evacuations at the beginning of May 2008 following advice from geologists observing the volcano. There was just one death from this eruption! On 2nd May the volcano erupted sending a plume of volcanic ash and steam into the atmosphere. Chaitén was evacuated along with other smaller communities in the area. The ash contaminated water supplies in the area and coated the local environment to a depth of 5–10 cms. Following the ash phase, pyroclastics and some lava emerged from the volcano and the ash was forced to 30 000 metres. The increasing force of the eruptions fused the double vent of Chaitén into one single 800 metre diameter vent and later blasted parasite cones onto its southern flanks. In February 2009 a lava dome in the vent collapsed sending pyroclastics flows to within 5 km of Chaitén.

Consequences of the eruption
Large areas of Chile and Argentina were coated with ash which had negative effects for agriculture in the short-term, but longer term it can make very fertile soil. The large amounts of ash also posed the risk of lahars (mud slides) which eventually caused flooding in Chaitén in May 2009 causing the destruction of a significant part of the town. The Chilean government responded to this by pledging to rebuild Chaitén 10 km north of its present site.

Eyjafjallajökull, Iceland 2010
Eyjafjallajökull is a stratovolcano with an ice cap which covers the caldera of the volcano. The volcano is situated in southern Iceland and straddles the Mid-Atlantic Ridge. It has two neighbouring volcanoes, to the north-east, which are also very active. Before its eruption in 2010 the volcano was last active in 1823.

Chronology of the eruption
In December 2009 seismic activity was detected with thousands of small quakes detected below the surface. By March 2010 approximately 3000 quakes per day were recorded as the magma chamber of the volcano began to fill. By mid March an eruption, in the form of a fissure vent, took place approximately 8 km from the top crater of the volcano delivering lava to the surface. In April the volcano resumed erupting and lava began to empty through the vent, causing a massive jökulhlaup (a sudden release of meltwater) to rush down rivers nearby. The resulting flood required 800 people to be evacuated to local emergency centres. This second, and increasingly explosive eruption, threw ash 8.5 km into the atmosphere, which led to the closure of airspace over many parts of Europe and resulted in air disruption for a six-day period in April with further disruptions in May.

Consequences of the eruption
- Airlines lost $200 million/day; 70 000 flights were cancelled at the height of the eruption; 6.8 million travellers were affected; £100 million in lost tourist business in London.
- BMW, LG, Nissan, Samsung, Honda suspended production as 'just-in-time' (JIT) production lines were halted.
- International trade was disrupted, e.g. Fed Ex and DHL cancelled flights of parcels and tonnes of New Zealand fresh salmon bound for Europe had to be destroyed.

Earthquakes

AQA	A2 U3
OCR	A2 U3
Edexcel	A2 U4
WJEC	AS UG1
CCEA	A2 U2

The point immediately above the focus on the Earth's surface is called the epicentre.

Earthquake intensity is measured on a seismograph. Two scales are used; Richter Scale measures the intensity of the energy released; Unercalli Scale is the amount of ground movement.

There are probably 500 000 earthquakes/year worldwide of which a hundred or so will do damage. These account for about 10% of natural disaster deaths per year.

Earthquakes are shock waves that are transmitted from a **focus**, which can lie anywhere from the surface to 700 km beneath the Earth's crust. The most damaging quakes have foci that are close to the surface and tend to arise along active plate boundaries. Earthquakes result from sudden movements along geological faults in response to convection currents in the deep mantle rocks (or because of volcanic eruptions). Earthquakes generate a variety of shock waves that have differing effects. Three types of seismic wave have been identified; **P (primary or push)** waves, **S (secondary or shear)** waves and **surface (or long)** waves. The speed of these waves varies according to the properties of the rocks through which they pass (P waves are the fastest and surface waves the slowest). P waves are compressional in nature and tend to cause the low rumbling associated with quakes and do damage to the area around the **epicentre**. They tend not to be as damaging as the S and surface waves. Shear waves cause the first shaking motion, vibration occurring at right angles to the direction of travel. Particle motion in surface waves is greatest at the surface and dies out at depth. Surface waves have the potential to shake large areas of the globe.

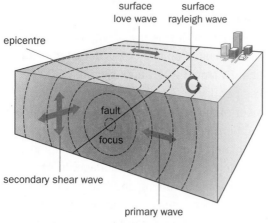

Earthquake magnitude

Magnitude
The magnitude and intensity of earthquakes depends on:
- the depth of shock origin
- the nature of the overlying material – hard rock absorbs; soft rock amplifies.

Effects
Earthquakes can cause:
- vertical or lateral displacements of the crust
- the raising and lowering of the sea floor
- landslides, liquefaction and tsunamis.

The impact of earthquakes

Impacts are often greater in LEDCs than in MEDCs as building regulations are poor, they are often remote and isolated and emergency planning is non-existent.

Case study: The impact of earthquakes

Bam, Iran, 26th December 2003 (LEDC)
The event: 7.7 on the Richter Scale. Epicentre 70 km south-west of Bam.
Issues: Densely populated; difficult to evacuate; weak; mud-brick used; no standard construction.
Effects: 26 271 deaths (many from cold after the event); 30 000 injured – worst earthquake in Iranian history. One-fifth of local teachers killed; half of local medical staff killed. 100 000 people made homeless. Mud-brick houses collapsed – walls first and heavy roofs fell on sleeping occupants. Tourism declined (Bam had received up to 10 000 tourists per year).
Short-term response: USA offered humanitarian aid which was eventually accepted. Red Cross, Crescent and the UN were also involved. Air lift of food aid into Iran. Iranians threatened to prosecute cowboy builders. 28 000 temporary homes built by Iranians.
Recovery/long-term response: Reconstruction to strict seismic guidelines. Bam held up as an example to the rest of the quake-prone Islamic world. Urban planning now always highlighted. Attempts have been made to improve public knowledge about earthquakes. The cost of reconstruction is estimated at $1 billion.

Haiti, 12th January 2010 (LEDC)
The event: 7 on the Richter Scale. Epicentre 25 km west of Port-au-Prince, resulting in a slip in crustal materials of about 1.8 metres.
Issues: Warnings were given of imminent earthquakes in 2006 and 2008. Construction standards were very low – there were no real building regulations. Many homes were built on stilts, which easily collapse. Port-au-Prince is the most populous part of Haiti due to recent massive rural to urban drift.
Effects: 230 000 deaths and 300 000 injured (morgues were overwhelmed). 1 million people were left homeless. 280 000 buildings collapsed. A tsunami killed three people. Lack of water. Riviere de Grande Goave dam was badly damaged. Communication damage (airport, port and roads damaged). Lack of leadership initially. Delays in aid distribution. Looting. Educational system totally collapsed.
Short-term response: The Dominican Republic was the first to react and send the Dominican Red Cross, doctors and mobile kitchens. Israel, Iceland, China, Qatar all sent help. UK search and rescue services mobilised. 100 000 people allowed into Canada and the USA without usual border handicaps. Adoptions expedited. USS Vinson led US navy task force. US marines/troops maintained law and order. Haitian government started process of moving population by boat to other points on the island. Many people returned to rural homes. UN appointed to take care of security on the island.
Recovery/long-term response: World Bank – debt payments waived for five years. Priority was to establish a working government, then work to sort out remaining refugees, roads and sanitation. Millions raised by individuals and governments around the world help Haiti's support for reconstruction.

Sichuan Province, China, 12th May 2008 (NIC)
The event: 8 on the Richter Scale. Epicentre 80 km north-west Chegdu.
Issues: Warning of earthquake in 2003 was ignored. The weight of water that filled the Zipingpu dam may have triggered the earthquake which lasted two minutes and had a shallow focus. In the hard ground of the central plateau lots of power develops in an earthquake. Many old buildings collapsed and those recently built were destroyed (it was suspected that regulations were not followed in the construction of the buildings).
Effects: 68 000 killed and 374 000 injured. Five million people were left homeless. Ranked the 21st worst earthquake of all time. 80% of buildings collapsed. An area of 596 km by 936 km in Sichuan was affected. All roads affected; schools and chemical plants destroyed. Copper prices rose (China major producer) and oil prices dropped (China is a major consumer). The mobile phone system all but collapsed. Zipingpu's HEP plant seriously damaged; 72 million stock animals killed; 34 earthquake lakes form behind landslides; 200 000 people moved from the Tangjiastan mountains as they are considered in danger from collapse of mud dams created by landslides. Damage estimated at $140 billion.
Short-term response: Within 90 minutes of the earthquake the Prime Minister visited and 50 000 troops were immediately sent to the area. Red Cross China and Taipei Fire department mobilised.

Taiwan offered further substantial help. China deployed 150 helicopters to the area. USA shared satellite imagery. Extensive rescue operation put into place. One-child family policy temporarily suspended. Quality of school construction criticised. UNICEF invited into China two days after the earthquake. Millions given to support the reconstruction.

Recovery/long-term response: China's government pledged $146.5 billion to rebuild Sichuan.

L'Aquila, Italy, 6 April 2009 (MEDC)

The event: Between 5.8 and 6.3 on the Richter Scale. Epicentre was near L'Aquila.

Issues: Giampaolo Giuliani, a local geologist, controversially predicted the earthquake having noted increased radon emissions. Building standards poor.

Effects: 308 deaths and 1500 injured; 3000–11 000 homes destroyed and 65 000 people made homeless. Earthquake proof buildings collapsed.

Short-term response: 40 000 find shelter in tented camps; 10 000 find shelter in hotels on the coast. Controversially those affected told by President Berlusconi to take the opportunity to take a holiday! The Italian government refuses EU or other cash help with the earthquake.

Recovery/long-term response: Poor building standards were criticised, criminal proceedings began immediately. Few convicted as many developments linked with Italy's underworld gangs. Italians pledged to re-build the area without outside help.

Responding to the earthquake hazards

Successful earthquake management depends upon a number of interacting variables.

Physical factors include:

- time of day
- depth of focus
- location of the epicentre
- base geology
- duration of the shake
- charting the recurrence interval
- identification of localities prone to liquefaction/folding or faulting.

> You must be able to describe a variety of management strategies and assess their effectiveness.

Human factors include:

- building style
- preparedness
- pinpointing the weaknesses in infrastructure.
- efficiency of the emergency services
- ability to cope and react

Earthquake engineering

Engineers attempt to match possible seismic events with stresses of such an event to the behaviour of buildings. They also try to estimate the consequences on built-up areas and its populations. The principle of such engineering is to control the vibrations and shock of earthquakes. Vibration control is used to:

- dissipate or spread earthquake wave energy within buildings
- disperse energy with a wider range of frequencies
- absorb portions of the seismic wave.

Examples of vibration control include:

- dry stone walling, e.g. the Aztec buildings in Machu Piccu in Peru use this method
- tuned dampers, a weight held high in a sky-scraper, e.g. in the world's second highest building Taipei 101, in Taipei, Taiwan
- rubber bearings which are built into a building's foundations
- pendulum bearings which allow displacement and lateral movements in the foundations, e.g. the San Francisco airport terminal
- building height control and profile adaptations, such as those in the Transamerica Pyramid skyscraper in San Francisco, and spring foundations.

Seismic construction involves the use of:

- reinforced masonry with ductile joints to allow some bending
- a light frame structure using wood, i.e. half timbering
- using concrete with a pre-stressed steel core
- cross bracing.

1.2 Rock types and weathering processes

LEARNING SUMMARY

After studying this section, you should be able to understand:

- the role and development that weathering and slope systems have on the landscape
- the effect that different geologies have on surface relief

Rocks and relief

AQA **A2 U3**
Edexcel **A2 U4**

Be able to draw these structures in cross-section.

(a) Material removed underground

(b) Appearance on the surface

The development of tors

Igneous rocks (begin as hot, fluid magma or lava)

Granite is a course-grained intrusive igneous rock consisting of three minerals: quartz (resistant), feldspar (susceptible to chemical weathering) and mica. The largest form of granite structure is **batholiths**, but it may occur in other forms such as **dykes** (vertical) and **sills** (horizontal).

According to Linton's theory of tor formation, granite batholiths formed about 6000 metres below the present-day surface of Dartmoor by the cooling of magma about 450 million years ago. Percolating groundwater caused chemical weathering by hydration and hydrolysis (see page 29), especially in zones which are jointed. Subsequent removal of the overlying sedimentary strata has exposed the unweathered remnants of the batholiths as **tors**, e.g. Hound Tor. The horizontal divisions between the blocks are pseudo-bedding planes; the result of pressure release as the overburden is reduced by denudation.

Sedimentary rocks (laid down in layers mostly under water)

Carboniferous limestone, such as at Malham in North Yorkshire, is composed of calcium carbonate. It displays horizontal bedding and vertical jointing and possesses secondary permeability. The best known feature of **karst** landscapes is **limestone pavement**, comprising **clints** (blocks) and **grykes** (vertical joints). The chemical weathering process of carbonation enlarges the grykes. Other karstic landforms include **sinkholes**, **caves** and **resurgent streams**.

Metamorphic rocks (sedimentary and igneous rocks changed by heat, pressure, fluids and strain)

Metamorphic rocks are formed from either igneous or sedimentary rocks by renewed heat and/or pressure. Examples include marble from limestone and slate from shale. All metamorphic rocks are impervious (they will not allow the infiltration of water).

Weathering: processes of disintegration

AQA	A2 U3
OCR	AS U1
WJEC	AS UG1/A2 UG3

Erosion is the mechanical or chemical breakdown of rocks and their transportation away from the site of breakdown.

Rock weathering is the disintegration and decomposition of rocks *in situ* (in one place), by natural agents at or near the Earth's surface. It is different from erosion, which requires moving agents.

Denudation is the general term given to the wearing away of the Earth's surface by weathering and erosion. There are three main types of weathering.

Physical weathering

Frost shattering (freeze-thaw) relies on water in cracks subjected to many freeze-thaw cycles (fluctuations in temperature around 0°C). Water expands by 9.6% of its volume as it freezes, exerting a stress on the rock of up to 2100 kg/cm^2 at 22°C. Though frost shattering has been observed happening at 14 kg/cm^2, strong rocks, if jointed, will split at 100 kg/cm^2. Frost action is at a maximum in rocks that do not drain freely. Chalk, with its pore spaces and moisture-retaining properties, disintegrates readily in freezing conditions. Mount Kenya in Africa has been dramatically affected by frost shattering!

Common in mountain area under glaciation.

Salt crystallisation is caused by the crystallisation of supersaturated solutions of salts (e.g. sodium chloride) occupying fissures and pore spaces within rocks. As the crystals grow, pressure causes surface scaling or granular disintegration. It also produces cavernous weathering (honeycombing of the surface) and contributes to the production of weathering pits and **tafoni**.

Common along coastlines.

Insolation weathering (exfoliation) occurs in desert environments with a large diurnal range in temperature (the difference between maximum daytime and minimum night-time temperatures). Because rock is a poor conductor of heat, the outer layers are subject to alternate expansion (due to intense daytime heating) and contraction (due to rapid cooling at night), while the inner layers remain cold. The outer layers eventually peel off. Water, in tiny amounts, plays an important part in the process. The effect of **exfoliation** is insignificant (0.5 mm/10000 yr).

Common in desert areas.

Pressure release (**dilatation**) occurs in many rocks, especially intrusive jointed granites which have developed at considerable pressure and depth. The confining pressure increases the strength of the rocks. If these rocks are exposed to the atmosphere at a later date (due to removal of the overburden by, e.g., glacial erosion) then there will be a substantial release of pressure (at right angles) which weakens the rock, allowing other agents to enter it and other processes to occur. Where cracks develop parallel to the surface, the process of sheeting causes the outer layers of the rock to peel away (along what are called pseudo-bedding planes). Pressure release is thought to be the dominant process in the formation of **inselbergs** and certainly perpetuates them. (Inselbergs are bare, rounded, steep-sided and dome-shaped features.)

Common in areas that are deglaciating.

Inselbergs are found in deserts.

Chemical weathering

This contributes to the disintegration of rocks by:

- weakening the coherence between minerals
- attacking the cements, e.g. in sandstone
- forming solutions, which are washed out by rain, making the rock porous and so ready to crumble by granular disintegration
- causing the formation of alteration products.

Hydration occurs in all rocks. Certain crystals grow in size due to the addition of water creating stresses in the rock, which may eventually crumble and disintegrate (granular disintegration).

> Chemical weathering is common in tropical areas.

Hydrolysis occurs in rocks containing the mineral feldspar (e.g. granite). Hydrolysis is a chemical reaction between water and the hydrogen ions in the rock. The feldspar in granite decomposes to form kaolin which is washed away. The laterites of the tropics, when rich in aluminium, give up bauxite.

Oxidation occurs in rocks containing iron compounds. Oxygen dissolved in water reacts with the iron to form oxides and hydroxides. This manifests itself in a brownish or yellowish staining of the rock surface which ultimately disintegrates.

Carbonation occurs in rocks containing calcium carbonate. Carbon dioxide in the atmosphere is absorbed by rain, creating a weak carbonic acid. This chemically converts the insoluble calcium carbonate (found in limestone and chalk) into soluble calcium bicarbonate, which is dissolved and carried away in solution.

Biological weathering

Plant roots widen cracks in rock faces; rotting plants create humic acid, which chemically decomposes the rock; animals produce ureic acid, which chemically decomposes the rock. In the UK there are 150 000 worms/acre, they move 10 to 15 tonnes of soil and other weathered rock fragments per year.

PROGRESS CHECK

1. What are the differences between igneous and metamorphic rocks?
2. What is the difference between weathering and erosion?
3. How does salt crystallisation operate?

Answers: 1. Igneous rocks are formed from molten rock and metamorphic rocks are formed under intense heat and pressure. 2. Erosion involves some transportation. 3. Salt crystals grow and split open the rock.

Slopes and slope processes

OCR **A2 U3**
Edexcel **AS U2**

Weathering processes produce debris, which is transported downslope by processes of mass movement. Mass movement processes may be sub-divided into rapid and slow, wet and dry. The type of process involved and its velocity depend on many factors including the angle of slope, the amount of water present in the waste material, the degree and type of vegetation, and human activity.

Types of movement

- **Soil creep** – this is very slow, does not require a lot of water and is the result of surface material being 'heaved' down slope.

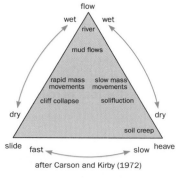

Slope processes

after Carson and Kirby (1972)

- **Solifluction** – occurs in periglacial climates. The ground tends to be wet (in the summer) and movements are relatively fast.
- **Cliff collapse** – this rapid slope movement occurs on steep slopes.
- **Mudflows** – when very wet this occurs very rapidly.

Slope movements on soft rocks occur rapidly and differ from the movements that occur on hard rock. The most common mass movements are those that occur through gravity on soft rocks. Most of these movements are rapid and occur over a clearly defined boundary, called a slip surface or shear plane.

Types of rapid movement include the following.

- **Translational slides** – the slip surface is parallel to the slope and they are not very deep. Heavy rainfall over a period of time can start them off.
- **Rotational slides** – large blocks of material rotate over a curved slip surface. The ground surface is left tilting back towards the main cliff. Often a number of blocks rotate down the hill giving a step-like slope profile.
- **Mudslides** – these usually occur in debris that has already slipped downslope. High water content means that they can flow at fast speeds for long periods.

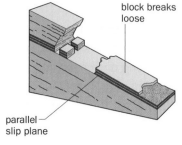

Translational slides

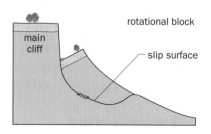

Rotational slides

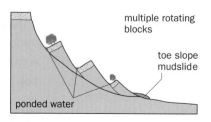

Multiple rotational slides

The loose debris at the foot of a slope is termed **talus**. The form (profile/angle) of a slope depends on the balance between subaerial supply of talus by weathering and mass movement processes and its removal at the cliff foot by, e.g., a river or the sea. Where subaerial supply exceeds basal removal, a low-angled, degraded profile develops. Where basal removal exceeds subaerial supply, a more vertical profile is maintained.

Case study: When slopes collapse

Hillside collapse in Oaxaca, Mexico, 2010

Hillside slopes around the town of Oaxaca collapsed after two major storms drenched the area in 2010. The huge rotational slips not only killed seven inhabitants, but also destroyed much of the town's infrastructure. It was estimated that 500 to 1000 people were buried as they slept in their homes.

Mudslides in China, 2010

Mudslides triggered by torrential rains hit a township in Gongshan County, located in a mountainous corner of Yunnan province, near Tibet. The Nu River swollen by days of rainfall flooded to a depth of 6 m. This water rushing across the area's slopes brought a torrent of mud burying a large part of the poor and remote town and killing more than 1200 people.

Slope movements may have to be controlled. They can be:

- **avoided**, by controlling the location, timing and nature of development
- **reduced**, by decreasing the angle
- **improved**, by draining and using retaining structures
- **protected**, through covering and compaction.

1.3 Tsunamis

LEARNING SUMMARY	After studying this section, you should be able to understand:
	• the causes, effects and ways of predicting tsunamis

Tsunamis

AQA **A2 U3**
Edexcel **AS U1/A2 U4**
WJEC **AS UG1**
CCEA **A2 U2**

All oceans of the world experience, or have experienced, tsunamis. The Pacific and Indian oceans are particularly prone.

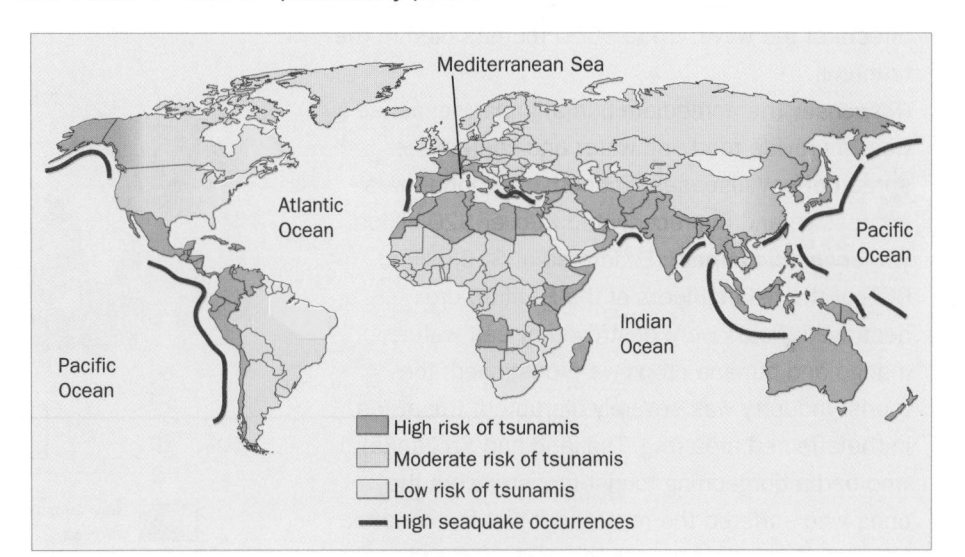

Worldwide risk of tsunamis

Tsunamis generally form after earthquake events or undersea landslides. The column of water displaced by these events contains all of its energy. The slope of most tsunamis at sea is quite low, so wave heights at sea are low. The length of a tsunami wave can be up to 1000 km and can move at up to 800 km/hr. Effectively, the energy from the tsunami is spread over the whole ocean area as it forms and travels forward. As it reaches shallow water its speed drops (to about 70 km/hr) and its amplitude (i.e. wave height) to heights of up to 30 m. The effect of a tsunami can last for several hours.

> Why are locally generated tsunamis so dangerous? Because there is no time for the population to escape!

Tsunamis can be detected once formed. The Pacific Tsunami Warning System is a group of buoys and transmitters that alert scientists to rapid and unusual changes in sea level. Because of the energy dissipation at sea the bouys and transmitters are notoriously unreliable, e.g. the 8.8 Chile earthquake of February 2010 sparked fear in the southern Pacific, but it eventually only produced a 1 m tsunami. It is thought that lateral movement during this event dampened the tsunami.

> Tsunamis usually demonstrate a phased approach, which involves a drawing down of the sea and then the arrival of the tsunami itself.

Case study: The Indian Ocean, Boxing Day tsunami, 2004

The Boxing Day tsunami in 2004 was caused by a 9.1–9.3 mega-thrust earthquake along the Sumatra-Andaman fault off the coast of Sumatra, Indonesia. The consequence was a displacement of sea 400 km by 100 km reaching a height of some 20 m, releasing energy equivalent to 1500 times that of the Hiroshima bomb. The 30 km^3 of water that was displaced by the earthquake formed the tsunami. The event itself is thought to have started three days earlier in an isolated part of New Zealand where an earthquake of 8.8 on the Richter scale was recorded. What is known for sure is that the eruptions of the volcanoes Leuser Mountain and Mount Talang, in Indonesia, are linked to the quake event.

Effects: Once generated the wave travelled at 500 km/hr to 1000 km/hr and struck the coastline near Banda Aceh just 15 minutes after it had formed, with a wave of over 17 m in height. Many countries around the Indian Ocean basin were affected, in particular those countries with a north to south aligned coastline, similar to that of the Sumatra-Andaman fault. Up to 275 000 died in the tsunami, one-third being children and four times as many women as men. It is generally thought the children and women were least able to resist the affects of the wave. Up to 9000 tourists died in the tsunami.

Response: The immediate humanitarian response was to provide food and water and contain the spread of any diseases. Economic response was impressive and unprecedented at over $20 billion.

The economic impact: Estimated at $10 billion; 66% of the fishing fleets of the Pacific were destroyed; fields were destroyed by salt water; mango and banana crops were destroyed; the tourist industry was severely disrupted. Countries in the affected area (e.g. Thailand and Sri Lanka) who had a burgeoning tourist industry were the ones who suffered the most – not just in economic terms, but also because these countries had cleared coastal mangrove swamps to make way for hotels.

Environmental impacts: Coastal ecosystems – mangroves, coral reefs, forests and wetlands were destroyed; massive water pollution; freshwater was contaminated.

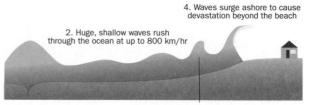

4. Waves surge ashore to cause devastation beyond the beach

2. Huge, shallow waves rush through the ocean at up to 800 km/hr

1. Undersea earthquake displaces large amounts if water in a sudden jolt

3. When they reach a gently sloping coastline, the waves slow and compress upwards

How tsunamis form

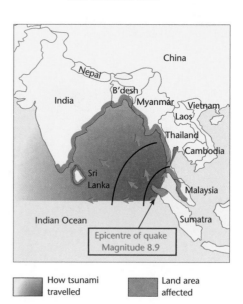

How tsunami travelled

Land area affected

Could it happen again? It will happen again, but the UN has now completed phase one of the Indian Ocean Warning Scheme, a tsunami warning system set up to provide warning to inhabitants of nations bordering the Indian Ocean of approaching tsunamis.

Case Study: The Papua New Guinea tsunami, 1998

The Papua New Guinea tsunami in 1998 was a 7.1 magnitude earthquake whose epicentre was estimated to be 30 km off the coastline of Aitape, on Papua New Guinea's west coast. This earthquake released a landslide that caused the tsunami. After the very first shock of the earthquake, there was a draw down which exposed the sea floor. Then the first tsunami wave arrived. A blast of air strong enough to knock people over preceded the wave. The second wave was larger and produced more damage. The waves uprooted trees, destroyed mangroves and piled debris as far as 500 metres inland from the coast. Wave heights have been estimated at between 10–15 metres. It is estimated that 2200 were killed and 9500 made homeless. Almost 500 people have never been found.

Why were there so many deaths in the Papua New Guinea tsunami?
- This was a surprise event – it had not been predicted.
- The population is concentrated on the coastline.

- At the time of the event the coastal roads and towns were busy with people.
- The siting of villages on vulnerable sand spits on the coast.
- The failure of residents to self-evacuate after feeling the earthquake.
- Substantial delays in mobilising a response to the tsunami.
- A lack of finance in Papua New Guinea means there is little work done on educating or protecting the nation.

1.4 Assessing hazard hotspots

LEARNING SUMMARY	After studying this section, you should be able to understand: • that specific points around the world are prone to hazard events • that not all hazards are confined to plate boundaries

Hazard hotspots

AQA	A2 U3
Edexcel	AS U1/A2 U4
CCEA	A2 U2

If you were to examine the relationship between natural hazards there is a clear link between national income levels and deaths. This seems to relate to the size of the hazard, the level of adjustment to the hazard, population density and the perception of the hazard threat. In wealthier countries the impact is more economic. But in poorer countries it is more to do with how disasters affect people.

The need to be well-prepared has received a necessary high profile in MEDCs, less so in LEDCs, but evidence suggests, e.g. in the Phillippines, that the situation is improving for the better.

Any location around the world that is at risk from more than one hazard is classified as a hotspot. There are a number of locations, countries and continents that might fall into this classification. Two are examined below.

Causes of multiple hazards

West Coast USA: California (and Washington and Oregon)
The most hazardous area of all the MEDCs is at the boundary of the Pacific and North American Plates, a mix of conservative and destructive boundaries. There are lots of big cities, such as Los Angeles and San Diego and also the states of Washington and Oregon.
- The Cascade Range contains many active volcanic peaks. The biggest are Mount Rainier and Mount St. Helens, which in the 1980 eruption killed 61 people and the resulting 25 km blast zone caused huge environmental and economic damage. Earthquakes along the San Andreas fault are generally of the medium scale and the huge conurbations of Los Angeles and San Francisco are also at earthquake risk.

The last earthquake in 1994 killed 60 people and caused $50 million of damage; a bigger quake is overdue.

- There were approximately 300 tsunami events recorded in the twentieth century on the west coast of the USA causing approximately $500 million damage to property. In 1964, 119 people died on the West Coast following an Alaskan earthquake which generated a tsunami that swept south over the Pacific. Much of the West Coast is at risk from tsunamis. The next tsunami to affect Los Angeles is expected to cost the state an estimated $42 billion.
- Landslides caused by earthquakes are induced on the loose sands and gravels. In 2003 a campsite landslide killed 16 people. Additional hazards include coastal flooding, flash flooding from intense winter rainfall, drought because of the Santa Anna (a dry offshore wind) and El Niño (a warm wind created by shifts in ocean currents– see page 39 for a full explanation) winds and wild fires (there were 915 deaths in 2003).

Issues

Many people have no insurance. Publicity and media work ensures that the perception of hazard issues is high; lots of up-to-date technology helps. Hazard zoning and mapping is important and reinforced buildings are common.

The Philippines

The Philippines is made up of 7100 islands in south-east Asia. The Philippines is an LEDC with rapidly growing population in established settlements and is increasingly industrialised. The previous two factors mean that the environment is degraded. Social and economic vulnerability only serve to increase hazard potential. The Philippines has 200 volcanoes, of which 17 are active.

- Located on two subduction zones, the Philippines fault runs through the islands. There are lots of earthquakes often leading to **liquefaction**, **landslides** and **volcanic activity**.
- *Worst earthquake:* 1990 Rizal City, Luzon; 1666 people died, 3500 people injured; over 2 million people were affected; £300 million of damage.
- *Worst volcanic event:* 1991 Pinatubo eruption; 800 people died, 100 000 made homeless; 15 million tons of sulphur dioxide discharged; 60 000 people were evacuated; 18 000 from USAF Clark were repatriated back to the US; 74 000 properties were damaged, mostly by ash dropped from the 34 km high ash cloud – which caused a 0.5°C drop in world temperatures for one year; conservative cost of the eruption was put at $0.5 billion.
- *Tsunami risk:* the islands of the Philippines are at risk from Pacific basin tsunamis. In 1994 Mindano, a 6 m wave, damaged a 40 km stretch of the coast; 78 people died, infrastructure was damaged and settlements and fish farms destroyed.
- *Landslides:* slope stability is low in the steep relief of the West Luzon Range; earthquakes cause increases in land-slipping as does deforestation; the Caraballo Mountains are especially prone to land-slipping.
- Additional hazards include *river and coastal flooding*, tropical storms, wild fires (1998 was worst year on record for this) and drought.

Issues

Hazard mapping is not common as there is a low risk perception (events are considered 'an act of God'), and there is a lack of technical help. There are few management groups, though PHIVOLCS has been effective and some land use planning and community training is in its infancy. There has been lots of donor help, from USA, UK and UN and many others, including NGOs.

Sample questions and model answers

(a) Study the photo, below, of downtown Los Angeles, California. There has been recent earthquake activity in the State of California.

(i) Describe the impacts of such events in this area. **[4]**

(ii) What impact do earthquakes have in LEDCs? **[4]**

(iii) 'If you understand how plate tectonics operates, then you can manage earthquakes appropriately.' Explain this statement for an area like that shown on the map. **[7]**

Talks about relative impact.

(i) At the epicentre of earthquakes the force is strongest. This would be where most buildings collapsed and perhaps where most lives are lost. However, in the built-up area of Los Angeles, where there is a much higher population density, and building density, more buildings and roads would be damaged. The presence of gas and electricity pipes would cause fires, and there may be more casualties here due to the overcrowding of the land. In areas further from the epicentre, such as Nevada, damage would be minimal and people would only feel a short, small shock.

(ii) When earthquakes occur in LEDCs, where there is less money available to the government. It is likely that the buildings would not be built to withstand earthquakes of this size - in shanty towns houses will fall down. Buildings, as well as infrastructure, would be destroyed, so rescue services would be less effective and the earthquake would have a worse effect, killing more people.

This student clearly understands plate tectonics and has studied pre-earthquake signs. Not much on theory.

(iii) Scientists have long studied plate tectonics, especially in areas such as California, where they have the problem of the very active San Andreas fault. By studying tectonic activity and how the earth changes before major earthquakes, it can help

Sample questions and model answers

This point really hones the focus of the question.

to predict a major earthquake when they see similar activities, for instance minor shocks become more frequent before an earthquake of high magnitude. This knowledge can be used to create hazard maps, which locate areas which are most at risk from an earthquake. In these areas, such as San Francisco and Los Angeles, they are then able to modify buildings to withstand pressure exerted by an earthquake, and also bridges and roads, such as the Golden Gate Bridge in San Francisco. In this way, scientists help to increase knowledge, so that preparedness increases and damage is kept to a minimum.

Part (iii) is well-written and purposeful – several management strategies that are clearly supported and balance the answer.

(b) Outline the nature, causes and consequences of geological/ geomorphological hazards. [15]

Weaker candidates might have begun with a list of hazards.

Geological hazards can be very varied, but the two most recognised hazards are earthquakes and volcanoes. Both of these hazards have been responsible for major disasters in recent years.

Demonstrates breadth of knowledge related to some good case study work.

An earthquake is caused by the movement of two plates together and occurs when energy builds up along a plate boundary and then the plates slip, thus resulting in a sudden release of energy which dissipates as shock waves. It is this release of energy, which causes the damage attributed to earthquakes. For humans, the hazards caused by an earthquake can vary greatly with the strength of an earthquake and the proximity of the epicentre and depth of the focus of the earthquake. In a strong earthquake, or where building quality is poor, a major hazard will collapse buildings, because of the shaking. In Agadir in 1960, a weak earthquake, 5.8 on the Richter scale, caused massive destruction and the death of 12 000 people. This was because of a high population density and poor building construction, which was not designed, or capable, of withstanding an earthquake. Many buildings collapsed trapping and killing those inside. Another earthquake hazard is the subsidence of buildings and liquefaction, both of which can cause the collapse of buildings. In an earthquake, hazards, such as collapsed electricity lines and explosions from ruptured gas pipes, also occur.

Clear understanding of the nature of the processes involved, some weaknesses on causes.

For a volcano the hazards depend greatly on the type of its magma. Lava from the volcano is usually a minor hazard because it is slow moving and lava flows can often be predicted. Lava flows are extremely destructive to houses and to vegetation. The greater hazard for volcanoes tends to come from pyroclastic flows which involve clouds of hot rocks and ash racing at high speeds down a volcano and destroying and killing everything in the way. Pyroclastic flows, or nuee ardentee, were responsible for the destruction of the city of St. Pierre in Martinique in 1902. Falling rocks thrown from the volcano and ash could also present a hazard - clouds of ash can often destroy crops and cause famine. Volcanoes erupt because of a build upon pressure within their core as magma and gases rise up from the mantle. As the pressure grows minor eruptions may occur until the pressure reaches a critical point when a major eruption may occur.

Concentrating on earthquakes and volcanoes can enable a top mark to be gained.

Both earthquakes and volcanoes can cause secondary hazards, such as landslides and avalanches. In California, because of tectonic activity, 95% of slopes are unstable. Volcanic eruptions can force water out of the crater, which will cause mudslides like those which wiped out villages around Nevada del Ruiz in Colombia in 1985. Gases from volcanoes can be toxic and will kill people and wildlife if they escape.

Geological hazards pose a hazard to life and also property. They can cause great economic damage and social damage. All of these can be caused by the range of geological hazards.

2 The challenges of the atmosphere

The following topics are covered in this chapter:

- Atmospheric processes
- Weather systems
- Climatic types
- Threats posed to, and by, the world's atmosphere
- Aspects of local climate

2.1 Atmospheric processes

LEARNING SUMMARY

After studying this section you should understand:

- that the atmosphere is an open system powered by and circulating energy
- that energy varies temporarily and spatially
- that the movement of air redistributes energy
- that air motion is a mix of forces
- that the circulation of air can be modelled
- that moisture in the air varies in time and space
- the basic mechanisms of condensation, lapse rates and forms of condensation

Systems and energy exchange

AQA	A2 U3
CCEA	AS U1

The atmosphere is an open system, with flows or movements of energy and materials between the different parts of the system. The sun is the driving force. The key to understanding the atmosphere is to understand why changes constantly occur in the atmospheric system.

Energy in the atmosphere

Solar radiation, or **insolation**, occurs as short-wave radiation and is the main source of external (heat) energy input into the Earth's atmosphere system, while the Earth's motions and gravitational pull (on air masses and moisture) provides a constant internal source of energy.

KEY POINT

The Earth's variable energy profile	
Global variations	**Temporal variations**
• Latitude determines the intensity of energy receipt. • Insolation decreases from the equator to poles.	• Seasonal shifts in radiation produce large latitudinal differences in hemispherical heating between January and July (seasonal anomaly maps display this information).
• Maximum solar radiation values are found in cloud-free areas such as the tropics.	• The distribution of heating can be influenced by the distributions of land and sea.
• Surplus radiation of low latitudes; heat transferred from low to high latitudes; deficit of radiation in high latitudes.	• Daily temperature changes can be related to diurnal changes/exchanges of radiation.
• Energy in the form of heat is transferred from the ground to the air.	• Reflective cloud, cooling wind and reflective snow modify daily temperatures.
• The unequal inputs and outputs of radiation (differences in heating and cooling) are the causes of all of our weather and climate variation.	• Temperature inversions may result when there is excessive loss of heat from the ground at night.

The atmosphere is reasonably 'transparent' to solar radiation, in that large amounts of energy are allowed to pass through to the ground surface. This stream of energy powers the atmosphere and the biosphere.

About 50% of the insolation received at the edge of the atmosphere is actually 'lost'. Most is scattered (by dust), reflected (by clouds) or absorbed (by clouds, dust and water vapour). Long-wave radiation emitted by terrestrial and atmospheric radiation is largely absorbed; this contrasts with solar radiation. The balance of the energy receipts is used to heat the ground and air by **conduction**, **convection**, **turbulence** and **evaporation**.

The atmosphere on the move

AQA	**A2 U3**
WJEC	**AS UG1**
CCEA	**AS U1**

The study of the forces that control movement is fundamental to our understanding of how energy (in the form of heat) is distributed around the globe by the global circulation.

Horizontal air movements are called **advection** and vertical movements are called **convection**. Air motion is initiated by a pressure gradient (air density variations) between places, with movement occurring between high and low pressure locations.

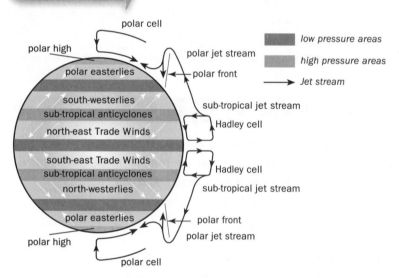

Convection cell model

The existence of large-scale convection models was recognised by George Hadley (in 1735) and William Ferrel (in 1889). Their postulations contributed to the equator-to-pole Convection Cell model (see diagram opposite).

Air circulates in a convectional fashion, but there are consequences for the atmosphere related to this movement. They involve the rate at which the air can rise or fall (the equation of state), the ability of air to expand or contract (the thermodynamic equation) and that air will continually circulate (the equation of continuity). Because we view movement from a moving platform, an apparent deflection of moving objects due to the Earth's rotation occurs. This is the so-called **Coriolis Effect**. In mid-latitudes the pressure gradient and Coriolis Effect are in balance. This leads to a **geostrophic wind** blowing not from high to low pressure areas, but between the two, parallel to the isobars.

Global air movements

Global air movements occur at a variety of scales. There is a close relationship between major winds and the world's pressure systems. Windbelts also contain the world's major weather systems, including hurricanes in the tropics and anticyclones and depressions in the mid latitudes (and at a smaller scale – tornadoes).

Mid-latitude movements

Found between the polar and tropical circulations is an area of complex upper air movements. The combination of the powerful Ferrel westerlies (the so-called circumpolar vortex) and the jet streams in the mid-latitudes, affect surface winds and pressure systems. These surface winds and weather belts are bound up in a series of waves known as **Rossby waves**. The link between the Rossby waves, jet streams and weather systems in the low and high pressure circulation is called the **index cycle**.

Influence of the oceans on atmospheric movement

Rossby waves form because of the Earth's thermal differences, the rotation of the Earth, and because of the destabilising effect of mountain ranges.

- The oceans provide water for the hydrological cycle.
- The oceans absorb and redistribute energy, e.g. the Gulf Stream and North Atlantic Drift benefit Western Europe by releasing heat and provides a milder climate than might otherwise be the case.
- Warm oceans and their currents can therefore affect local and distant locations.

Case Study: El Niño Southern Oscillation (ENSO)

El Niño Southern Oscillation (ENSO) occurrences are global climatic events linked to climatic anomalies. Drought in particular has a strong teleconnection with ENSO. Every two to seven years ocean currents and winds shift off the coast of western South America bringing warm water westward. In recent decades, scientists recognised that El Niño, as it is called, is linked with shifts in global weather patterns. Its effects last between 14 and 22 months. The southern oscillation, 'a seesaw of atmospheric pressure between the East Pacific and Indo-Australia', is closely linked to El Niño. Generally, El Niño refers to the oceanic properties and southern oscillation refers to the atmospheric component. See diagrams below.

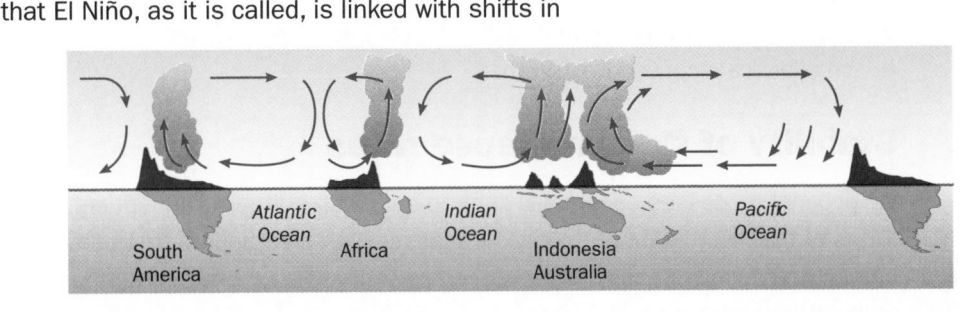

Normal circulation with southern oscillation

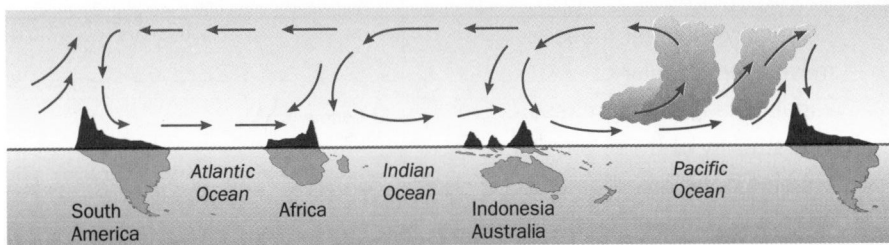

ENSO event: no southern oscillation

The first signs of an El Niño are:

- Rise in surface pressure over the Indian Ocean, Indonesia and Australia.
- Fall in air pressure over Tahiti and the rest of the central and eastern Pacific Ocean.
- Trade winds in the south Pacific weaken or head east.
- Warm air rises near Peru, causing rain in the northern Peruvian deserts.
- Warm water spreads from the west Pacific and the Indian Ocean to the east Pacific. It takes the rain with it, causing extensive drought in the western Pacific and rainfall in the normally dry eastern Pacific.

La Niña, is the opposite of El Niño. It refers to a period of cold surface water in the Pacific. The two temperature cycles oscillate back and forth. La Niña's effect is as devastating as El Niño. Both El Niño and La Niña are thought to have oscillated back and forth through the period 2007 to mid–2010 bringing snow to Canada in 2007, floods to Malaysia in 2008, drought to Singapore in 2010 and no snow for the Vancouver Olympics in 2010.

Moisture in the atmosphere

AQA	A2 U3
CCEA	AS U1

Remember warm air holds more moisture than cold air.

You need to be aware of these two methods of precipitation formation: Bergeron Findelson and Coalescence theory.

The **Bergeron–Findeison** process in the formation of precipitation by ice crystal growth.

Water droplets are lifted rapidly by turbulent air currents. Rapid freezing occurs and water vapour turns instantly to ice through sublimation. As ice falls from the clouds it melts and turns to rain. (Cloud seeding uses the same method! Super-cooled clouds are seeded with dry ice and silver iodide, etc.)

Coalescence theory

Water droplets grow in turbulent air. Large clouds form and droplets grow and eventually fall. Collision in the clouds cause the drops to grow.

Types of precipitation include:

· rain   · sleet
· drizzle   · fog
· hail   · fog drip
· snow   · dew.

To understand weather you must understand this.

Remember: Adiabatic changes occur in a vertical direction and are due to internal pressure change, not heat exchange. Temperature varies with height, i.e.:
ELR = 6.5°C/km climbed
DALR = 10°C/km climbed
SALR = 6°C/km climbed
(These 'rates' vary at the same rate up or down!)

The amount of moisture in the atmosphere is relatively small, but highly varied. When air is holding the maximum amount of water vapour possible, it is said to be saturated. The water vapour content of the air can be expressed in terms of its absolute humidity or, more commonly, in terms of its relative humidity i.e. percentage ratio between the actual amount of water and the maximum amount that the air can hold at that temperature.

Evaporation and condensation are two important phase changes in the hydrological cycle and are accompanied by the absorption and liberation respectively of latent heat.

The main forms of condensation are clouds, mostly at high levels, and fogs, at or near the ground. These forms of condensation occur when air is brought to saturation point or its dew-point temperature. Condensation is assisted in the atmosphere by condensation nuclei around which water can form. Condensation does not always lead to precipitation; gravity has to overcome the ability of air currents to keep the water vapour buoyant before this can occur.

Most saturation and condensation in the atmosphere takes place as a result of air cooling, principally by the vertical ascent of air. Air is forced to rise and cool by:
● orographic uplift
● frontal uplift
● large-scale convergence and ascent in low pressure systems
● smaller-scale convective currents.

Stability of the air – lapse rates

Advectional processes continually seek to establish a stable atmosphere, in terms of the distribution of energy. Air masses with their uniform temperature and humidity typify this stability. However, within air masses pockets of air must move vertically (or adiabatically), especially when air masses move over irregular or variably heated land masses. The stability of these smaller bubbles of air has dramatic effects in terms of local weather.

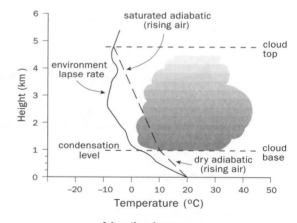

Advectional processes

In general, temperature falls as you move higher into the atmosphere. This is known as the **environmental lapse rate** (ELR). As pockets of rising air gain height they cool, expansion occurs due to decreasing air pressure, further energy (kinetic) is lost and the pocket of air cools still further. So long as it is not saturated it cools at a fixed rate, the **dry adiabatic lapse rate** (DALR). Once the rising air reaches the condensation level it becomes saturated. Latent heat is produced and water vapour turns to water. The rate of cooling is now reduced and the air now cools at a slower rate than previously. This rate is known as the **saturated adiabatic lapse rate** (SALR).

Unstable air occurs when a vertically displaced warm air parcel is encouraged to rise. At a sufficient height it cools and condenses into clouds. In this case, the ELR is greater than the DALR or SALR.

> This will happen when the air is cooler than surrounding air.

With **stable air**, a parcel of air displaced vertically upwards or downwards in the atmosphere will tend to return to its original position. Here the ELR is less than the DALR or SALR.

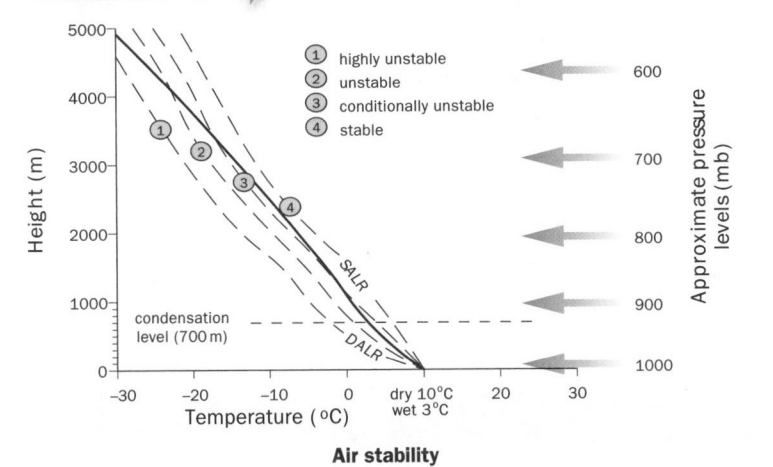

① highly unstable
② unstable
③ conditionally unstable
④ stable

Air stability

Air stability is an important concept as it determines the buoyancy of the air and thus development of cloud and fog. Cumulus clouds, often of large vertical extent, are characteristic of unstable air, whereas horizontally developed stratus clouds tend to form in more stable air.

> Unsaturated air that becomes saturated is said to be conditionally unstable.
>
> ELR<DALR>SALR

PROGRESS CHECK

1. Why does energy vary globally?
2. How is air heated?
3. What is a geostrophic wind?
4. Describe movements of air in a Hadley (convection) cell.
5. What is El Niño?
6. What is the environmental lapse rate?

Answers: 1. Latitude determines intensity, max solar radiation is found at the tropics. 2. By conduction, convection, turbulence and evaporation. 3. Air moving between high and low pressure systems parallel to isobars. 4. Rising hot air from the tropics cools as it rises, clouds form and air becomes cold/heavy heading towards the ground warming as it is compressed 5. Every two to seven years currents and winds shift off the coast of the western South America. 6. ELR is the general decrease in air temperature with height not related to movement, cooling at 6.5°C/km.

2.2 Weather systems

LEARNING SUMMARY

After studying this section, you should be able to understand:

- the mid-latitude circulation, air masses and fronts
- the causes and effects of extreme weather
- how man influences and affects the atmosphere

Britain and Western Europe's weather and climate

AQA **A2 U3**
CCEA **AS U1**

Britain has a cool temperate humid (maritime) climate.

- A maritime climate is typically found in the mid-latitudes 40° and 60° N (or S).
- Marine influences are predominant and can influence areas far inland.
- Eastward moving air bringing depressions (cyclonic conditions) with characteristics of temperature and humidity derived from the oceans.

- The weather can be extremely changeable and variable.
- Winters are mild, the ocean's effect is marked; summers are warm, 13–18°C is common in July (the oceans can cool summer temperatures too).

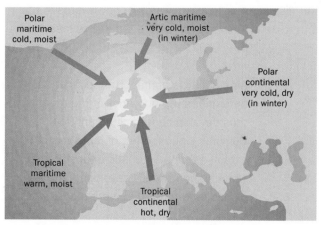

Features of the UK climate

- Drought conditions are rare, but the ingress of North African air can bring spells of prolonged hot weather.
- Rain occurs throughout the year, but varies from place to place based on the relief (it chiefly runs in on cyclonic fronts and decreases to the east).
- Depressions (cyclonic conditions, lows) are less common in summer than in winter and generally bring warm, but wet, conditions.
- Anticyclones bring hot, sunny weather in the summer and fog, snow, cold and still conditions in the winter.

Local weather processes and patterns

AQA	A2 U3
CCEA	AS U1

Low pressure	High pressure
What is it?	**What is it?**
Low pressure is a zone of rising air. Depressions are low pressure systems.	High pressure is a zone of sinking air. Anticyclones are high pressure systems.

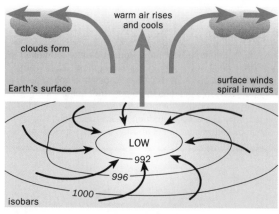

Causes of low pressure

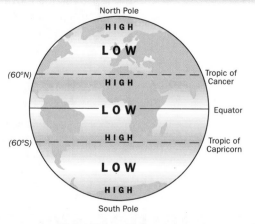

Global high and low pressures

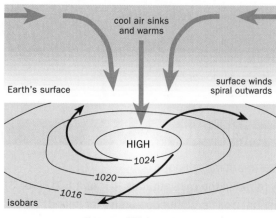

Causes of high pressure

What are the weather effects of low pressure?

Unstable rising air causes unsettled weather. Typically depressions bring:

- cloudy skies
- low levels of sunshine
- wet weather
- temperatures that are mild for the time of year
- windy conditions
- changeable weather.

What are the weather effects of high pressure?

Stable sinking air causes settled weather conditions:

- clear skies
- sunshine
- dry weather
- high day and low night temperatures
- calm weather
- dew and frost
- fog and mist
- thunderstorms
- snow in winter.

High pressure

The 'synoptic' winter and summer anticyclone

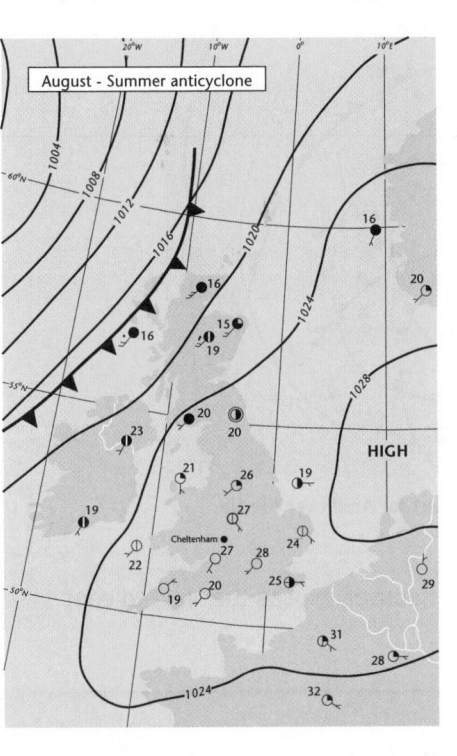

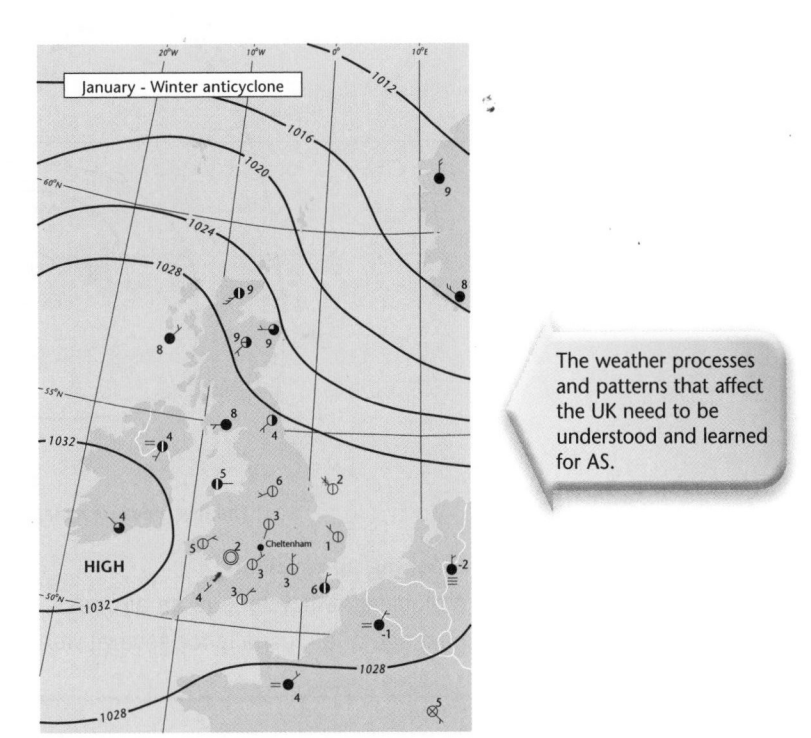

> The weather processes and patterns that affect the UK need to be understood and learned for AS.

Low pressure

How fronts form and the typical passage of weather

Above   Side

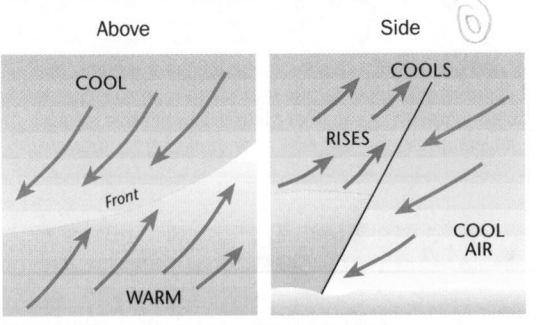

Stage 1: Warm and cold air meet - warm air rises

Stage 2: Low pressure develops

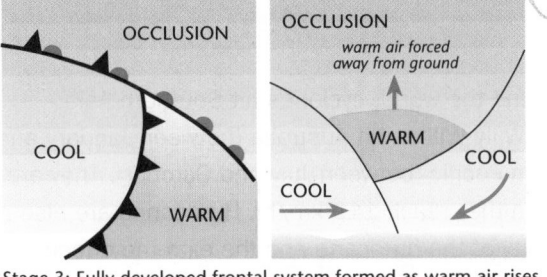

Stage 3: Fully developed frontal system formed as warm air rises

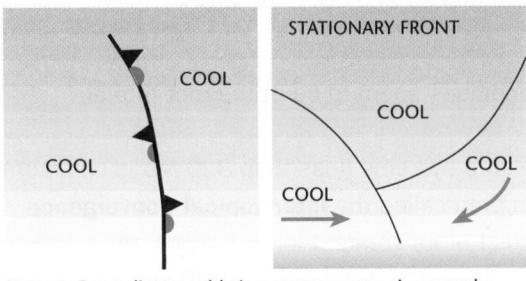

Stage 4: Front dies as cold air squeezes warm air upwards

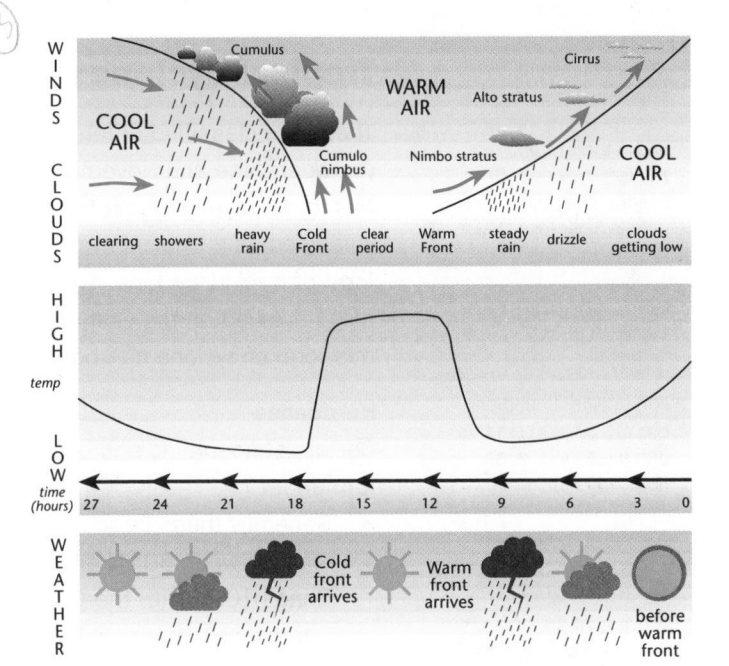

43

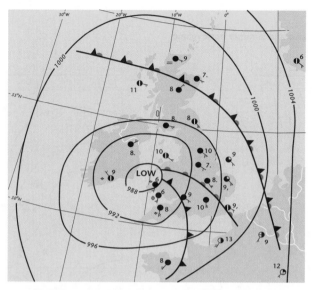

The low pressure synoptic chart for April over the UK

Blocking anticyclone

A blocking anticyclone is an air mass that effectively blocks the west to east passage of air, usually for several weeks.

PROGRESS CHECK

1. What are the features of the cool temperate humid climate?
2. What are the causes of low pressure?
3. What is the typical passage of a low pressure system like?

Answers: 1. Mid-latitude in position, eastward moving air, changeable weather, mild winters and summers that are warm 2. Unstable rising air 3. Clear and warm – drizzle and part cloud – heavy rain and thunder and lightning possible – sunshine - heavy rain and thunder and lightning possible – drizzle and part cloud – sunshine again.

Severe weather and related phenomena

AQA	**A2 U3**
OCR	**A2 U3**
Edexcel	**AS U2**
WJEC	**A2 UG3**
CCEA	**AS U1**

Extremes of precipitation and temperature cause most of our weather hazards. Extremes of air movement surprisingly create more minor effects.

Tropical cyclones

18% of the world's population live on coastal areas that are under threat from tropical cyclones. On average 20 000 people/year lose their lives to tropical cyclones.

These are classified differently around the world; cyclones hit Southern Asia (between June and November), the Willy Willies hit Australia (between January and March) and typhoons hit the western Pacific (between July and October). They are classified as a tropical cyclone when they reach 119 km/hr. Hurricanes are also a form of tropical cyclone. The intensity of the hurricane and the extreme effects they have had on the Gulf coast of the eastern seaboard of the USA ensure that this form of cyclone has been well researched and documented.

Hurricanes

Wind spins around an eye of 25 to 50 km this decreases in size as hurricanes speed up.

Certain atmospheric and oceanic conditions seem to be at the root of their formation.

The ITCZ is where convergent air meets near the equator. This area is intermittent and can vary its position.

- Generally these intense low pressure systems (less than 920 Mb is common) form between 30° N and S, in an area called the **Intertropical Convergence Zone** (ITCZ). With the Coriolis Effect at its greatest in the tropics, the systems rotation can be easily initiated (winds in the order of 119–300 km/hr are common).

- The sea in this area is well heated, and exceeds the minimum 26–27°C necessary (between August and October) for latent heat to be released to strengthen the hurricane. The ocean then provides the initiating and sustaining energy for hurricanes. They quickly 'die' once they move onto the land. The high levels of moisture held also point to a sea origin (typically in the order of 10–25 cm/day).

- Towering (15 km is common) cumulo-nimbus clouds form around the central eye in highly unstable conditions. These clouds further fuel the hurricane as latent heat energy exchanges moisture from gas to liquid. Spinning weather sub-systems develop all around the main hurricane mass.

Case study: Hurricane Katrina

Using Hurricane Katrina as the example the effects of hurricanes involve:

- storm surges – water was forced through the Lower Mississippi and New Orleans levee system up to 20 km inland forcing buildings and cars inland

- deaths – 1836 died during the Katrina event

- damage to property – over 1 million properties were damaged or destroyed (34 000 trailers replaced these properties)

- huge insurance claims – the average claim was $94 000 (most companies now no longer insure the Gulf Coast!); the cost of the whole event is put at $105 000 billion; damage was done to the oil, forestry and farming industry which could bring the final bill to about $150 billion

- refugees – 1 million from the Gulf Coast were dispersed across America; by 2006, one year on, Louisiana's population was still 5% down

- spoilt landscapes – 233 000 km^2 was left uninhabitable, equal to the size of the UK

- tourism was the mainstay of New Orleans's economy pre-Katrina, providing $5.5 billion/year to the industry and 40% of City taxes. Almost 85 000 were employed in the tourism industry.

> Between 1900 and 1991 there were 152 direct hurricane hits on the USA.

> Examiners look for a good mix of process and factual understanding. Simple diagrams like this convey information very rapidly.

> Newspapers are a good source of up-to-date information. AS questions frequently use this topical information.

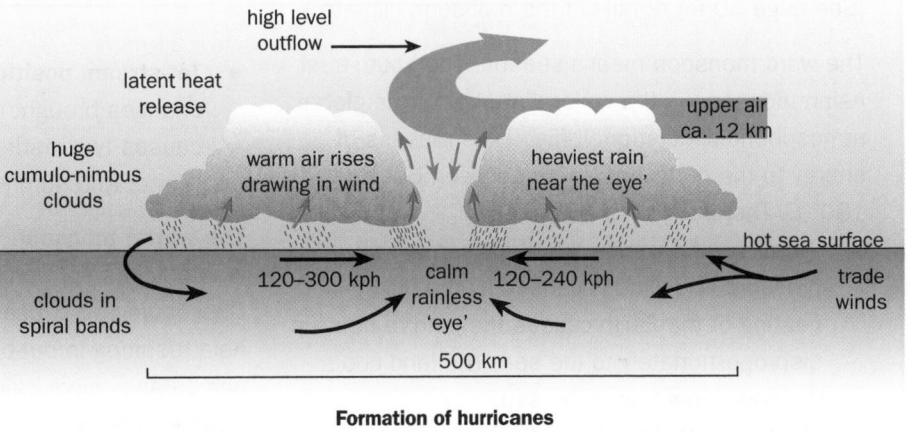

Formation of hurricanes

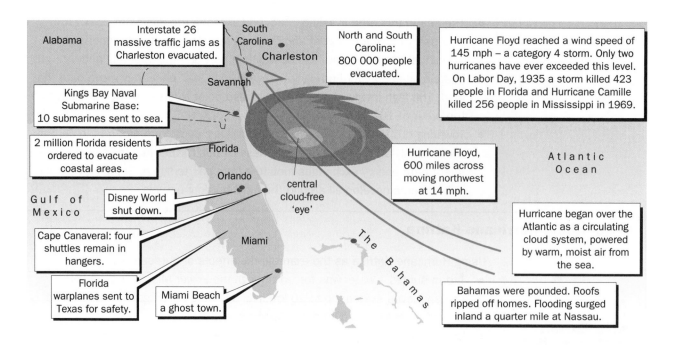

Hurricane Floyd menaces Georgia and the Carolinas, September 1999

Tornadoes

> A tornado is a thin funnel of spinning air and forms because of pressure drops, cooling and consequent saturation. Tornadoes are measured on the Fujita F1 to 5 scale.

Tornadoes are a concentration of cyclonically spinning air (about 3 km across), found overland, rather than over water. Most often a visible cloud forms from a large cumulo-nimbus origin. The tremendous rotations are initiated by rapid convergence at the base of the cumulo-nimbus cloud as rapid updraughts develop. Tornadoes are common over the mid-west and southern states of the USA during the heat of early summer afternoons, e.g. in Yazoo County, Central Mississippi, April 2010, 17 people died in winds that reached 240 km/h.

Case study: The south-east Asian monsoon
(See page 50 for details of the monsoon climate.)

The word monsoon means season. The south-east Asian monsoon is the major disturber of the global atmospheric circulation. It displaces masses of energy to the north and south between 60° and 180° E. There are several causes.

- **Pressure and wind** – the displacement of pressure and wind is caused by the direct heating of the Earth causes the land to heat up disproportionately to the sea. The land cools disproportionately to the sea.
- **ITD (inter tropical discontinuity)** – seasonal shifts of the ITD and its windy belts.
- **Mountain influence** – without the mountains, patterns of rainfall would be different in India.
- **Concentrations of carbon dioxide** – reduced levels over India because the effect of the uplift over the Himalayas causes less heat to be absorbed, which leads to cooling.

- **Jet-stream positions** – in India the seasonal changes brought about by the jet-stream causes two markedly different monsoon periods (winter and summer).

The winter monsoon
Westerly jets split around the Tibetan Plateau. Descending air causes high pressure over central Asia resulting in out-blowing north-east winds across Asia, clear skies and little rain. Sunny weather results.

The summer monsoon
Beginning in the spring and finishing in about June, the phases of the summer monsoon are below:
- In March the northern jet dominates the area.
- The sun is directly over India.
- The ITD moves to an area south of the Himalayas.

- Low pressure moves northwards – the monsoon trough.
- One part of the trough moves west over India, the other part to the east over the Bay of Bengal. They meet and join over northern India in June.
- Low pressure continues to deepen and develop over the area and the north Indian Ocean. Under the intense isolation of the area's clear skies, warm moist air starts the cycle of the monsoon rains

- Lots of warm moist air is drawn in over India from the north Indian Ocean: intense rainfall results.

The monsoon brings drought breaking rainfall and wind and invariably initiates devastating subtropical cyclones e.g. Orissa, India 1999.

Impacts

Agricultural yields increase during the 'good' monsoon years. This has to be balanced against flooding and loss of life.

Case study: Severe weather

Cyclone 05B, Orissa, India 1999
This was India's deadliest storm since the 1970s; wind's were recorded at 260 km/ph winds; 45 000 took shelter; 15 000 killed; pressure dropped to 912 mbars; the cyclone stalled over Orissa, hence the damage; 17 100 km^2 of crops were destroyed; $5 billion damage; starvation was an issue after the event.

Typhoon Ondoy, North Luzon Philippines, 2001
In 6 hours 410 mm of rainfall fell; floods in parts of Luzon exceeded 2.5 m; more rain fell than during the Hurricane Katrina event; 250 people were killed and hundreds were missing when the typhoon passed.

PROGRESS CHECK

1. Why do hurricanes increase in intensity over the sea?
2. The primary impact of a hurricane is death and destruction. What are the secondary and tertiary effects?
3. What are the causes of the monsoon?

Answers: 1. Latent heat is released. 2. Secondary impact is the cost of the rescue and the tertiary impact is the long-term economic effects. 3. Pressure and wind movements, position of the ITD and jet streams

Weather: human aspects

OCR	**A2 U3**
Edexcel	**AS U2**
CCEA	**AS U1**

Every aspect of our lives is affected by the weather. The table below summarises the effects on various areas of economic and social activity.

Effects on economic and social activity

KEY POINT

Agriculture	Livestock gets distressed in excessive heat or cold, particularly young animals. Fodder and grazing need to be available at all time for animals; this is difficult in winter. Few crops grow above 60° N and S. Most crops can be affected by variations in day-to-day weather fluctuations. A late frost affects the coffee crop. Droughts are a threat to crops. Pests tend to be weather dependent (locusts).
Fishing	Fishing is completely weather dependent. Icing of the superstructure of ships leads to disasters, as do storm conditions.

KEY POINT

Construction	Concrete laying and groundwork cannot take place in extreme cold or wet periods. Tenders are often adjusted to allow for weather risks. Using the services of weather forecasters saves £120 million/year for the construction industry in the UK.
Transport	Adverse weather causes delays and accidents. In the UK airlines can save millions of pounds per year, local highway divisions can save by only gritting where necessary, and shipping can save fuel and time, all by using forecasting to help them be more efficient.
Power	There is a strong and obvious link between weather and fuel used. Most utilities estimate, using past weather records, when greatest demand will be.
Television	Reception is affected by rapid changes in temperature and humidity. Anticyclonic conditions in particular affect reception.
Business and Retail	Most businesses would be unwise to ignore the weather, e.g. in the UK summer supermarkets stock up on BBQ fuels and foods, salad products and fruit. Cinemas lose their attraction. DIY enthusiasts buy equipment and gardeners buy plants. Ice cream sellers do a roaring trade. In winter, stocks of antifreeze and batteries for cars sell quickly. Plumbers attend more leaks and bursts. Pharmacists dispense more medicine. Wet weather equates to more business for retail outlets in tourist areas, and so on.
Leisure and sport	By using forecasting we can plan our activities for maximum enjoyment and safety. In the UK indoor parks (Centre Parcs) and covered stadia (Millennium Stadium in Wales, and Wembley Stadium Complex) are all responses, perhaps, to our unreliable weather.
Health	Housing, dress and way of life all relate to the weather and climate of our respective home area. Our comfort (addressed by air conditioning or heating), activity (i.e. cold restricts activity) and health (asthma, heat-stroke, hypothermia and hayfever) are all affected.

2.3 Climatic types

LEARNING SUMMARY

After studying this section, you should be able to understand:

- global differences in wind speed, average temperature and average rainfall
- the monsoon, equatorial, savanna, desert and western margin climate

Climate

AQA   A2 U3

'Climate' is a very general term that has a variety of closely related meanings. Usually, climate refers to the average, or typical, weather conditions observed over a long period of time for a given area.

Climates vary from place to place and countries around the world do not fit into neat categories of climate and weather. Many factors influence a location's climate, such as distance from the equator or the local topography (natural features). Coastal areas often have a different climate to those inland. Climates can be classified according to the average and typical ranges of different variables, most commonly temperature and precipitation. The most commonly used classification scheme is the one originally developed by Wladimir Köppen. The **Köppen system** is based on the idea that vegetation type is the best expression of climate. So climate boundaries are selected with vegetation type in mind.

There are five classes in the Köppen system:

- Class A: Tropical
- Class B: Dry
- Class C: Temperate
- Class D: Continental
- Class E: Polar

Each class has further sub-classes.

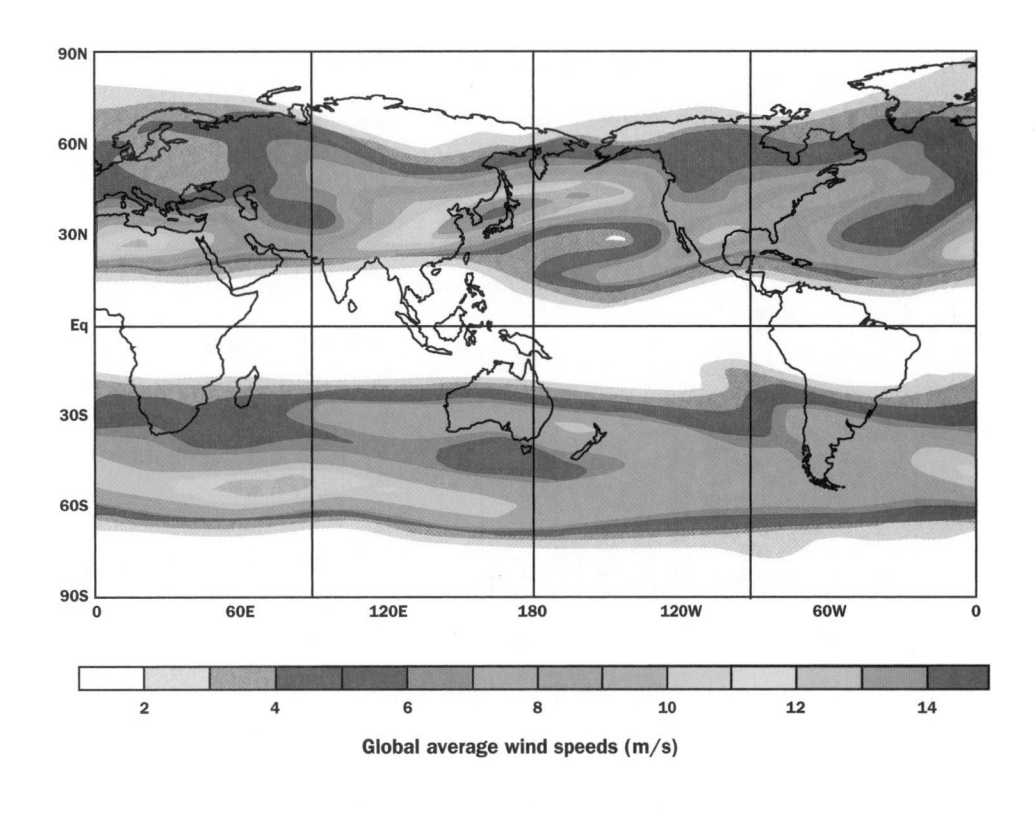

Global average wind speeds (m/s)

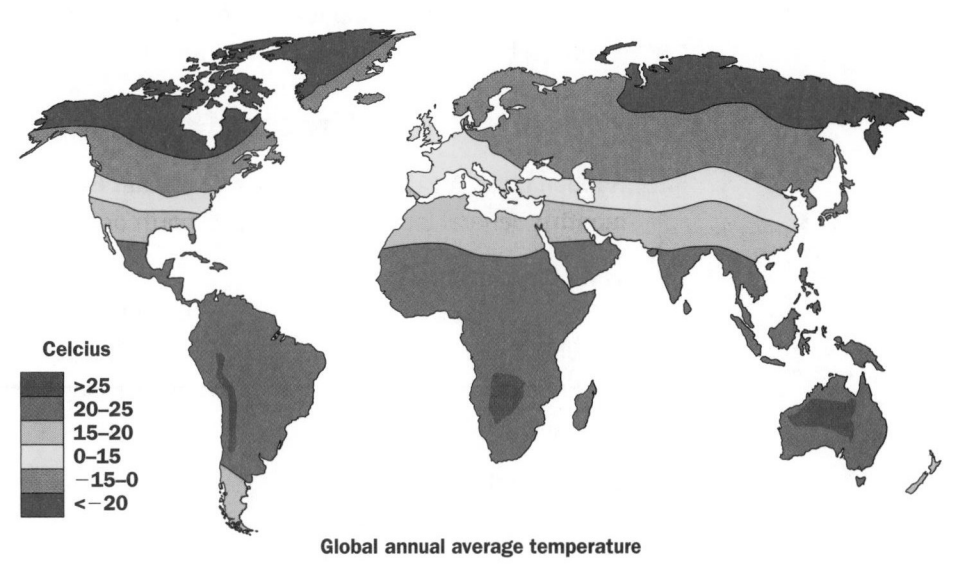

Global annual average temperature

You must know global differences in wind speed, average temperature and average rainfall.

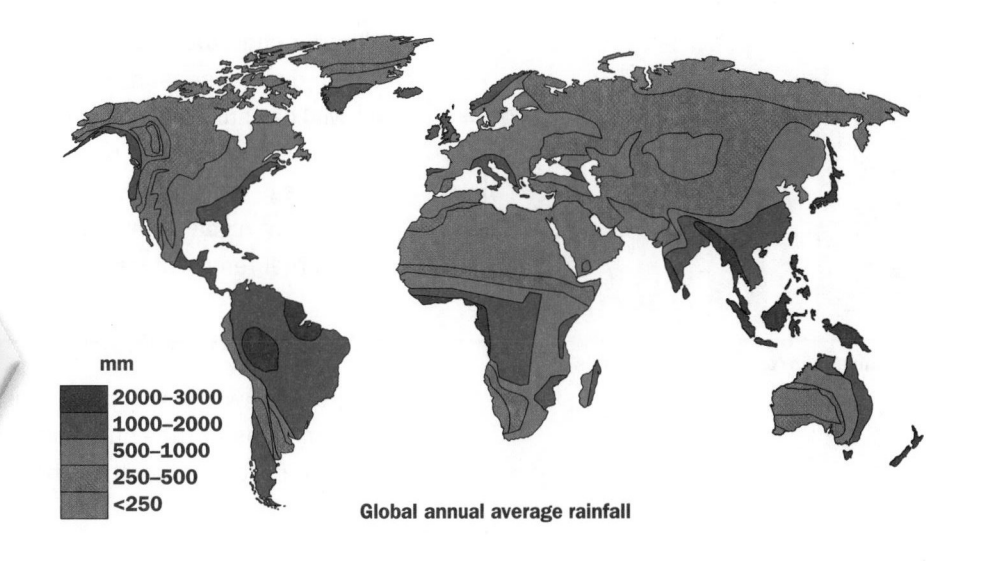

Global annual average rainfall

The monsoon climate

Köppen class A (Am – monsoon type): Found along the coastal areas of India, Sri Lanka, Bangladesh, Columbia Mexico and Burma.

Where in the world?
- Villahermosa, Mexico
- Medellín, Columbia
- Mangalore, India
- San Juan, Puerto Rico
- Yangon, Burma
- Monrovia, Liberia

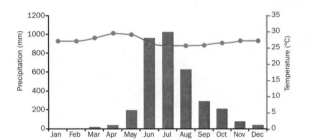

Annual rain/temperature in Mangalore, India

General features
- Controlling factor is the monsoon circulation.
- The monsoon is a seasonal change in wind direction (onshore during the summer and offshore during the winter).
- The changes in wind direction relate to the differences in the way the land and sea heats up.
- Monsoons can be found in areas around the world, not just in south-east Asia.

Climatic characteristics
- Temperatures remain high all year at about 27°C (the range is about 3.6°C).
- Lower temperatures occur during the monsoon as clouds block the sun.
- Seasonal rainfall is concentrated in the 'high-sun' season.
- The Himalayas aid the formation of clouds through the orographic effect.
- Drought is often a feature during the 'low-sun' season.

The equatorial/tropical climate

Köppen class A (Af – no dry season, at least 60 mm of rain in the dryest month): General location is up to 12° north or south of the Equator.

Where in the world?
- Apia, Samoa
- Biak, Indonesia
- Entebbe, Uganda
- Iquitos, Peru
- Malé, Maldives
- Singapore
- Penang, Malaysia

General features
- All months have precipitation.
- The ITCZ dominates this climate.
- There is no obvious summer and winter season.

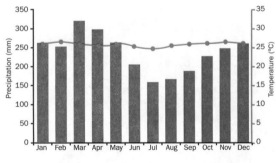

Annual rain/temperature in Iquitos, Peru

Climatic characteristics
- No distinct seasons.
- Mid-day sun is nearly vertically above.
- Heavy convectional rainfall is an everyday feature.
- Average daily temperatures are about 28°C (clouds prevent higher temperatures).
- Diurnal temperatures range between 2°C and 5°C.

The savanna climate

Köppen A (Aw – distinct dry season): Location on the outer margins of the equatorial zone at about 5°–10° and 15°–20° north and south of the Equator.

Where in the world?
- Dar es Salaam, Tanzania
- Darwin, Australia
- Dakar, Senegal
- Mombasa, Kenya
- Ibadan, Nigeria
- Manila, Philippines
- Mumbai, India

General features
- Covers a quarter of the world's land surface.
- Usually found between the tropical rain forest and the deserts of the world.
- Hot all year (c20°C) with a long dry season and marked wet season.

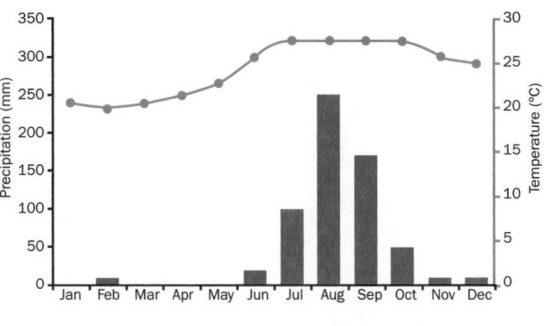

Annual rain/temperature in Dakar, Senegal

Climate characteristics
- Large annual temperature range.
- The mean monthly temperature is about 10°C$^+$ (to over 25°C).
- Maximum temperatures occur at the onset of the rains.
- Diurnal temperatures range between 10°C to about 15°C.
- Seasonality of rainfall is related to the position of the ITD (about 86 cms to 150 cms is the usual range).

The desert climate

Köppen B (BW – desert): Typically found in continental interiors in the sub-tropics.

Where in the world?
- Phoenix, Arizona
- Muscat, Oman
- Cairo, Egypt
- Karachi, Pakistan
- Tindouf, Algeria
- Mendoza, Argentina

General features
- Rainfall is sporadic and absent completely in many years.
- The sub-tropical high pressure and continentality ensure deserts stay dry.
- Cool seas, leeward locations in rainshadow are also influential.
- Desert areas are described as arid, that is, annual precipitation is less than half the annual potential evapotranspiration.

Climate characteristics
- No month has average temperatures below 18°C and may reach 50°C – all related to high sun angles.
- Skies are cloud-free, because of the high pressure.
- Massive diurnal ranges (e.g. 50°) because cloudless skies allow heat to escape at night.
- Rainfall is irregular and unreliable; less than 25 cms/year (relative humidity is less than 10%).
- The subsiding air leads to very stable atmospheric conditions.

Annual rain/temperature in Tindouf, Algeria

The western margin maritime climate

Köppen C (Cfb – at least 30 mm precipitation in the driest month): Found on west coasts of middle latitude countries.

Where in the world?
- London, UK
- Copenhagen, Denmark
- Hobart, Australia
- Paris, France
- Rize, Turkey
- Sofia, Bulgaria
- Dublin, Ireland
- Valdivia, Chile

General features

- Typically cool dry summers and warm winters and humid most of the year.
- Polar maritime air masses play a big part in this climate – bringing mild temperatures and high humidity.
- Sea currents may moderate temperatures.

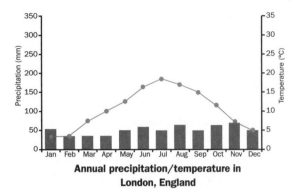

Annual precipitation/temperature in London, England

Climate characteristics

- Mild summer and winter temperatures. Small annual temperature range. In the UK 14°C is the average temperature.
- Affected by westerly winds.
- Heavy cloud cover is a feature, linked to cyclonic activity; this can bring a lot of precipitation over 2000 mm is not uncommon whereas 300 mm is the average.

PROGRESS CHECK

1. What are the effects of weather and climate on the economy?
2. What are the features of the savanna climate?

Answers: 1. In agriculture there are extreme effects, such as the effect of hot and cold weather on animals and humans, pest infestations related to differing weather conditions; in transport bad weather causes delays, more power used in cold weather to keep warm and avoid ill health. 2. Hot all year (c20°C). ITCZ affects it, found between tropical rain forest and desert areas.

2.4 Threats posed to, and by, the world's atmosphere

LEARNING SUMMARY	**After studying this section, you should be able to understand:**
	- how a variety of threats to the atmosphere threaten us at a variety of different levels

Climate change: Causes and effects

AQA	**A2 U3**
OCR	**A2 U3**
Edexcel	**AS U2**
WJEC	**A2 UG3**
CCEA	**AS U1**

Chlorine is released readily in the cold (below −80°C) sunless skies of the Poles.

Ozone

Ozone is a thin layer of gas found 10–50 km in the stratosphere. Being very unstable, ozone is easily broken down by high energy ultraviolet (UV) light. The process of breakdown causes absorption. This blocking effect is beneficial to use because high intensity UV can cause skin cancer.

In recent years ozone was being depleted by use of chlorofluorocarbons (CFCs) in industry, as a coolant in refrigerators and in aerosol propellants. Once released, CFCs rise high in the atmosphere, releasing chlorine, which in turn destroys the ozone. CFCs has since been largely phased out.

For the past 30 years during the southern hemisphere spring, chemical reactions involving chlorine and bromine caused ozone in the southern polar region to be destroyed rapidly and severely. This depleted region is known as the 'ozone hole' though the hole is now gradually becoming smaller.

The effects of decreasing amounts of ozone in the upper atmosphere can cause different types of cancer. Decreased amounts of ozone can have an effect on the ability of plants to retain nitrogen. Reduced ozone can also allow global warming to increase.

Smog

> You should be able to write about various air pollution control technologies, land use planning strategies and about efforts to reduce pollution including using primary regulation or the law, fuel efficiency and conversion to cleaner fuels.

The smogs we experience today are different to those of the 1950s, e.g. London was known as 'Old Smokey' or the 'Smoke' because of the thick smoky fogs that would often envelop the city. These only disappeared in the early 1960s when legislation and smokeless fuels ended the polluting effects of bituminous coal. Today we have photochemical smog, produced by the exhaust gases of vehicles and industry. Small amounts at ground level are lethal, affecting breathing, causing conjunctivitis-type irritations and affecting plant and animal tissue.

Mexico City has the world's worst levels of smog. As a result, children and the elderly are advised not to live in the city. The heavy smog levels have had serious health effects in the city, though many poorer people who need city jobs have to live there and suffer from the smog. Smog is not only a city problem. As smog levels increase, winds are carrying smog away from urban areas and harming people and ecosystems far away from the cities. Agriculture is also affected by smog. Soybeans, wheat, tomatoes, peanuts, lettuce, and cotton are all subject to infection when exposed to smog. Smog and its constituent chemicals costs $1 billion damage in the US each year.

The greenhouse effect and global warming

> We add 10 tonnes/person/year of CO_2 to the atmosphere. Current values are 240 parts per million (ppm). During the last ice age it was 200 ppm.

Warming near the Earth's surface results in the atmosphere trapping the Sun's heat – without the greenhouse effect the Earth would be 33°C cooler. Carbon dioxide (CO_2) is a major contributor to the greenhouse effect and with water vapour it absorbs vast amounts of heat. Since the Industrial Revolution carbon dioxide has been released into the atmosphere. Once emitted, a carbon dioxide molecule stays active in the atmosphere for about 200 years. Carbon dioxide from human activity is therefore a major cause of global warming, along with methane from termite colonies, flatulent cattle and rice paddy fields. Some computer models estimate a 1.5–4.5°C increase in temperature over the next two centuries because of global warming.

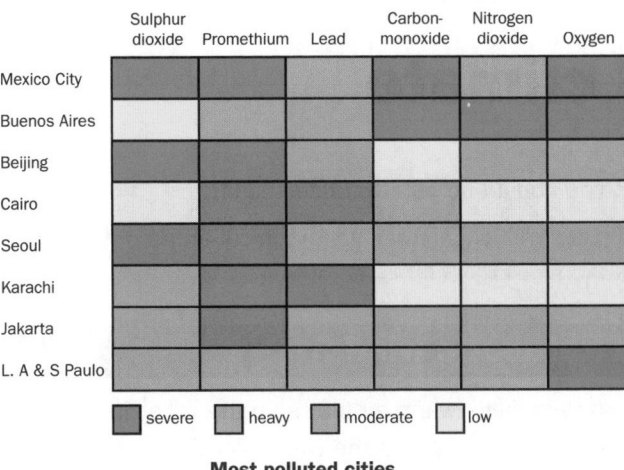

Most polluted cities

It should be remembered that the climate varies naturally and that data that supports the theory is difficult to come by and verify. Certainly some areas in the world have been hotter and drier and river flows have varied. However, many people believe that the data suggests a new ice age is upon us! Best scientific evidence supports a significant increase in the Earth's temperature over the last 100 years (this is well within normal averages/fluctuations of the Earth's atmosphere). This increase will affect sea levels, e.g. the Marshall Islands, Maldives and Kiribati islands and lower areas of the Nile and Bengal deltas could all be threatened by increases in sea level.

The developed world is attempting to reduce levels of carbon dioxide, e.g. by using energy efficient lights carbon dioxide, sulphur dioxide (SO_2) and nitrogen dioxide (NO_2) emissions might be reduced. Burning leaner and cleaner fuels and the implementation of energy efficiency campaigns will reduce the amounts of greenhouse gases released into the atmosphere.

Global dimming is another effect of global warming. There is 10+% less sunlight today than 30 years ago. This is because of industrial activity. This may off-set global warming, but will have a negative effect on solar power production, and a lack of vitamin D may affect human health.

The effects of global warming

The most dangerous effects of global warming are outlined below.

- **Melting polar ice caps:** This will raise sea levels. If all glaciers melted today the seas would rise about 70 m. With so much freshwater being delivered into the sea, ocean currents could be shut down. Animals would die, unable to cope with their changing landscape. The world will continue to heat up as the albedo (reflectivity) effect of the ice and snow is lost.
- **Economic consequence:** Most of the more extreme weather events and outbreaks of disease have been linked with global warming. These events cause billions of dollars of damage and in the treatment and control of outbreaks of disease.
- **More droughts and heat waves:** Some areas of the Earth will become wetter due to global warming, while others will experience serious droughts and heat waves. Severe droughts can be expected in Europe. Water is already a rare commodity in parts of Africa – global warming will exacerbate the conditions and is already leading to conflicts.
- **More extreme weather events:** As global temperatures rise, the oceans accumulate and retain energy, so the probability of more frequent and stronger tropical cyclones is a certainty, e.g. Katrina in 2005.
- **Spread of disease:** As northern continents and countries get warmer, disease carrying insects migrate north, such as malaria-carrying mosquitoes.

Acid rain

Quantities of sulphur dioxide and oxides of nitrogen (N) are emitted into the atmosphere from industrial activity and vehicles. In the atmosphere chemical changes turn rain into weak acids (in the order of pH 4.3). In the late 1960s lakes became acidic and fish populations plummeted in Scandinavia, caused by pollution from Western Europe. Germany has been particularly affected – in the mid 1980s 58% of trees showed some signs of damage. Since 1990 Germany has spent £300 million, attempting to sort out its problem forests.

Acid rain also 'rots' buildings e.g. the Houses of Parliament in the UK. Various clean air acts in the westernised countries have reduced the problem significantly. Remedially, lakes can be limed to neutralise acids.

2.5 Aspects of local climate

LEARNING SUMMARY	After studying this section, you should be able to understand:
	• how local weather effects can influence small geographical areas

Air flow in mountains (anabatic and katabatic winds)

AQA	A2 U3
OCR	A2 U3
WJEC	A2 UG3
CCEA	AS U1

During the day when conditions are calm, warm air blows up the valley in response to the heating of the air in contact with the valley slopes (anabatic). At night the reverse happens (katabatic) completing the circulation. Accumulations of cold air in valley bottoms can cause frost hollows to form.

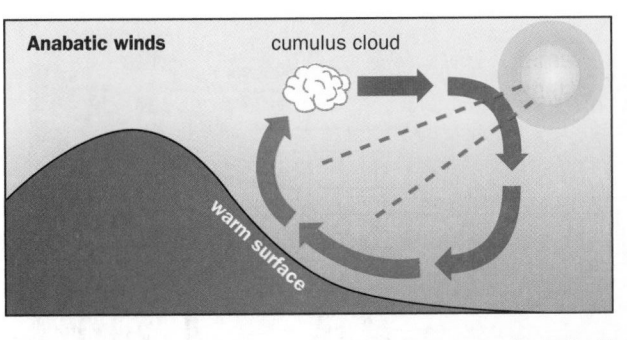

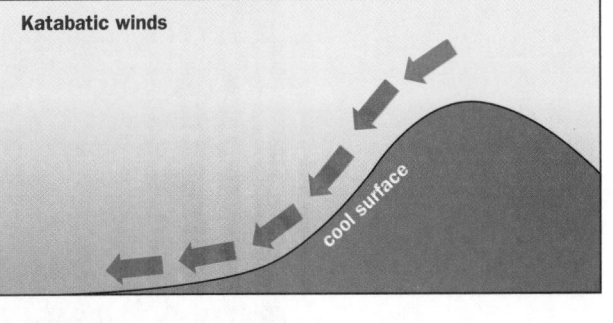

Anabatic/Katabatic winds

Land and sea breezes

AQA   A2 U3

On warm days air over the land heats up and expands. This tilts the pressure gradient. The result is that a small landward blowing circulatory system develops, with compensating out-blowing wind aloft. The reverse happens at night. These breezes have a marked effect on coastal climates.

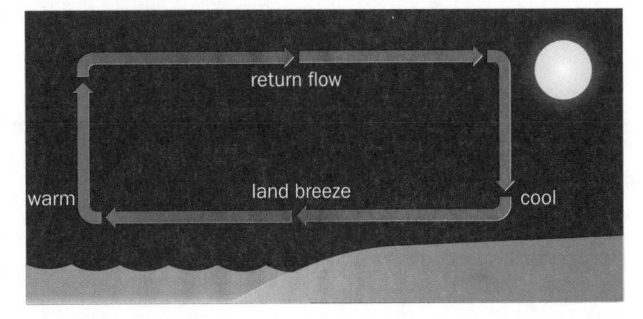

Land and sea breezes

Urban climates

AQA	A2 U3
OCR	A2 U3
WJEC	A2 UG3
CCEA	AS U1

Urban areas can modify climates by affecting temperature, wind direction, speed, precipitation and pollution.

The urban heat island

Urban areas tend to be warmer than their rural neighbours (1–2°C). The intensity of the heat island effect does vary, e.g. on hot summer days urban and rural areas heat up at about the same rate, but at night the urban areas retain their heat (late afternoon to early evening being the best time to assess the heat island effect).

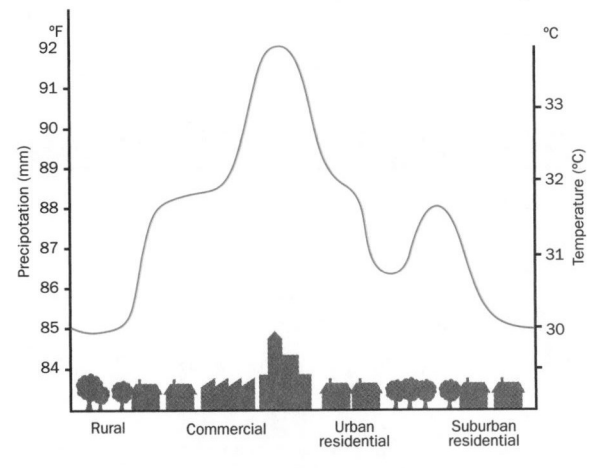

Urban heat island profile

Heat islands form because of the combined effects of:

- The **type of building** materials used – the urban fabric heats up very quickly. The reinforced concrete that covers so many cities heats much more rapidly than soil for instance. The colour of buildings also has an effect on the albedo (the reflectivity of the surface). Light buildings reflect heat away and dark ones absorb it.
- The **height and pattern/density** of buildings – in the planned jumble of heights that cities are heat gets trapped, tall buildings and the spaces between them holding the heat in and re-reflecting it around the city.

0 = dark
1 = bright

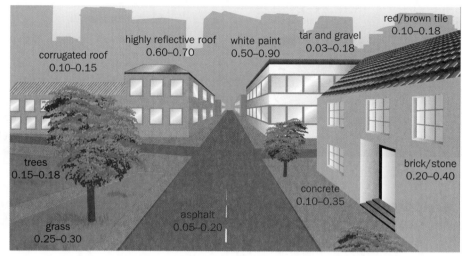

red/brown tile
0.10–0.18

highly reflective roof
0.60–0.70

white paint
0.50–0.90

tar and gravel
0.03–0.18

corrugated roof
0.10–0.15

trees
0.15–0.18

brick/stone
0.20–0.40

concrete
0.10–0.35

asphalt
0.05–0.20

grass
0.25–0.30

Urban environment albedos

- **Human activity** releases heat, e.g. from vehicles moving about the city or from air conditioning and heating.
- The presence of **water in the city** should not be under-estimated – more water and the city cools, less water and heat is released into the city atmosphere.
- **Pollution** is a feature in all cities. Dust and other pollutants can limit the amount of heat that can penetrate the cities' atmosphere. In Mexico City light levels are reduced by pollution, so-called **photochemical pollution** or **smog**.

Wind in cities

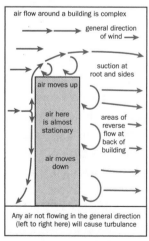

buildings tend to get taller towards the city centre

urban 'canyon'

streets may get narrower towards centre

winds are squeezed into increasingly restricted streets producing the Venturi effect

wind speeds increase

(a) Wind in cities

air flow around a building is complex

general direction of wind →

suction at root and sides

air moves up

air here is almost stationary

areas of reverse flow at back of building

air moves down

Any air not flowing in the general direction (left to right here) will cause turbulance

(b) Wind in cities

Wind in urban areas generally moves at slower speeds because of the frictional effect of the buildings. However, the layout of the city, with its tall buildings means that air is funnelled through parts of it. This **funnelling** effect increases wind speed – the **Venturi effect**. Tall buildings also have the effect of calm air movement at pavement level.

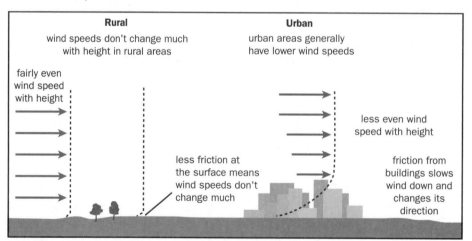

Rural
wind speeds don't change much with height in rural areas

Urban
urban areas generally have lower wind speeds

fairly even wind speed with height

less even wind speed with height

less friction at the surface means wind speeds don't change much

friction from buildings slows wind down and changes its direction

Wind in cities (c)

PROGRESS CHECK

1. What are the threats to the atmosphere?
2. What is the difference between anabatic and katabatic winds?
3. What is the Venturi effect?

Answers: 1. Smog has increased and ozone has decreased; the greenhouse effect; acid rain. 2. Anabatic wind moves up the valley sides and katabatic wind moves down the valley side. 3. Funnelling of wind in cities.

Exam practice questions

The diagram below shows the position of areas of high and low pressure and the direction of the surface winds blowing between them.

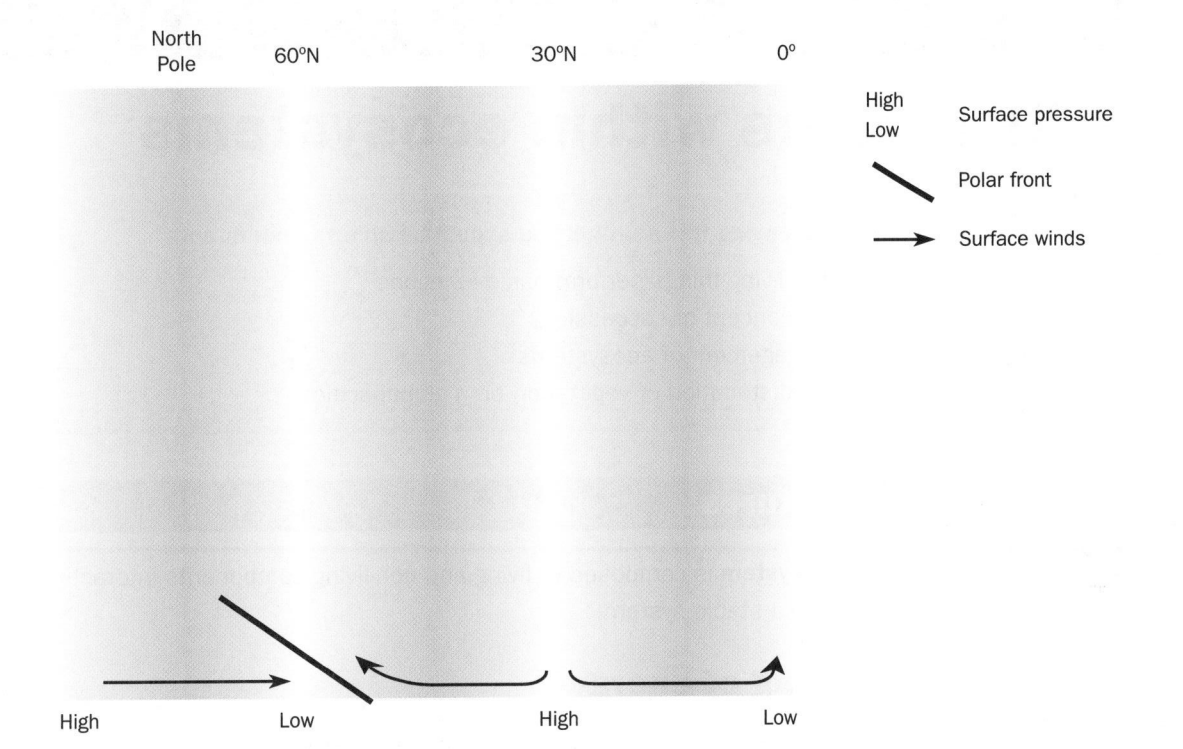

North Pole   60°N   30°N   0°

High
Low   Surface pressure

Polar front

Surface winds

High   Low   High   Low

1 Complete the above diagram.

   (a) Indicate a Hadley Cell.   **[2]**

   (b) Account for the size, shape and directions of air movements drawn.   **[2]**

   (c) Outline the differences between the area of low pressure around the Equator and those at the polar front using the following three headings:

      – reasons for the air rising;

      – type of cloud that usually forms;

      – precipitation characteristics.   **[6]**

   (d) How can El Niño affect the weather and economics of the tropical and subtropical zones?   **[3]**

   (e) Explain how the weather associated with periods of long winter cold affect humans.   **[8]**

2 Describe and explain the differences between temperate depressions and tropical cyclones.   **[20]**

3 Ecosystems

The following topics are covered in this chapter:

- **Structures within ecosystems**
- **The distribution of biomes**
- **Threats to fragile biomes**
- **Management of fragile biomes**

3.1 Structures within ecosystems

LEARNING SUMMARY

After studying this section, you should be able to understand:

- that living things perform basic functions
- the concept of succession
- management of ecosystems
- the distribution of vegetation on a global scale

Basic ecosystem structures

AQA **A2 U3**   CCEA **AS U1**
OCR **A2 U3**

An **ecosystem** is composed of living and non-living components interacting to produce a stable system.

Cycling of energy

GPP is the total amount of energy fixed by green plants. NPP contributes to the production of new plant tissue. NPP = GPP – Respiration

This is perhaps the most obvious process in an ecosystem. Energy cannot be created or destroyed, but it can be transferred. Primary production via photosynthesis fixes energy from the Sun into plants. The rate of energy fixing is expressed in $kg/m^2/year$ (or equivalent) and is either fixed as **gross primary production** (GPP) or as **net primary production** (NPP).

The trophic level is the position that an organism occupies in a food chain – what it eats and what eats it.

Energy is lost/released through respiration in the photosynthetic process and is moved or transferred in **food chains**. Whenever energy is exchanged in a food chain, a 'stage' or **trophic level** is reached. There are few food chains with more than five trophic levels because of this energy loss or exchange. The numbers of individuals, biomass and productivity decreases as trophic levels four and five are reached.

Producers = autotrophic

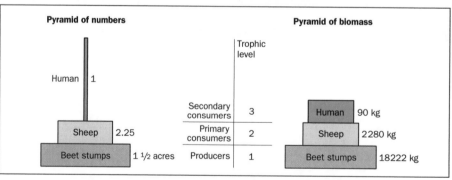

Cycling of energy

Simple food chains seldom exist, it is more realistic to display energy flows in a **food web**.

Cycling of nutrients

This is the second of the processes in an ecosystem. Nutrient cycling moves chemical elements out of the environment to organisms and then back again to

nutrients returned to litter
store from dead and decaying
plants and animals

precipitation

BIOMASS

fallout
pathway

LITTER

growth and
uptake pathways

inorganic substances
converted into organic
living biomass

decay
pathway

run off

SOIL ← weathering

decomposition allows
inorganic material
to return to the soil

leaching

Cycling of nutrients

the environment. Nutrients are stored in compartments and are recycled via pathways, as shown in the diagram opposite. Stores are not in a static condition. As nutrient amounts increase or decrease, store sizes vary. [N.B. they can be drawn any shape!] Humans can affect the cycling of nutrients in ecosystems by misusing or overusing the land.

Abiotic means the non-living, chemical and physical components of the ecosystem and includes climate, soil characteristics, parent rock, relief of the land and drainage characteristics. As with any system there are inputs, outputs, stores and flows. Ecosystems are **open systems** because energy and living matter can both enter and leave the system.

The most important **input** is energy from the Sun, which drives photosynthesis and so enables plants to grow. Other inputs include animals that arrive from elsewhere and water transported into the ecosystem by precipitation or rivers.

Nutrients are transferred out (**outputs**) of the system in a number of ways. Animals can physically move and water can move out of the ecosystem in rivers and by **evapotranspiration**, through flow and groundwater flow.

Flows within an ecosystem include nutrients which can be transferred from one store to another, e.g. from the soil to the vegetation through capillary uptake by plant roots.

Succession and climax: the changing ecosystem

| AQA | **A2 U3** | CCEA | **AS U1** |
| OCR | **A2 U3** | | |

Primary succession starts from bare ground. Secondary succession arises on land that had previously been covered in vegetation.

Ecosystems are dynamic and continually evolving. After colonisation, and the growth of **pioneer** species, begins the predictable pattern of vegetational change called **succession**. Each stage is called a **sere** (the chain of vegetational changes that are gone through is called a **prisere**). Each stage is an advance on the previous one, offering increased protection and shelter, better soil conditions and an increased nutrient stock for the plants that follow. Ultimately, when the vegetation reaches its most complex stage, **climax** is reached. Four types of succession exists. Two are xeroseres (successions in dry conditions) – the **lithosere** (on rock) and **psammoseres** (on sand), and two are **hydroseres** (successions in water) – the hydrosere (in freshwater) and the **halosere** (in salt water).

The development of a psammosere in West Wales

Ynyslas Dunes in West Wales are a typical example of a psammosere.

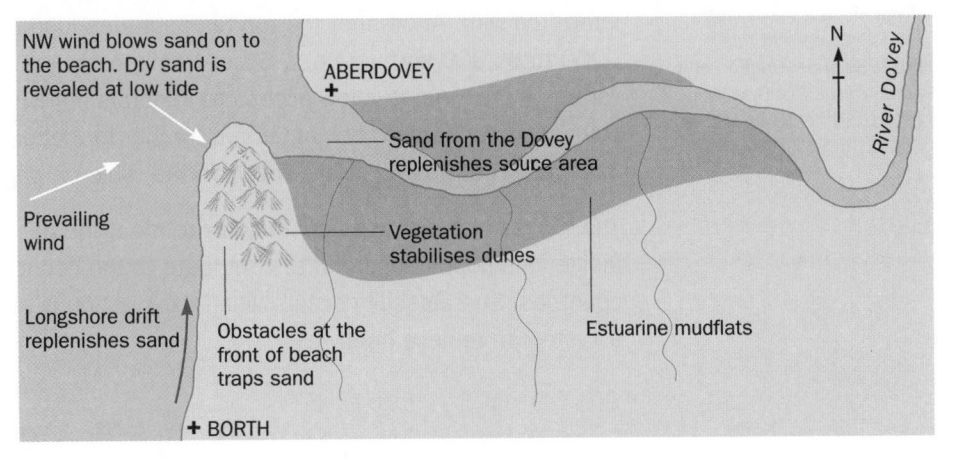

Embryo dunes – is an extremely difficult environment. Sand is dry, lacking in nutrients, has poor water retention, high alkaline pH and high salt content. Pioneer plants have to be **xerophytic** (drought tolerant) and **halophytic** (salt tolerant). Typical plants include sea rocket and sea couch. Few animals colonise the embryo dunes.

Mobile dunes – are areas of exposed sand reshaped by the wind. **Marram grass** is the dominant vegetational type. It is tolerant of the extremely dry conditions and begins to stabilise dune systems. Groundsel and sea spurge are also common. There is still no real colonisation by animals.

Semi-fixed dunes – have better conditions for plants. Soils begin to develop, mosses and lichens start to colonise the area and a range of flowering plants appear. Increased nitrogen fixation aids plant development. Animals are common.

Fixed dunes – are the most stable of the dunes and are covered in vegetation. Plant growth is at its maximum as conditions are ideal. Humus incorporation into the soil begins.

Dune slacks – are hollows formed by erosion by wind until the water table is revealed. These can be rich in different varieties of vegetation.

Management of ecosystems

| AQA | **A2 U3** | CCEA | **AS U1** |
| OCR | **A2 U3** | WJEC | **A2 UG3** |

Management at the local scale

Ecosystems may need to be managed at a local scale. Dunes perform an important role in protecting the coastline, but their soils and vegetation are fragile: trampling by humans and the burrowing and grazing by animals destroy the thin mantle of regolith revealing sand. This can lead to blowouts, massive deflation hollows on dune crests. Management techniques include:
- moveable board walks, zig-zagged to deter motorbikes and other vehicles
- low fencing to deter 'wanderers' and to funnel visitors
- advice boards and directions to help visitors make the most of their tour
- wardens, tolls and admission fees to control numbers of visitors
- designated car parking
- 'sacrificing' small areas to visitors in some dune systems
- controlling animals by culling, poisoning and shooting.

Management at the global scale

The world's **tropical rainforests** (TRF) can deteriorate at a phenomenal rate when interfered with. At present the top ten countries lose rainforest at the rate of 3000 to 35 000 km^2/yr; i.e. 1.5% of total forest area per year. It is cleared for a variety of reasons: to allow access to minerals, for HEP production, to allow cultivation, for cattle ranching and for timber for paper, furniture and energy. The impacts of deforestation are shown in the table on page 61.

The TRF can be managed sustainably to meet the needs of plants, animals and indigenous populations and to contribute to the economic development of countries. Sustainability occurs when exploitation is not greater than the ability of a system to replace itself.

Plants and animals	The water cycle	Landscape	Climate	Population
• Soils become nutrient impoverished as the nutrient cycle is breached. • Species are lost and the genetic stock is lost. • Niches are lost. • Fertile soil is lost. • Seeds don't germinate.	• Greater sediment load causes silting and flooding. • Run-off increases with no vegetation cover. • Quality of the water is affected.	• Soil is lost. • Gullying and sheet wash is more common. • Duricrust forms an infertile hardpan.	• Carbon release accelerates global warming. • Rainfall decreases. • Humidity drops. • Daily temperatures change. • Albedo effects on bare soil.	• Indigenous people lose their homes and way of life. • Exposure to outside influences allows disease to spread.

KEY POINT

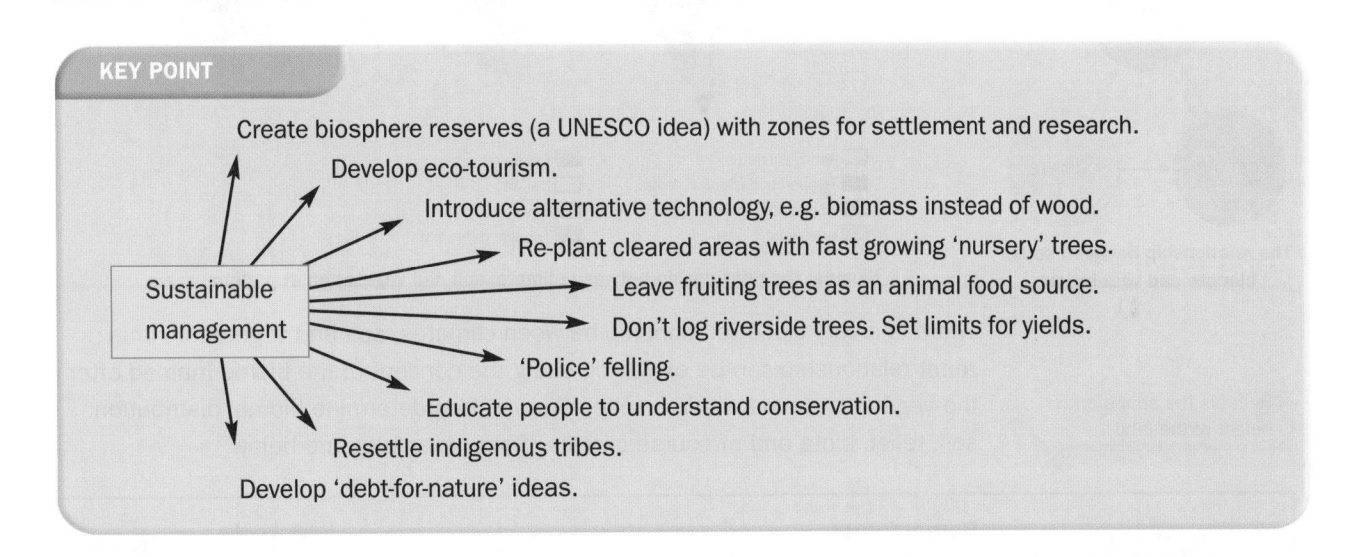

Sustainable management

- Create biosphere reserves (a UNESCO idea) with zones for settlement and research.
- Develop eco-tourism.
- Introduce alternative technology, e.g. biomass instead of wood.
- Re-plant cleared areas with fast growing 'nursery' trees.
- Leave fruiting trees as an animal food source.
- Don't log riverside trees. Set limits for yields.
- 'Police' felling.
- Educate people to understand conservation.
- Resettle indigenous tribes.
- Develop 'debt-for-nature' ideas.

Small-scale ecosystems in the UK

AQA	**A2 U3**	WJEC	**A2 UG3**
OCR	**A2 U3**	CCEA	**AS U1**

The broadleaf temperate deciduous forest (BTDF)

The Community Forest Organisation in the UK

In the 1990s, 3500 hectares of trees were planted in twelve community forests in the UK, reviving areas of damaged and derelict land, mostly on the edges of cities. All Community Forest Organisations (CFOs) have prepared detailed plans that will take them into the middle of this century.

As an example, the Greenwood Community Forest (incorporating Sherwood Forest) in Nottinghamshire, provides opportunities for commercial forestry, access, community involvement and education. The New Forest in Hampshire is an example of a managed plagioclimax community. It has been maintained for over 1000 years using grazing, careful tree planting and drainage.

Present plant species in the **broadleaf temperate deciduous forest** (BTDF) date, on the whole, from the last ice age. Since its first appearance, the BTDF has been markedly depleted, especially near the major cities in the UK. Globally 19% of the biomass by biome is BTDF. Characteristic of BTDF include:

- leaf loss, a form of dormancy in response to the cold winter months when there is little soil water and low sunlight/energy
- a growing season of about seven months
- extreme sensitivity to temperature changes
- a definite tree line, above which it will not grow.

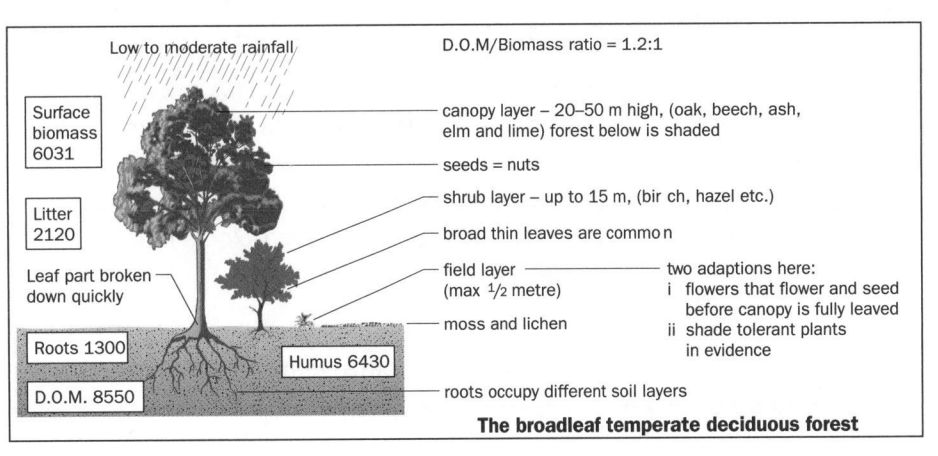

Low to moderate rainfall

D.O.M/Biomass ratio = 1.2:1

Surface biomass 6031

Litter 2120

Leaf part broken down quickly

Roots 1300

Humus 6430

D.O.M. 8550

- canopy layer – 20–50 m high, (oak, beech, ash, elm and lime) forest below is shaded
- seeds = nuts
- shrub layer – up to 15 m, (bir ch, hazel etc.)
- broad thin leaves are commo n
- field layer (max ½ metre) — two adaptions here:
  i flowers that flower and seed before canopy is fully leaved
  ii shade tolerant plants in evidence
- moss and lichen
- roots occupy different soil layers

The broadleaf temperate deciduous forest

Principle soils of the BTDF are the brown earths. Significant uses of the BTDF include game management, timber production, as shelter-belts and recreation. Most conservationists would advocate that this multi-purpose use is, if managed properly, the best way to use BTDF.

Biomes: the distribution of vegetation on a global scale

| AQA | **A2 U3** | WJEC | **A2 UG3** |
| OCR | **A2 U3** | CCEA | **AS U1** |

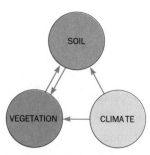

The relationship between soils, climate and vegetation

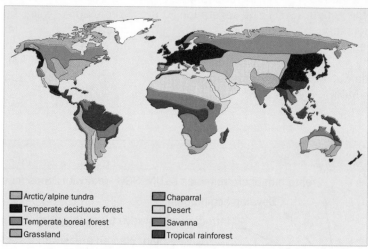

Arctic/alpine tundra
Temperate deciduous forest
Temperate boreal forest
Grassland
Chaparral
Desert
Savanna
Tropical rainforest

Biomes: the relationship between climate, soil and vegetation at a global scale

There is a very close relationship between climates, soil and vegetation. This **zonal** relationship can be explained using the concept of the biome (named after the vegetation found within a zone). Four factors determine biomal distribution: soil, relief, biota and of course climate. Some examples are below.

> Relief is the shape and height of the land.

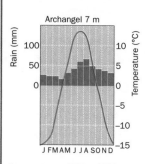

Temperature: –15° to +12°C
Precipitation total: 525 mm

Boreal forest

Evergreen trees. Needles instead of leaves, these are waxy, resisting drought and frozen conditions. Thick bark. Snow is easily shed by the branches. Growing season is about three months. Ground vegetation is sparse. **Productivity: 800 g/m^2/yr**

Soils – dominant water movement is downward. Low temperatures mean decomposition is slow. Acid in nature.

Human use – boreal forest is highly productive, because of its large biomass. Is now generally commercially managed.

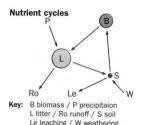

Nutrient cycles

Key: B biomass / P precipitaion
L litter / Ro runoff / S soil
Le leaching / W weathering

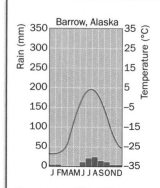

Temperature: –35° – 15°C
Precipitation total: 150–250mm

Tundra

Almost no trees due to a short growing season and permafrost; lichens, mosses, grasses, sedges, shrubs.
Productivity: 32 to 780 g/m^2/yr

Soils – with soil moisture frozen for much of the year chemical alteration of minerals is low. Parent material is mechanically broken particles. Lots of raw humus and peat because of slow plant decomposition. No distinctive layers, but a mix of sandy clay and raw humus. Colours range from dark to grey through the profile. Often loose on the surface.

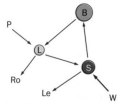

Key: B biomass / P precipitaion
L litter / Ro runoff / S soil
Le leaching / W weathering

Human use – discovery of oil in the tundra areas is a serious threat to this very fragile ecosystem both in terms of pollution and in terms of the permafrost melt.

Tashkent 478 m

Temperature: -1° to +26°C
Precipitation total: 375 mm

Temperate grassland

Natural grassland, can be tall and frequently has extremely long roots. **Productivity: 1600 g/m²/yr**

Soils – active soil animals ensure that decomposing material is rapidly incorporated into the A horizon. There is a strong crumb structure. Mild leaching in evidence.

Human use – grasslands are used extensively for agriculture. They also support cattle in ranching operations. Soil erosion can be a problem.

Key: B biomass / P precipitaion
L litter / Ro runoff / S soil
Le leaching / W weathering

Kisangani 418 m

Temperature: 24° to 26°C
Precipitation total: 1740 mm

Tropical rainforest

Trees are large and the leaves are waxy. Trunks are covered in thin bark. The crowns of the trees form a canopy. There are a great number of tree species/ha. Buttress roots improve stability. **Productivity: 2200 g/m²/yr**

Soils – ferrallitic soils that are heavily weathered. Translocation is dominant, as is leaching. This soil is a poor nutrient retainer.

Human use – the TRF is being rapidly cleared caused by population pressure. Removal causes soil erosion, loss of gene pool, climate change and affects the water table. A tourist attraction is lost.

Key: B biomass / P precipitaion
L litter / Ro runoff / S soil
Le leaching / W weathering

PROGRESS CHECK

1. GPP – Respiration = ?
2. What is a sere?
3. Why is marram grass so important in the psammosere succession?
4. What are the four factors that determine biomal distribution?

Answers: 1 NPP (net primary production). 2. A stage in the successional sequence. 3. It is able to tolerate dryness and salt and it is the principal vegetation type holding and stabilising dunes. 4. Soil, relief, biota and climate.

3.2 The distribution of biomes

LEARNING SUMMARY	**After studying this section, you should be able to understand:**
	• the global significance and fragility of each biome
	• the distribution and characteristics of each biome
	• the problems of destruction and management of the biomes

The tropical rainforest biome

AQA **A2 U3** CCEA **A2 U2**
OCR **A2 U3**

Distribution and characteristics

The TRF is a complex and rich ecosystem that contains the Earth's biggest gene pool, thought to be the result of mutations and the subsequent evolution of species, because of the vast amounts of UV-B radiation absorbed. Ultraviolet radiation occurs in three types – A, B and C. UV-B is the strongest of the ultraviolet rays. The TRF is also the world's most prodigious terrestrial ecosystem, with primary production and productivity being in the order of 1000 to 3500 g/m^2/yr (the result of high rainfall and temperatures). Additionally, the TRF plays a significant role in maintaining stability in the atmosphere – it is the world's most finely balanced system.

Distribution

The TRF has a world biomass productivity of 765 000 tonnes/ha.

The TRF is the world's oldest major ecosystem.

The main areas for TRFs include:

- the Amazon Basin and into Central America – 58% total area
- the Zaire (Congo) Basin in Africa – 19% total area
- the Indo-Malay/Indonesia to Australia – 23% total area.

All these areas lie between 10°N and S of the Equator and cover 718 million ha, or 5% of the land surface of the Earth.

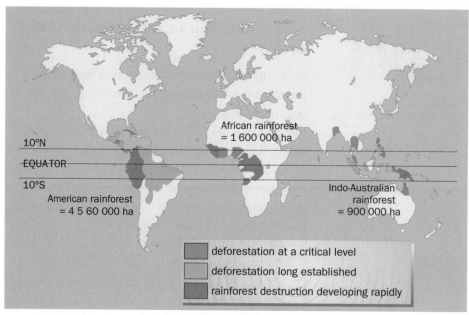

African rainforest = 1 600 000 ha

10°N

EQUATOR

10°S

American rainforest = 4 5 60 000 ha

Indo-Australian rainforest = 900 000 ha

☐ deforestation at a critical level
☐ deforestation long established
☐ rainforest destruction developing rapidly

The global distribution of TRF

Vegetation

Plants which exist in the TRF do not, as is generally thought, grow continually.

TRFs occur in similar climatic zones: they are the **climatic climax community**. All TRFs resemble one another in structure and habitats because they experience similar environmental variables; in ecological terms they are a **convergent community**. Variations reflect differences in climate, relief and altitude and **edaphic** (soil) conditions.

Climate

Köppen identified a tropical climate as one with temperatures greater than 18°C and daily differences greater than seasonal ones. The best students will recognise and appreciate Köppen's findings.

TRFs occur in the humid tropical areas with rainfall exceeding 2000 mm/yr. The seasonal temperature averages 27°C, though daily temperature swings can range between 6°C to 12°C. High humidity and low wind speeds are typical features, but TRFs are not dripping wet, as many are led to believe. They are well-drained and are often completely dry on the forest floor. Changes in altitude and micro-climate can cause a 'seasonal' effect in the forest. There is a **daily rhythm** to tropical rain weather, hot, wet and predictable.

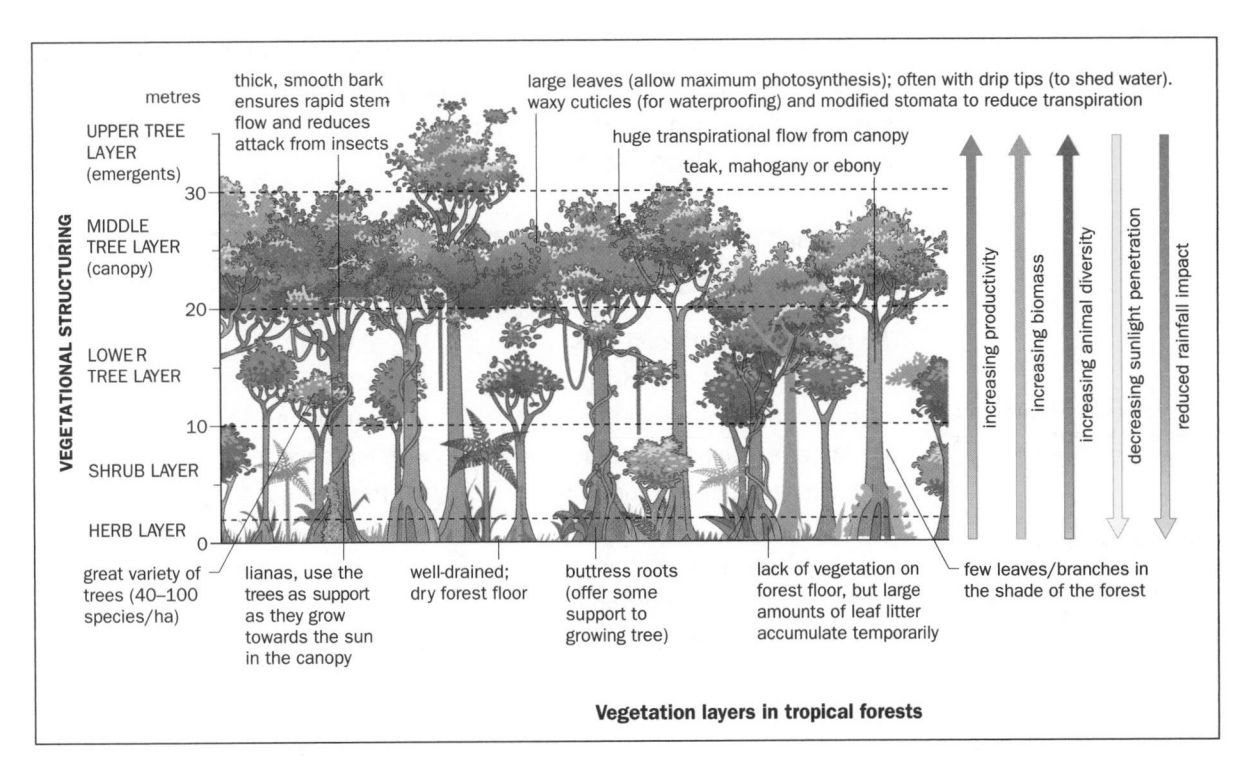

Vegetation layers in tropical forests

Tropical latosol soils

Typically these soils are:

- up to 30 m deep
- very well weathered (in places turned into kaolinite, a type of clay)
- acidic and highly leached; this is a dominant process
- rich in iron and aluminium in the upper layers; hence their red colour
- low in organic matter because of high rotting rates; they are humus poor
- downward drained, typically the soils receive more precipitation than is evaporated
- highly translocated; lots of material is moved through them by water, a process known as **ferrallitisation**.

> The best candidates will always show an understanding of soil processes and realise the results of even slight changes to the system.

> Students should be able to describe ecosystems in detail with reference to vegetation, climate and soil.

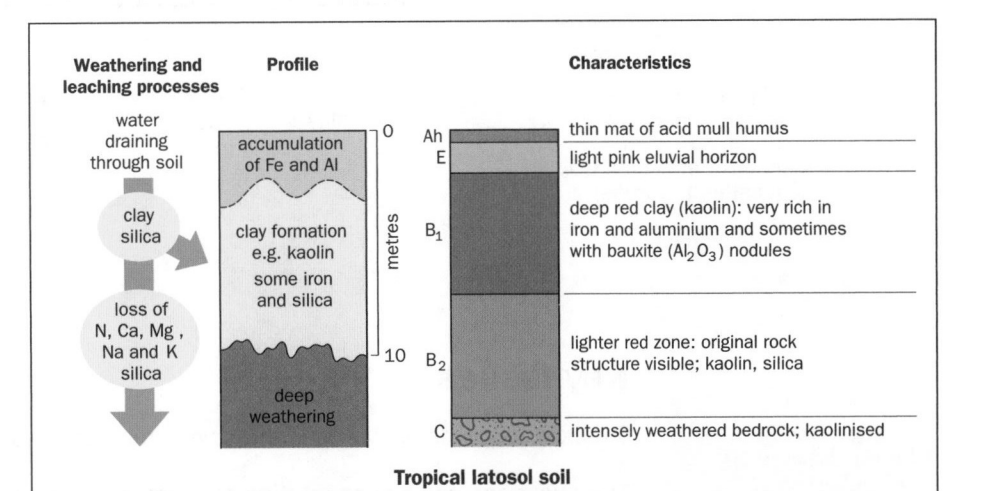

Tropical latosol soil

Management of rainforests

AQA **A2 U3** CCEA **A2 U2**
OCR **A2 U3**

The use of the TRF is influenced by its productivity and by the distribution of its nutrient reserves. Most of these are held within the timber, the living biomass. Massive forest clearances, and even small disturbances, because of fuel, ambitious development plans and the pressure of burgeoning populations, cause an immediate loss of nutrient reserves along with the formation of laterites, soil erosion and leaching. Such areas therefore have to be managed if the forests are to provide for the longer term future.

The cause and consequences of rainforest loss and possible solutions

The causes of rainforest loss:

Complex interrelationships: understand them!

Fifty species of plant and animal become extinct every day.

- economic and material resources, e.g. pharmaceuticals
- timber extraction, e.g. for buildings, fuel and furniture
- HEP dam and reservoir construction (Brazil has plans for 31 dams in the Amazon by 2010)
- pressure of population

- wood-pulp for paper
- plantations
- clearance for ranching
- the need to 'develop'
- colonisation programmes
- mineral extraction
- tourism
- over-fishing.

KEY POINT

Consequences of rainforest loss: As forest clearance increases, the ecosystem becomes increasingly susceptible and fragile.

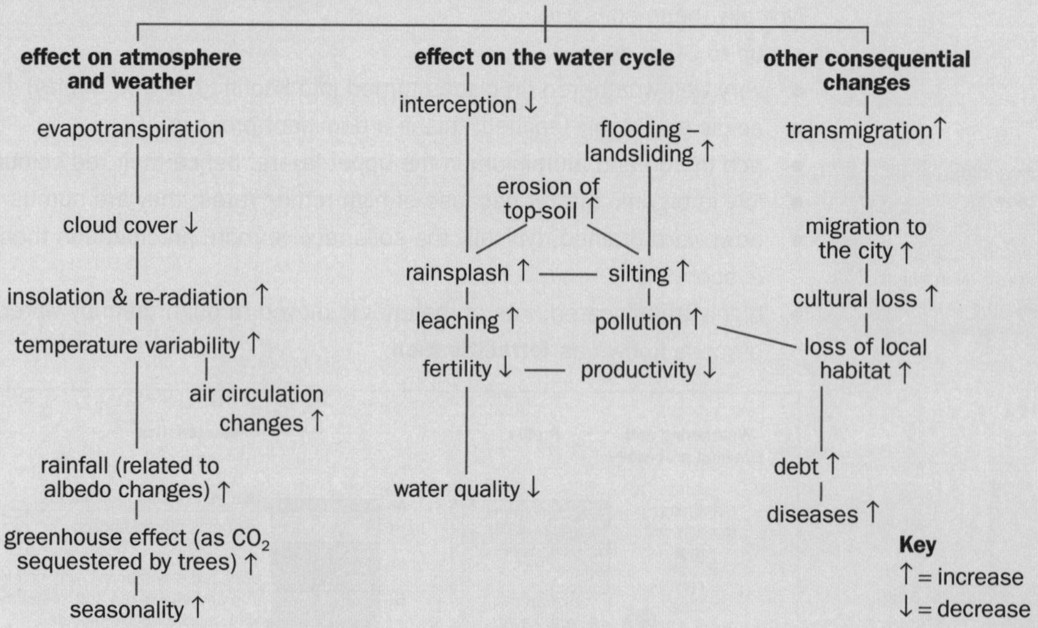

effect on atmosphere and weather

evapotranspiration
|
cloud cover ↓
|
insolation & re-radiation ↑
|
temperature variability ↑
|
air circulation changes ↑
|
rainfall (related to albedo changes) ↑
|
greenhouse effect (as CO_2 sequestered by trees) ↑
|
seasonality ↑

effect on the water cycle

interception ↓
|
flooding – landsliding ↑
erosion of top-soil ↑
rainsplash ↑ —— silting ↑
leaching ↑   pollution ↑
fertility ↓ —— productivity ↓

water quality ↓

other consequential changes

transmigration ↑
|
migration to the city ↑
|
cultural loss ↑
|
loss of local habitat ↑
|
debt ↑
|
diseases ↑

Key
↑ = increase
↓ = decrease

Key Issues

Burning of the TRF sends 2 billion tonnes of CO_2/yr into the air.

The flow diagrams above raise three very important issues related to the TRF.

The questions and issues relating to deforestation and development

Deforestation is viewed as an LEDC problem. Some LEDC countries fell timber commercially to generate incomes. Another way of viewing deforestation is to relate the rates of deforestation to the process of development creating new roads, new markets and pioneer settlements. The clearing of the forests could be seen as development in its own right?

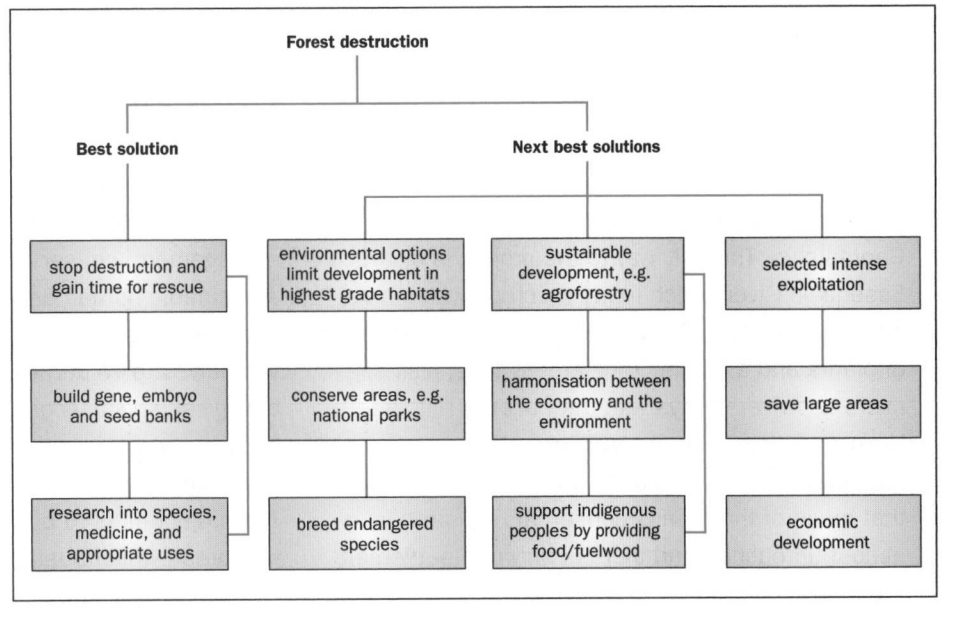

Tropical rainforests have to be managed sustainably if they are to survive. Many of these solutions are sustainable. Sustainability occurs when exploitation is not greater than the ability of a system to replace itself.

Global repercussions and eco-catastrophe

Destruction of the TRF damages resources. Their genetic materials support modern agriculture, medicine and industry. The burning of the rainforest is also known to contribute to the stock of atmospheric carbon dioxide. Will TRF clearance constitute an eco-catastrophe?

Management and sustainable development

Any activity or schemes that improve (e.g. thinning) or preserve (e.g. through revenue enhancement from leisure activity) can be termed **management**. Should the management schemes enable the TRF to replace itself at a greater rate than it is destroyed, then the system is said to be **sustainable**.

Strategies to protect the rainforest

- **Labelling** – timber certification enables consumers to choose a product that meets specific environmental, social and other criteria.
- **Guidelines** – where timber is extracted management should increasingly follow an established set of sustainable guidelines, related to yield and felling needs.
- **People power** – has resulted in important steps, such as the mahogany moratorium in Brazil and the World Bank's forestry policy, prohibiting logging in primary tropical forests.
- **Controls** – to change and halt destructive loans by multilateral development banks (MDBs) and to promote sustainable forest management.
- **International agreements** – e.g. the Kyoto Protocol (1997) put limits on totals of greenhouse gas emissions.
- **Debt swopping/or debt-for-nature** – involves writing off debt if the country involves itself in conservation and sustainable development.
- **Creation of extractive reserves and protected areas** – so that the productivity of the TRF area is utilised, but only in designated areas.
- **Creation of biosphere reserves** – these are zoned protecting reserve areas, where conservation is the key activity. In areas around the protected area limited eco-tourism and other activity might take place.

Case study: The Iwokrama Project, Guyana

The 360 000 hectare Iwokrama Project began in 1989 when the President decided to set aside a rainforest site in Guyana as a 'living laboratory'. Iwokrama is in central Guyana, 300 km south of Georgetown. The only way to the project is by the Essequibo River, which forms much of Iwokrama's eastern boundary. The area consists of a vast range of plants and animals, living among hills, plains, rivers and river valleys with the Pakatau Hills and Iwokrama Mountains in the centre. High temperatures and rainfall of up to 2500 mm mean that much of the area is dense TRF. The area is almost untouched and there is only one settlement within it – Kurupukari (Fairview). Only a small numbers of visitors are allowed in and this is in controlled conditions.

The United Nations Development Programme (UNDP) granted $3 million to help set up the project. In 1994 training and research began. The Iwokrama Project focuses on five areas: sustainable management; conservation and use of biodiversity; sustainable human development; forest research; information and communication. It is managed as two halves: a Wilderness Preserve to preserve the biological, social, economic and cultural relations between humans and the area; and a Sustainable Utilisation Area (SUA), for sustainable development. The project aims to preserve the biodiversity of the rainforest, whilst sustainably developing its economic potential.

The tropical grassland biome

AQA **A2 U3** CCEA **A2 U2**
OCR **A2 U3**

Distribution and characteristics

> Savanna is from the Hispanicised Amerind term for 'plain'.

The term savanna is generally used to describe the tropical grassland biome which ranges from virtually treeless grassland to those areas with drought resistant trees or shrubs. The obvious dominant species are the perennial, tussocky and **xerophytic** (resilient to drought) grasses and sedges. Savanna develops in regions where the climax community should be a form of seasonal woodland, but edaphic (soil) and other disturbances prevent the climax community from forming.

Vegetation of the savanna biome

A transect across the savanna grassland shows how the natural vegetation changes in response to the latitude and variation in rainfall (climate).

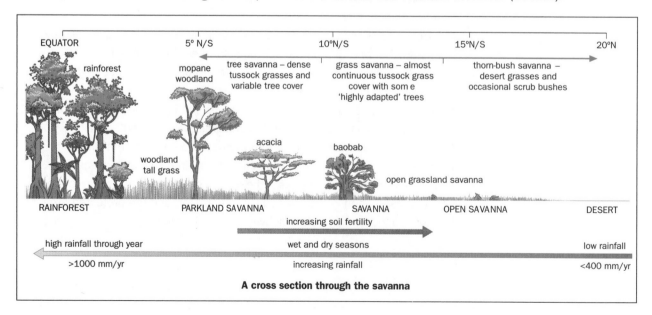

A cross section through the savanna

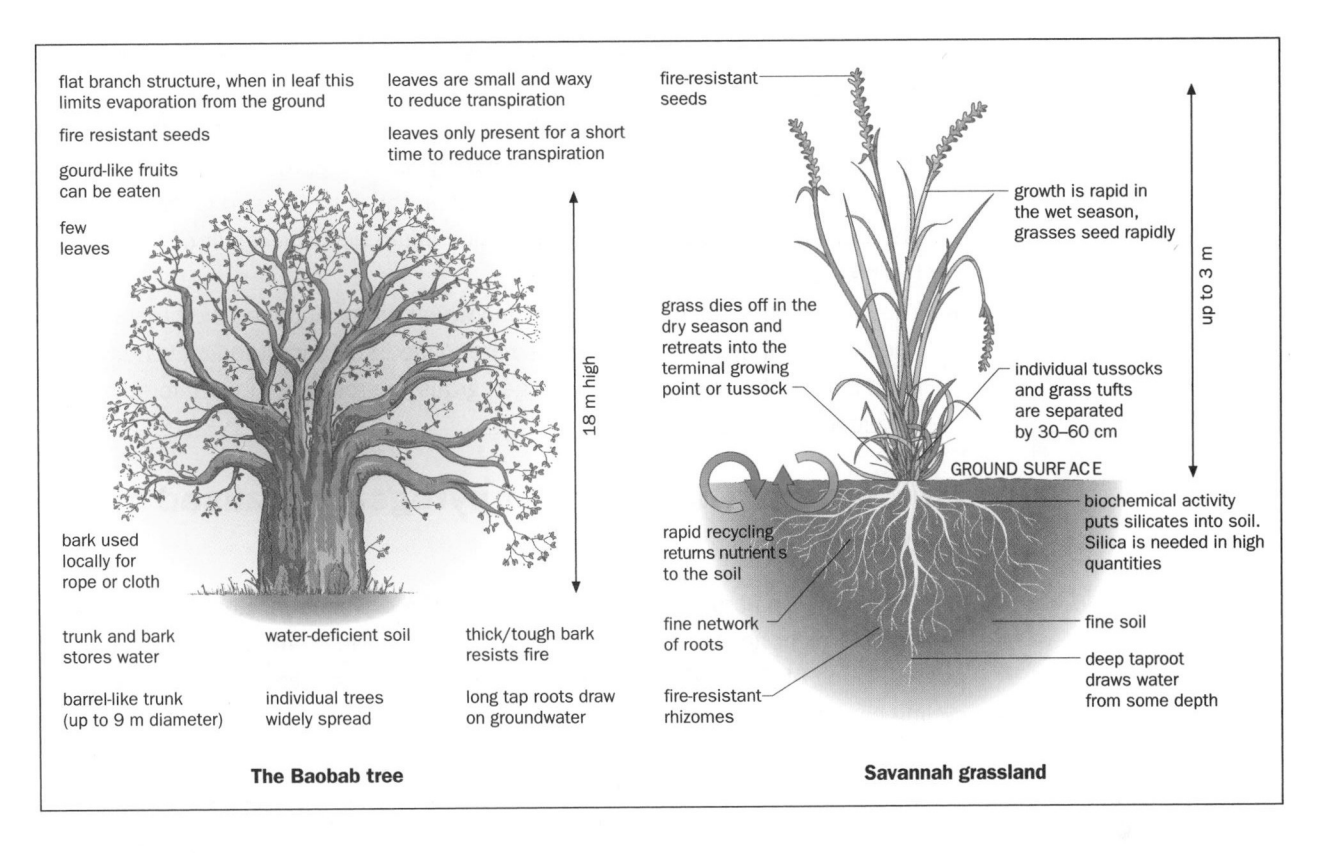

flat branch structure, when in leaf this limits evaporation from the ground

fire resistant seeds

gourd-like fruits can be eaten

few leaves

bark used locally for rope or cloth

trunk and bark stores water

barrel-like trunk (up to 9 m diameter)

leaves are small and waxy to reduce transpiration

leaves only present for a short time to reduce transpiration

18 m high

water-deficient soil

individual trees widely spread

thick/tough bark resists fire

long tap roots draw on groundwater

The Baobab tree

fire-resistant seeds

growth is rapid in the wet season, grasses seed rapidly

grass dies off in the dry season and retreats into the terminal growing point or tussock

individual tussocks and grass tufts are separated by 30–60 cm

up to 3 m

GROUND SURFACE

biochemical activity puts silicates into soil. Silica is needed in high quantities

rapid recycling returns nutrients to the soil

fine network of roots

fire-resistant rhizomes

fine soil

deep taproot draws water from some depth

Savannah grassland

Causes of the savannah biome

One would expect, given the distribution of this form of vegetation that it might fit into a pattern based on climate alone. In reality this is not the case, the highly variable savanna biome seems to be the result of responses to a number of quite variable and separate controls.

> The savanna is associated with Köppen's wet and dry climate type.

Climate

The two season savanna climate, otherwise known as the **wet and dry climate**, seems to have a dramatic effect on this grassland biome.

- The **wet season** is caused by north-east and south-east Trade Winds blowing in towards the equator, meeting in the low-pressure area called the doldrums (the ITCZ or Inter Tropical Convergence Zone), which is over the Tropic of Cancer in June. This area experiences massive heating as air rises and cools, clouds form and it rains. Moisture laden winds blowing in from the sea aggravate the situation.

- The **dry season** occurs when the ITCZ follows the Sun south to the Tropic of Capricorn in December. At this time the Trade Winds blow away from the land and no rain falls. Annual rainfall is between 700 mm and 1200 mm. Temperatures typically are 19°C–27°C.

JOS 9°54'N, 8°53'E

Rainfall (mm)

Temperature (°C)

J F M A M J J A S O N D

Savanna climate

Soil (edaphic) controls

In general **laterisation** is the dominant soil-forming process and low-fertility **oxisols** can also be expected. This is all very similar to that of the TRF. However, lower rainfall and the marked dry season means that the soil is less intensely weathered and leached, organic matter accumulations are higher, and (uniquely) silicon is not easily moved. But, perhaps the most marked control relates to the amount of soil water available and that many savanna areas cover old landform and forest surfaces, most of them depleted of mineral nutrients and composed of hard **duricrust**. In effect, the savanna can cause **edaphic savanna sub-climaxes**. There are several examples.

- **The East African Savanna** (of Zimbabwe, Botswana and Namibia) developed on droughty, but nutrient-rich volcanic sandy soils, e.g. the Serengeti Plains in Tanzania (though these are largely controlled today by fire and grazing).
- The **Cerrado of Brazil** which tolerates aluminium rich (the result of laterisation), low nutrient conditions.
- The **Llanos of Venezuela and Colombia** which are maintained by the annual flood of the Orinoco, which creates waterlogged soil and standing water for a good part of the year. This encourages grass to grow, but inhibits forest development.

Fire: natural (lightning) or pyrogenic (caused by people)

Fires often sweep across the savannas during the dry season. Bush fires destroy litter which might have enriched the soil, but also provide nutrients such as potash. Firing of the landscape appears to have a rejuvenating effect; nutrient cycling can come to a halt if grasses/trees are not periodically burnt off.

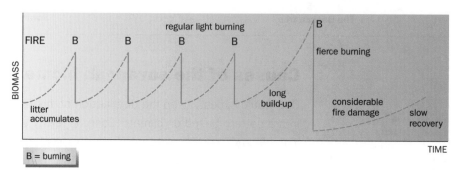

The fire and recovery sequence

Generally, grasses are better adapted to fire than trees.

Many areas on the savanna, particularly those under pressure from population growth, are deliberately and seasonally put to the torch, to encourage the growth of grass species that will have high nutritive value for domesticated grazing animals. The savannas of south-east Asia are generally thought to have been 'man-made'. The considerable pressure on the land of Africa will ensure that fire is used as an aid to hunting and the clearing of land for some considerable time to come.

Grazing

It is possible to view the savanna as a grazing sub-climax. In the natural world large mammals debark and knock over trees, grazers eat and trample seeds and inhibit the growth of the larger plants. Overgrazing has contributed to both the savannas' spread and to its demise in some areas (see the section on desertification below). Growing populations and the restrictions placed on nomadic tribes (e.g. the Fulani/Tuareg) mean that grazing areas are limited and over-used. The result is that the grassland never has a chance to recover, seedlings don't grow and the grass does not develop fully. Bare ground starts to appear, evaporation increases and very rapidly a dry microhabitat develops, e.g. in Ethiopia.

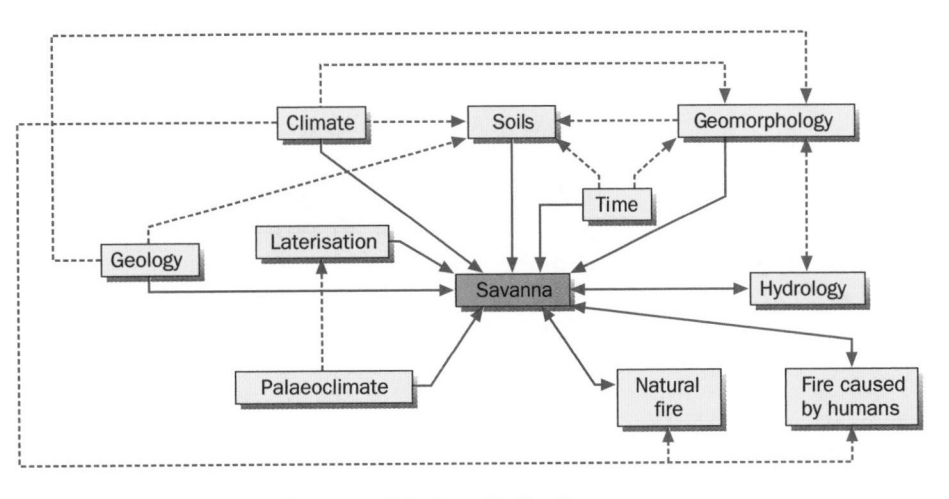

Summary of factors affecting the savanna

The desert biome

AQA	AS U1
OCR	AS U1
Edexcel	A2 U3
WJEC	A2 UG3
CCEA	A2 U2

Savanna areas can be very productive, but are also extremely precariously placed. According to the UN, desertification results from various factors, including climatic variations and human activities.

There is little doubt that in recent years the world climate has fluctuated; in Africa average and above average rainfall has followed long periods of drought. Over the last 30 years the precipitation falling onto the continent has been up to 48% lower in the central savanna belt. Causes of these climatic fluctuations could include **El Niño**, global warming (affecting the position of the ITCZ) or a general global shift in weather patterns. However, it is generally assumed that it is humans that bring imbalance to the savanna biome through their mismanagement.

There is little doubt desertification is occurring and that its effects continue to creep insidiously forward. However, there are a range of ways in which this desertification can be managed to increase the savanna biome's resilience and recovery rates, to enable it to continue providing for indigenous tribes and to maintain the natural flora and fauna. See the table below.

Reduce overcultivation	Reduce overgrazing	Reduce deforestation	Improve soil conditions	Alter social and economic conditions
• Use fertilisers to increase yields. • Use high yielding/drought resistant crop varieties. • Use crop rotations. • Use irrigation in the dry season.	• Improve grazing through controlled burning of grasslands. • Introduce new breeds. • Improve the medical care of the stock. • Rotate grazing areas. • Introduce game ranching.	• Use agroforestry (a combination of agriculture and forestry). • Tree planting schemes (Majjia Valley Project Niger). • Introduce alternative sources of fuels and building materials.	• Use natural mulches or plastic sheets to trap moisture. • Build low stone or earth walls parallel to the contours to trap sediment and surface run-off (Burkino Faso). • Plant leguminous crops to add nitrogen. • Reduce the effects of salinisation. • Add fertilisers and organic material.	• Decrease the dependence on subsistence agriculture either by introducing commercial agriculture or developing craft agricultural based industries. • Extend tourism. • Provide loans and grants.

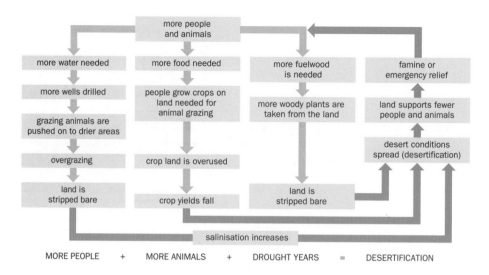

'People push the savanna system out of balance'

Case study: The Sahel, North Africa

The Sahel is the transitional zone in Africa between the Sahara desert to the north, and the savanna and tropical forest to the south. This dry belt stretches across Africa.

The environmental emergency that is the Sahel is the result of drought and localised environmental degradation, which means that agricultural production and livestock numbers have dropped. There has been migration from the area and some urban growth, but the region is still mostly rural and dependent upon rain fed agriculture and animal rearing.

Three huge and long-lasting droughts struck the region in the late twentieth century, killing 100 000 people and wreaking havoc on the land. A once prosperous and self-contained savanna, people almost overnight became vulnerable. But things are changing for the region. Reaction to the Sahelian problem has focused on:

- famine early warning systems developed by USAID
- soil and water conservation through agro-forestry and other cheap sustainable methods
- more expensive production techniques, i.e. high yielding rice, drought resistant crops, irrigation
- repositioning and restructuring the Sahel's place in Africa; funded by foreign aid, investment has occurred in basic industry, agricultural exports and transport.

These responses have also accounted for the fact that the population will double by 2040 (this problem needs to be addressed too!). Future development must recognise the complex social and ecological structures. Progress has been made through the intervention of foreign governments and has enabled the local democratic governments to ensure a productive and lasting future for the inhabitants and the environment.

PROGRESS CHECK

1. Name two countries where deforestation has reached critical levels.
2. The TRF is a climatic climax community. What does this mean?
3. Why is the temperature constant in tropical areas?
4. What process translocates, using water, material through the soil in the TRF?
5. Many plants in the savanna grassland areas have to be xerophytic. What does this enable plants to do?
6. What are the causes of the savanna biome?
7. Why is fire seen as a good thing for the savanna?
8. What two factors really work together to make the savanna one of the world's most fragile environments?

Answers: 1. For example Honduras, Guatemala, north-west Congo, Thailand, Malaysia. 2. That in the tropical climate the TRF is the 'top' vegetational type 3. Temperatures remain at a constant level as the sun is virtually overhead at all times. 4. Ferrallitasation. 5. Such plants are resilient to drought. 6. Climate, soil, fire (and latterly overgrazing). 7. Firing the savanna has a rejuvenating effect – providing nutrients and encouraging growth of grasses. 8. A lack of reliable, or variability, of rainfall and the presence of more people and animals on the savanna.

The temperate grassland biome

AQA	A2 U3
OCR	A2 U3
Edexcel	A2 U3
WJEC	A2 UG3
CCEA	AS U1

Distribution

Temperate grassland biomes are found in the Prairies in central north USA and Canada, the Pampas in Argentina, the Veldt in South Africa, the Steppes in the Ukraine and smaller areas, such as the Canterbury Plain of New Zealand and the Downs of south-east Australia.

temperate grassland
grass litter
mull humus
fauna mix upper layers of A horizon giving rich dark brown/black colour
capillary rising enhances calcification (in summer)
increasing alkalinity
limited breakdown of organic inputs lower in the horizon
there is often no discernible A/B horizon
calcic layer (illuviation)
nodules of lime CaCo₃
soil biota active
krotovinas (burrows)

- temperature range = – 5°C to 25°C rainfall = 200 mm to 500 mm evaporation = +800 mm
- found in lower-mid latitudes
- very fertile agricultural soil
- climate = temperate semi-arid

lime – rich parent material

Chernozem soil type

Climate, NPP and soil type

Temperate grasslands occur in both **cool temperate** and **warm temperate** climatic zones. Rainfall ranges from 500 mm to 750 mm. NPP averages about 600 g/m²/year. Most precipitation (75%) falls in the summer months, coinciding with the highest evapotranspiration rates. Winter temperatures are usually substantially sub-zero, particularly in continental interiors. Soils are usually a **chernozem type**.

The vegetation and animals

The grassland vegetation consists of species such as grama, buffalo grass (50 cm), feather grass and tussock (or tufted) grass (2 m). Trees are prevented from growing by the tightly-knit 'sod' that the grasses form and by the grazing pressures of herbivores. The main carnivores include wolves, coyote and predatory birds such as eagles and hawks. Grass roots extend down up to 2 m to the water table. This helps to bind the soil and reduce soil erosion. The grasses die back to ground level in autumn in which seeds lie dormant until the snowmelt, spring rains and higher spring temperatures stimulate germination. This leads to rapid growth in the summer. The largest store of nutrients is the soil and includes the rhizomes and roots of the grasses.

The soil profile

Horizon

Decomposed & part undecayed material	O	
The top-soil		
Well mixed organic and inorganic material	A	shedding is at its height here
The eluvial zone		
The sub-soil		
This horizon receives material, washed from above	B	accumulation important in this zone
The illuvial zone		
Weathered parent material	C	

Case study: Threats to prairie biodiversity

North Dakota: There was once a range of breeding grounds for migratory birds and a variety of environmental niches for other animals related to climate, relief and soil. But, 98% of the animal habitats have been destroyed and native plants have been outcompeted by alien species. Other impacts of human activity in the Prairies include:

- modern cereal farming, which causes soil erosion of up to 1 m of soil being lost from some fields
- damming major rivers, irrigation and flood control mean prairie grasslands have disappeared under the water behind dams and under flood and irrigation schemes
- hunting and trapping, which has resulted in 60 million bison lost from the prairies since the seventeenth century
- mineral extraction – little environmental consideration is given to mining projects.

The 1930s 'Dust Bowl': In the 1930s, the prairie lands of south-east Colorado, south-west Kansas and parts of Oklahoma and New Mexico, Texas and New Mexico were turned to the 'Dust Bowl', due to severe drought, overgrazing and attempts to plough and cultivate former grassland areas. It led to economic ruin for many farmers. Out migration took place from the region in the Great Depression. Soil erosion remains a problem in the area.

Cattle ranching: Cattle ranching and beef production are an important part of the economy in the USA. Ranching should work well on the prairies, the turf should remain unploughed and the cattle merely replace wild animals such as bison. But, the carrying capacity of the Prairies has probably been reached as stock levels have risen – in many cases at the request of the government! Production of red meat is not sustainable and does not feed great numbers of people.

Reduced plant cover leads to less leaf litter, higher evaporation and lower interception rates. This leads to faster run-off with a greater risk of flood and a lack of ground cover which accelerates soil erosion. Soil that gets into rivers silts them up and contributes to eutrophication. Cattle also damage river banks (the so-called riparian zone) and invertebrate habitats. They drink more than native herbivores and rely on irrigated crops such as alfalfa, which has increased demand for water too. At least half of the available forage on the prairies is eaten by cattle. Grass-fed ranging cattle require much lower levels of inputs, limiting wildfire hazards.

The tundra biome

AQA **AS U1**   Edexcel **A2 U3**
OCR **AS U1**   WJEC   **A2 UG3**

Characteristics of tundra include:
- a short season of growth and reproduction
- low biotic diversity
- a simple vegetation structure
- an extremely cold climate
- energy and nutrients in the form of dead organic material
- large population oscillations
- limitation of drainage.

The tundra is the coldest of all the biomes. Tundra comes from the Finnish word *tunturi* (treeless plain). The tundra landscape is moulded by frost; it experiences extremely low temperatures, little precipitation, poor nutrients and has a short growing seasons. Dead organic material provides the nutrient pool. Tundra is separated into two types.

Arctic tundra

The **Arctic tundra** circles the North Pole and extends to the coniferous forests (the taiga). The tundra region experiences cold, desert-like conditions. Only about 60 days have conditions suitable to grow crops in. Temperatures range from −35° C to 15° C and rainfall is 15–25 cm. The soil that forms sits on a layer of permanently frozen subsoil called **permafrost** (mostly of gravel and finer material). In the summer the surface is often boggy.

Even in this extreme biome some 1700 small shrubs, sedges, mosses, liverworts and grasses still survive. Animals have also to handle the cold conditions if they are to survive. Young are raised quickly. Many mammals and birds have large amounts of fat to insulate themselves against the cold. Many animals hibernate during the winter because food is not abundant; others migrate towards the South.

Plants are adapted to the cold winds; being short and grouping together to resist the cold. They are able to photosynthesise at a low temperature and low light intensity. Most reproduce asexually.

Alpine tundra

The **Alpine tundra** is located on mountains throughout the world at high altitude above the tree-line. The growing season is approximately 180 days. The night time temperature is usually below freezing. Unlike the Arctic tundra, the soil in the Alpine is well drained. Plants and animals are well adapted to the Alpine tundra.

There are several threats to the tundra.
- Oil and gas development contributes to global warming. As the permafrost melts, because of oil field development, the tundra ecosystems collapse. As permafrost begins to decay they release more carbon dioxide and this accelerates global warming.
- Ozone depletion at the North and South Poles; stronger UV rays destroy the tundra.
- Air pollution poisons lichen, which in turn feeds many animals.
- Buildings and roads draws heat into the permafrost and causes it to melt.
- Invasive species, such as Japanese knotweed, push aside native vegetation and reduce the diversity of plant cover.
- Oil spills can kill wildlife and significantly damage tundra ecosystems.

3.3 Threats to fragile biomes

LEARNING SUMMARY

After studying this section, you should be able to understand:
- the importance of biodiversity and its management to the efficient functioning of the planet
- that threats to biodiversity threaten mankind

The threat to ecosystems, biomes and biodiversity

AQA   AS U1/A2 U3
OCR   A2 U3
Edexcel A2 U3
CCEA   AS U1/A2 U2
WJEC   A2 UG3

Biodiversity describes the range of genes, species and ecosystems in an area. When an ecosystem is healthy it means there are high levels of biodiversity. The biosphere is very important to the Earth – it converts carbon dioxide into oxygen, provides rare and unusual medicines, recycles chemicals and contains all the systems (air, water, heat, land) that support life on Earth.

What threatens biodiversity?

There are few pristine ecosystems left in the world. As populations increase and resource boundaries are extended, so humans destroy and disrupt biodiversity. This is done through rapid tourism development, over-fishing, mining, alien species invasions, eutrophication, deforestation, desertification, climate change and farmland encroachment, especially when located in marginal areas. There are few areas in the world that are totally natural. These few areas have to be protected and are known as 'biodiversity hotspots'. These areas that face threats are unique and have thousands of species present (generally we are talking about tropical areas and some mountains and islands).

Attitudes to biodiversity around the globe

In MEDCs people have a positive attitude towards ecosystems and want them protected, conserved and restored. Importantly MEDCs can pay for this. In newly industrialised countries (NICs) and recently industrialised countries (RICs) ecosystems are used as a resource in economic development. In LEDCs the ecosystem in many cases has yet to be completely and irretrievably exploited.

Biodiversity hotspots

AQA	AS U1/A2 U3
OCR	A2 U3
Edexcel	A2 U3
WJEC	A2 UG3
CCEA	AS U1/A2 U2

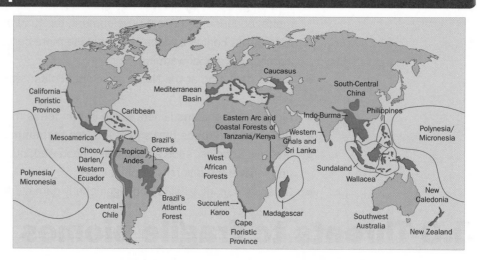

Biodiversity hotspots for conservation priorities

What influences biodiversity?

- The amount of sunlight and rainfall received.
- Quantity and quality of nutrient cycling.
- How often humans interfere with the ecosystem.
- Where complex food chains are supported biodiversity tends to be high.
- Where evolution is allowed to proceed unaffected by humans, biodiversity tends to be high.
- Where there are a range of ecological zones biodiversity tends to be highest (altitude and distance from the coast can affect the amounts of biodiversity).

Why preserve biodiversity?

Ecosystems have 'different' values – some are intangible, some are financial and some are essential for life on earth!

- Biodiversity acts as a regulator (controlling floods and the atmosphere's gases).
- Biodiversity is a supplier (of food, fuel and wood for building).
- Biodiversity is important in intangible ways (for tourism and for education for instance).

The management of biodiversity

The world's population cannot continue to use and abuse the global ecosystems. All ecosystems have a level at which no long-term decline in the system will occur. This level is called the **maximum sustainable yield level**. What is important to realise is that this level leaves no room for changes brought about by global warming for instance. What is clear is that many interested parties have different agendas with regards biologically diverse areas. Some see ecosystems as resources to be exploited while others see them as areas to be conserved at all costs. Others look only to conserve and fight, e.g. the World Wide Fund for Nature to preserve species such as the panda. Others work to seal the concept of **keystone species** such as the bee. Then there are organisations, such as the World Trade Organisation (WTO), intent on conserving huge **eco-regions** (e.g. the Amazon) or small biologically **diverse hotspots** (such as island ecosystems). Ecosystems and biodiversity can be managed in many ways. At one end of the scale are exploitative practices (e.g. zoos) and at the other end are protective practices (e.g. the limited entry to Antarctica). Conservation management has increasingly gone down the biosphere reserve model route and in many locations ranging from the Galapagos Islands to Devon!

Case study: The North Devon UNESCO Biosphere Reserve – small scale

The North Devon coast Area of Outstanding Natural Beauty (AOB) was designated in 1959 and covers approximately 171 km sq of coastal landscape. It is an important designation for the Biosphere Reserve. It includes the North Devon Heritage Coast and the core area of the UNESCO Biosphere Reserve at Braunton Burrows. The Biosphere Reserve is home to a number of rare and characteristic habitats and species. These include:

- A culm grasslands habitat that is unique to south-west England and increasingly rare. It is home to otters, glow worms and the marsh fritillary butterfly. The grasslands help prevent flooding by soaking up water, holding on to it and releasing it gradually into the rivers.
- The Taw/Torridge Estuary is home to pea crabs, lugworms and hydrobia snails. The mudflats support wading birds such as the Curlew and Redshank.

- Braunton Burrows (and Northam Burrows) is internationally recognised as one of the finest dune systems in the northern hemisphere.
- Lundy Island, England's first Marine Conservation Area, is home to seals and other marine species, as well as endemic coral.
- Western oak woodlands – the clean damp air blowing in off the sea and across the reserve provides ideal growing conditions.

Furthermore, farming is encouraged and generally follows the Biosphere Reserve needs that farming should support the continued health of wildlife and the environment's capacity to deliver ecosystem services (e.g. reduced carbon emissions or flood prevention) and also to preserve and enhance the landscape.

What future for the world's biodiversity?

> 2010 was the UN International Year of Biodiversity.

Four possible futures were identified in the *Global Environment Outlook: environment for development (GEO-4) Report* published in 2007. The report assessed the state of the global atmosphere, land, water and biodiversity and identifies priorities for action. It identified four 'futures' for the world.

- A **markets first world** where profit drives continued the degradation of biodiversity.
- A **policy first world** where humans are the primary concern, with ecosystems protected when possible.
- A **security first world** where wealth remains with the rich MEDCs and the environment continues to be exploited.
- A **sustainability first world** where humans and ecosystems are given equal weighting and ecological well-being is given high importance.

Case study: Northwest Forest Plan (NWFP), USA – landscape and large scale

The Northwest Forest Plan (NWFP) consists of policies that determine land use in 24 million acres of the north-west region of the USA. It was developed in 1994 to protect the northern spotted owl habitat and now aims to protect rivers, wildlife and forests. There are National Parks, National Wildlife Reserves and military bases in the area. The aims of the plan were to:

- protect and restore watersheds to provide high-quality water
- protect and restore forests to store carbon and help fight climate change
- restore ecosystems and conserve biodiversity
- restore and embrace natural processes in order to create and maintain diverse ecosystems
- provide appropriate access to nature and

recreation opportunities to reconnect people and communities to nature
- involve the public in forest planning.

The NWFP has been successful in achieving its goals. It has developed into an integrated conservation strategy and it has conserved old-growth forests and all the forests have now become dynamic ecosystems. But the plan has been controversial as timber yields within National Forests have decreased, with job losses both locally and outside the area.

3.4 Management of biomes

LEARNING SUMMARY

After studying this section, you should be able to understand:
- that even small ecosystems can contribute to the urban landscape
- that small biomes are rich sources of biodiversity and need managing and preserving

Ecosystems in urban areas

AQA	A2 U3
OCR	A2 U3
Edexcel	A2 U3
CCEA	A2 U2

Urban ecosystems are familiar to all of us. They cover 4% of the world's land surface. An urban ecosystem is simply the community of plants, animals, and humans that inhabit the urban environment in parks, yards, street plantings, greenways, urban streams, commercial landscaping, disused buildings and which provide the living heart of the urban ecosystem. These communities might seem separate and fragmented, but they are not. Trees in urban areas are a good example. When viewed from above these trees form a 'forest' that is easy to see snaking through many urban areas.

Urban ecosystems are of course highly disrupted systems. Buildings and roads form a covering for the soil that affects how water flows through the urban landscape and what can survive in it. The plants in urban areas are non-native species and weeds. Even in the city's more natural areas, such as parks, the vegetation is often highly altered, e.g. in wooded areas. The ground layer is often replaced by shade-tolerant grass, disrupting the natural soil forming processes.

The development of urban parks

It was the Victorians who came up with idea of the formal urban park. The idea was to have a pleasant green space fringed with big, desirable houses which would sell at a high price to the wealthy giving the investors a good return for their money. These parks, however, would also be close enough for the populous as a whole to enjoy them in their leisure time. The Victorians had a great interest in trees and many new species were introduced into these parks and even the relatively small suburban gardens contained exotic trees and shrubs. Many of the original Victorian parks and squares are still in place today.

In spite of harsh conditions (high levels of air pollutants, road salts, disease, poor soil quality, frequent drought, reduced sunlight), city green spaces are full of life, e.g. Washington, D.C. was found to have over 100 bird species and Chicago was found to have the best remaining fragments of tall-grass prairie and oak savanna in the USA.

Case study: Urban forests, UK

Hampstead Heath in London was once part of the common lands of the manor of Hampstead, while most of the rest was the parkland/farmland of Kenwood House. The Hampstead Heath Society was formed in 1897 to protect and conserve it. The landscape is varied with woodland, meadows, heathland and ponds. Long-established features like hedgerows and old trees provide links with the past. The range of wildlife is extraordinary for an urban site and includes kingfishers and reed warblers. Over 300 species of fungi have been recorded and it is also home to a number of species of bats.

Epping Forest is an area of ancient woodland on the Greater London and Essex border. It is a former Royal Forest and is managed by the City of London Corporation. It contains areas of woodland, grassland, heath, rivers, ponds and bogs and has numerous scientifically important sites. It covers approximately 72 km^2. Lying between the rivers Lea and Roding the gravelly soil was unsuitable for agriculture, but ideal for forestry.

Other urban green spaces

Cemeteries: In the early 1800s private entrepreneurs in many UK cities started to create private landscaped burial parks independent of the church. In 1832 Parliament passed a bill encouraging the development of seven private cemeteries around London. The cemeteries increased burial spaces, but also proved popular for people to walk or ride through.

Wasteland: This includes brownfield sites, demolition sites, brick pits and disused and under-used allotment sites. The particular value of these sites are the micro-habitats that are created. Secondary successions are frequently seen on wasteland areas, colonising old car-parks and walls.

Threats to urban green spaces

Today the rate of re-development is faster than the rate of creation of urban green spaces. There are three main threats to urban green spaces.

- The main threat is the housing shortage. Government policy in the UK had been to put at least 60% of all new housing on brownfield sites by 2008. By 2005 in London 73% of new housing was built on brownfield sites. Around the world **two billion homes** will be added to urban areas by 2030.
- There is a lack of awareness of the importance of these sites to urban biodiversity. Wasteland sites often look unattractive, are vandalised and often are used for antisocial and illegal purposes.
- These areas are frequently subjected to a cycle of fly tipping and 'tidying' (tidying wasteland actually removes habitats that are best left with the bulk of their rubbish) that is detrimental to wildlife.

Case study: The Thames Gateway, 'England's little rainforest'

The Thames Gateway and West Thurrock marshes (covering 68 acres) contain a large expanse of brownfield land, coupled with nationally important populations of endangered insect and invertebrate species. One derelict oil terminal beside a giant superstore at West Thurrock has been found to have more wildlife per square foot than any national nature reserve. The Thames Gateway has also been identified as Europe's largest regeneration area and will be subject to major house building over the next 10–20 years. A campaign to save the site has begun with the government being asked to take a much broader perspective on the impact of development to this area.

Urban green spaces contribute essential controls to urban areas.

- **Shade and temperature control:** Green spaces help to control urban temperatures. Trees provide shade and also transpire large amounts of water that, when evaporated, provide a cooling effect. The shading effect of trees can reduce energy use and planting more trees can reduce air conditioning use and costs.
- **Air filtering:** Trees can filter out as much as 85% of air pollution and particulates. In Baltimore, Washington it is estimated that tree stands (a continuous area of trees) remove over 17 000 tons of air pollutants/year.
- **Noise reduction:** Trees and shrubs can filter noise pollution.
- **Storm and flood water control:** The permeable nature of green spaces are able to soak up excess water more effectively than hard tarmaced surfaces.
- **Biodiversity and wildlife habitat:** Cities support a variety of plants and animals.
- **Recreation:** Parks and green spaces provide space for recreation.
- **Food production:** Around the world people grow food in their gardens and on roadside verges, etc. In Kenya and Tanzania, two out of three urban families are engaged in farming. In Taiwan, half of all urban families are members of farming associations. In Cuba urban agriculture produces nearly one million tons of organic produce every year. A million or more brownfields exist in cities worldwide and offer the chance to create new green spaces, lessen congestion and ease development pressure on the remaining green areas. If well managed, urban green spaces can really add to the already proven health and education benefits of urban ecosystems.

PROGRESS CHECK

1. Why are the chernozems of the temperate grasslands so fertile?
2. How does alpine and arctic tundra differ?
3. How do urban ecosystems contribute to controlling the microclimates of urban areas?

Answers: 1. Rapid recycling of nutrients. 2. Alpine tundra is found in highland areas and is well-drained. 3. They offer shade and temperature control, filter air and aid noise reduction.

Sample question and model answer

1 Describe some of the factors that could account for the changes in vegetation in an area over time. **[20]**

Candidate roots answer successfully in a suitable case study!

The changes that occur over time in the vegetation of an area are known as succession. A good example of where this can be seen is Holkham Beach in Norfolk, on the sand dunes. On any new patch of land a new species will appear and grow, for instance on the embryo and fore dunes at Holkham sea, rocket and marram grass (both resistant to salty and dry conditions) appear. These are called the pioneer species.

Answer uses the language of the topic.

These pioneer plants have the effect of stabilising the sand with their roots. This allows water to be trapped along with nutrients. Gradually conditions improve enough for new species to colonise the area, which perhaps would not have survived there before. This succession continues with each new species that comes along replacing another as conditions continue to improve. Looking at the dunes at Holkham, the succession of different species can easily be observed by taking a transect back from the embryo dunes to the mature dunes and slacks. The further you head back the more growing conditions improve and diversity of species can be seen. Different species add new nutrients to the soil and eventually certain wildlife can also be found to colonise the area, such as rabbits that also add nutrients to the soil through their faeces, as they survive on the existing vegetation.

An easy read – it flows well.

Eventually there becomes a point when succession begins to cease and the vegetation stabilises to equilibrium with the environment. The final type of vegetation is known as the climax vegetation and can be dependent upon many things. Many geographers believe that the principal factor in determining an area's climax vegetation is the climate. For this reason the vegetation would be known as climatic climax vegetation. Unusually in Norfolk, however, the climax vegetation at Holkham is an area of extensive pine forest. This does not fit the idea that it is entirely dependent on climate, as coniferous trees are usually found in colder environments with very high rainfall, such as the north western side of the UK. The reasons for this climax vegetation must therefore be put down to other things.

Other factors that exist are things like geology and human impact. At Holkham the geology may be the reason and if so the climax vegetation would be known as geo-climax vegetation. The area's geology is such that it would provide the bases for the acid soils, which would suit the growth of conifers.

Maintains a 'local', and small scale focus.

Humans are very likely to have an impact in the vegetation type at Holkham. This would result in a plagioclimax vegetation. Holkham is a very heavily managed area, controlled by English Nature, as the sand dunes are literally being worn away. For this reason many paths have been installed in the dunes and trees planted. It is at this stage that one might reasonably assume that the pine trees had in fact been planted and are in fact a plagioclimax community.

Overall: competent, organised, reasonably accurate and well supported. Consistently relevant to the theme.

Holkham just gives one example of how vegetation can change over time and the many reasons that can account for it. Other examples would easily be seen in the ecosystems of lakes and perhaps mountains.

Exam practice question

1 Explain briefly how man has interrupted and disturbed the nutrient cycle. **[25]**

4 Hydrological and fluvial challenges

The following topics are covered in this chapter:

- The hydrological cycle and drainage basin
- Fluvial processes and landforms
- Water management

4.1 The hydrological cycle and drainage basin

LEARNING SUMMARY

After studying this section you should be able to understand:

- that water is continually entering and leaving the river system
- that this open system coexists within the closed hydrological system
- that drainage basins are more or less self-contained and are therefore a convenient means of considering the effect and action of running water
- that most streams and rivers have quite simple origins

Rivers as open systems

AQA   **AS U1**
OCR   **AS U1**
WJEC   **AS UG1**
CCEA   **AS U1**

Rivers dominate the physical landscape, producing widespread changes. Rivers remove rock and material from mass wasting, particularly – but not exclusively – in the humid areas of the world. Rivers move under gravity in a channel, transferring both water and sediment downstream. Rivers therefore act as open systems. Stores and transfers vary in size according to changes in inputs of water. If all sections of the 'system' balance it is said to be in **dynamic equilibrium**.

> Look out for this question. It is often set using the approach of 'fill in the missing word'.

The hydrological cycle

What happens in the river is part of a cycle of events called the hydrological cycle (below).

> You need to learn this drainage basin terminology.

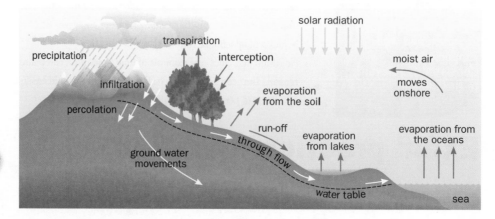

It is possible to compile graphs of precipitation, potential and actual evapotranspiration. These are called soil moisture budget graphs (below).

The soil moisture budget for Norfolk.

Rainfall in mm

soil moisture deficit

rainfall recharge of soil

soil full

soil at its driest

J F M A M J J A S O N D

------ precipitation

·············· actual evapotranspiration

——— potential evapotranspiration

A knowledge of the soil moisture budget and potential evapotranspiration allows irrigation 'need' to be calculated.

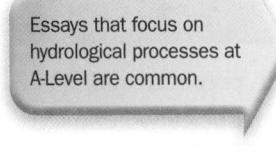

Essays that focus on hydrological processes at A-Level are common.

KEY POINT

Evaporation: Water that is warmed, usually by the sun, so that it changes into a gas (water vapour).

Overland flow: When water flows over the surface of the ground. This occurs for a number of reasons. The soil may be saturated and therefore unable to absorb any more water or the underlying rock may be impermeable or the ground may be frozen.

Percolation: The movement downward of water through the soil or underlying porous rock. It collects as groundwater.

Precipitation: Water falling in any form. Rain, sleet, hail, snow and dew are all forms of precipitation.

Throughflow: The movement of water within the soil sideways towards the river.

Transpiration: The water loss from vegetation into the atmosphere.

Water table: The top of the zone of saturation.

The river represents the flow of water overland, but two further key 'hidden' aspects are of significance – infiltration and groundwater.

Infiltration

This is probably the most significant and fundamental portion of the drainage basin system, contributing as it does to groundwater storage, throughflow and stream channel recharge.

Infiltration rates are affected by the following.

- The **nature of precipitation**; its duration, total, the area covered, the size of raindrops, frequency and its chemical composition.
- **Vegetation interception** depends on the nature of the vegetation, e.g. pine forests take out as much as 96% of low-intensity rainfall, leaving only 4% to infiltrate.
- The effect of **depression storage**, the geological structure and man's agricultural activity. Water captured in the landscape or plough-furrows infiltrates!
- **Evapotranspiration** – available heat, wind, and the texture and depth of soils affect how much water returns to the atmosphere from exposed water surfaces and from vegetation.
- The **nature of the soils**, their permeability, porosity and texture.
- **Slope length and angle**, determines whether run-off or infiltration is the dominant process.

divides of sub-basins

source

confluence

tributary

catchment area

mouth

watershed

Groundwater

Groundwater appears and collects in permeable rocks, known as **aquifers**, which are saturated as a result of infiltration by rainfall. Water flows between the small grains of rock and soil that make up the aquifer. Variations in the level of the water table reflect the surface topography, i.e. the water table is near the surface in valleys and deep down in the hills. It is because of the topography of the landscape that groundwater 'flows'. The velocity of flow is proportional to the **hydraulic gradient** (simply the slope of the water table induced by the topography of

It is important that you know and understand the nomenclature of the drainage basin.

the landscape); speeds are typically between 1m/year to 1m/day. Most rivers gain their water/flow from a combination of surface run-off and groundwater discharge. Groundwater contributes most water in the summer and autumn in the UK, surface run-off contributing in the wetter winter and spring months, e.g. Edwards Aquifer, San Antonio, USA and the chalk aquifers of southern England.

PROGRESS CHECK

1. What is infiltration?
2. What is permeability?
3. What is throughflow?
4. What is percolation?
5. What is the water table?
6. What is groundwater flow?
7. What is overland flow (surface run-off)?

Answers: 1. Water soaking into soil. 2. Water seeping through soil. 3. Movement of water downhill through soil. 4. Seepage into the water table. 5. Top level of saturated rock. 6. Flow below water table through permeable rock. 7. Water flowing over the land.

4.2 Fluvial processes and landforms

LEARNING SUMMARY

After studying this section you should be able to understand:

- that erosion, transport and deposition determine fluvial landforms

Fluvial processes

AQA	AS U1
OCR	AS U1
WJEC	AS UG1
CCEA	AS U1

As water flows potential energy (the energy of position) is converted to kinetic energy (the energy of motion).

River channels not only transfer water and sediment, but the **potential** and **kinetic** energy within the water does the work of erosion, transportation and deposition within the river. Flow type affects the efficiency of a river to erode, transport or deposit. Two principle flow types are recognised: **turbulent flow** and **laminar flow**. Both are affected by friction.

Energy and flow

The ability of a river to perform processes of erosion, transportation and deposition is determined by the amount of energy it possesses.

Energy flow relationships

Faster rivers have more energy so more work is done. A real examiners' favourite at AS level is to be able to read and interpret Hjulström graphs.

During periods of low flow, below bank stage flow or when rivers are at base flow, low energy conditions are experienced and little 'work' is done. Channel adjustments are most likely to occur during the rising phase of the storm hydrograph, when there is a lot of water in the system; it is at this point that erosion is at its height. Deposition occurs as discharge and energy declines on the falling limb of the storm hydrograph. A relationship exists between volume and velocity and individual particle erosion, transportation and deposition. This is displayed on the **Hjulström curve** (see opposite).

The Hjulström curve shows the relationship between the size of sediment and the velocity required to erode, transport and deposit it. The critical erosion curve shows the **minimum** velocity required to lift a particle of a certain size. The critical deposition curve shows the **maximum** velocity at which a river can be flowing

1000 | 100 | 10 | 1
Velocity (cm/s)

Erosion

Transportation

Deposition

0.01   0.1   1.0   10   100
Diameter of particles (mm)

Hjulström curve

before a particle of a certain size is deposited. The zone in-between is the zone of transport. Note that the velocities for transport are lower than that for erosion, because it takes much more energy to lift sediment than to maintain it in transport. The other strange pattern is that it takes more energy to erode some of the smallest particles. This is because they are clay particles which are strongly bonded together and therefore require a lot of energy to be eroded.

Competence is the maximum size of load that a river can carry and this is largely determined by velocity. The **capacity** is slightly different in that this is the total amount of load carried.

In summary, river landforms and landscapes go through long periods of stability, interspersed with short, but rapid, periods of change, related to variations in energy input and material surges.

The energy available to a river enables it to accomplish three main types of work.
- The land surface over which the river flows is **eroded**.
- Eroded material is **transported** away.
- The river **deposits** the material that has been transported.

> Process questions are frequently used at the start of structured questions.

KEY POINT

Increased energy/water input increases channel capacity and efficiency. Width and depth increase and velocity and throughput of water increases. Erosion and transportation are at their most efficient.
Decreased energy/input of water subsides and so discharge and velocity decreases. Erosion ceases and transportation slows, deposition begins and the river becomes narrower and shallower.

Erosion

To understand the important work that erosion completes in a river, three areas have to be further explored: **volume**, **velocity** and **load**. The latter is important as a river 'charged' with sediment is able to wear the land surface away more effectively than one that is merely dissolving it away.

Volume
Most streams and rivers obtain their water from rainfall or any of the other forms of precipitation. This precipitation evaporates, soaks in or contributes to the run-off or drainage of the land surface. As rivers flow from high (source areas) to lower areas (the mouth, usually the sea, unless the river has entered an arid basin), their volume increases as contributions from other parts of the drainage basin via tributaries are added. There can be variations in a river's volume relating to seasonality of rainfall (in monsoon areas), the contribution of snowmelt and of springs and groundwater.

> Discharge is defined as the volume of water passing a particular point in a river in a unit of time, expressed as m^3/s^{-1}, or in cumecs.
> $Q = A \times V$ (A = Cross Sectional Area (Width × Depth) and V = Velocity

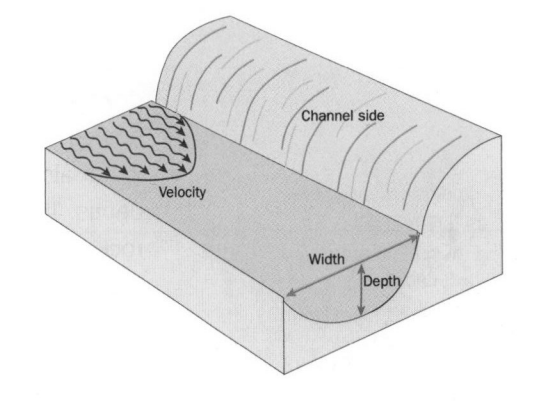

Channel side

Velocity

Width

Depth

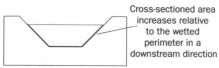

Cross-sectioned area increases relative to the wetted perimeter in a downstream direction

A hydraulic radius is a cross-sectional area divided by wetted perimeter. The higher the number, the more efficient the river is at transporting the water through it. The most efficient parts of the river are in the lower course.

Know this for AS. Remember the Hjulström curve?

Velocity

Velocity is more or less constant along the length of the river. It is true that steeper slopes do encourage higher velocities, but the larger channels of the lower course exert relatively less friction than the small channels of the upper course, causing an increase in velocity and allowing the river to become much more efficient.

Load

Small streams carry a greater quantity of fine material than coarse material. Conversely, large rivers with more available energy carry larger/coarser material.

The **erosive power** of a river is for the most part determined by the **charge** of debris it carries. Running water has restricted erosional ability. The proviso being that with increased load a river is more likely to begin to aggrade, or deposit.

> **KEY POINT**
>
> Volume of water carried + velocity of this water = energy of the river.
> Energy availability determines the capacity (total load) and calibre (the weight/size dimensions of individual particles).

The mechanism of erosion

For a particle to be used in the erosional process it has to be removed from the bed and banks of the river by erosion and entrained. This entrained material is acquired in three ways.

- **Vertical erosion** deepens channels, aided by weathering mass movement and soil creep. Characteristics of channels undergoing vertical erosion include a large bed load comprising coarse hard particles. **Potholes** are common, as are deep narrow gorges.
- **Lateral erosion** increases a river's width. A large sediment load has to be entrained for this process to work most effectively. It is responsible in conjunction with the processes of slope transport and mass movement for valley widening, meander migration and river cliff formation.
- **Headward erosion** increases the length of a river. This process is most active in the source area of a river or where a bed is locally steep. It causes accelerated erosion and is commonly associated with **waterfall** formation.

Actual erosional work is then carried out by a number of processes.

- **Corrasion** and **attrition** are two processes which rely upon the load of the river to achieve their effects. Corrasion occurs most often during periods of higher river flow, bed load being used as an abrasive agent, scratching and scraping at the solid bedrock. A correlation exists between this process and accelerated vertical erosion. The debris that results from the corrasive processes is free to collide and bash into itself. This process is known as **attrition** (or **communition**) and causes a reduction in particle size in a downstream direction. Smaller particles ensure that the channel remains smooth and that friction does not compromise efficiency in the lower course.

Erosion = wearing away and movement.

Erosion + weathering = denudation.

Weathering break down rock *in situ*, little movement is involved.

- Water alone is not as effective an agent of erosion as a river 'charged' with debris. The process **hydraulic action** is least effective in areas of hard base rock. In the middle and lower courses, where the bed and banks are likely to be composed of incohesive sediments and where there is a degree of sinuosity, bed and bank scour is effective in removing material vertically and laterally. An extreme form of hydraulic action is **cavitation**. The sudden and violent collapse of bubbles created by this process shatters banks extremely rapidly.

- **Corrosion** is a process where the solvent action of water dissolves soluble materials and carries them away in solutions.

Transport

For sediment to be moved the following must occur.
- Resisting forces have to be overcome.
- When drag and embedded particle **inertia** is overcome and the particle begins to move, this is called the **critical tractive force**.
- **Competent velocity** has to be achieved. This is the lowest velocity at which particles of a particular size are set in motion, i.e. the bigger the particle the greater the velocity needed to move it.

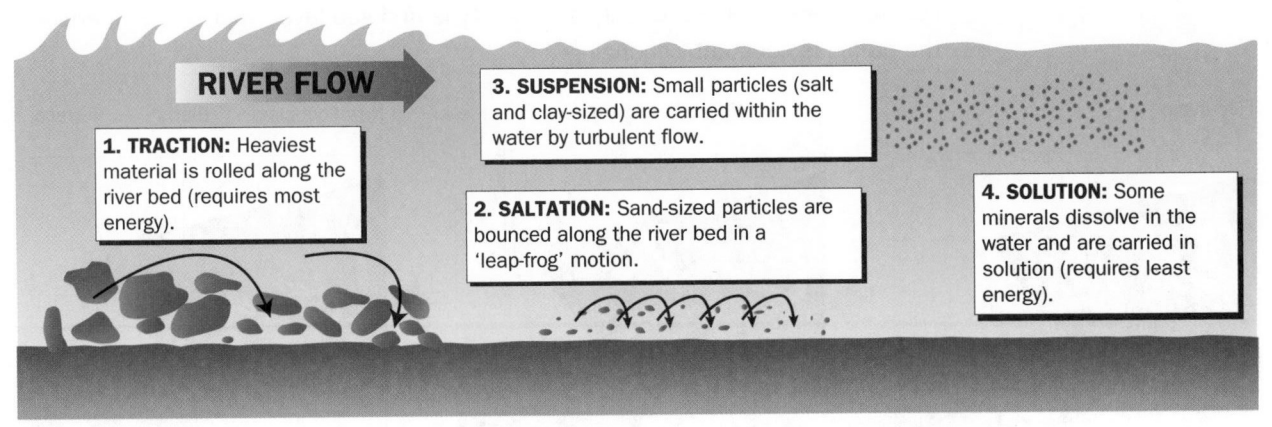

Methods of sediment transport in rivers

Downstream changes in sediment
- Amounts of material moved increase in a downstream direction, as weathered material is input and as tributaries add material.
- Material gets progressively smaller and rounder on its journey downstream.

A stream which is 2 m wide by 10 cm deep can move 1½ tonnes of alluvium daily to the sea.

Deposition

Depositional processes occur when the carrying capacity of a river is reduced. Several factors can affect the capacity of a river to retain its transported load.
- **Water velocity** changes due to changes in gradient or a break in slope. This is usually caused by variations in geology.
- **Geology** can, by changing the chemical composition of river water, cause rapid vegetational growth. This slows water and causes deposition.
- **Evaporation** (as on the Nile) or over-abstraction (as on the Colorado) can reduce flow and consequently bring on deposition.
- Additional **debris and water volume from tributaries** can result in the slowing down of water and deposition.

Thus, deposition is not confined to the lower reaches of rivers: it can occur at almost any point along the river's course. This localised deposition can, during periods of high flow, cause localised flooding.

Change and grade

Knowing these changes might be useful in your enquiry work.

Different parts, or sub-sections, of a river's course have different characteristics. The increasing efficiency is achieved through changes in the variables that influence the channel shape and form, as shown on page 88.

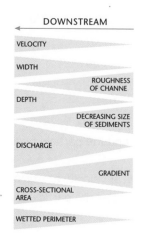

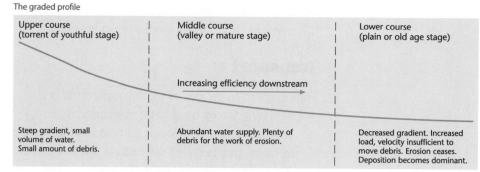

The graded profile

Upper course (torrent of youthful stage)	Middle course (valley or mature stage)	Lower course (plain or old age stage)
Steep gradient, small volume of water. Small amount of debris.	Increasing efficiency downstream. Abundant water supply. Plenty of debris for the work of erosion.	Decreased gradient. Increased load, velocity insufficient to move debris. Erosion ceases. Deposition becomes dominant.

The graded profile, or state of equilibrium, is achieved when the river's course is as efficient as it can be from source to mouth. Do remember though – a river course is rarely graded! Changes in rock type and sea level (**rejuvenation**) ensure the profile is constantly changing!

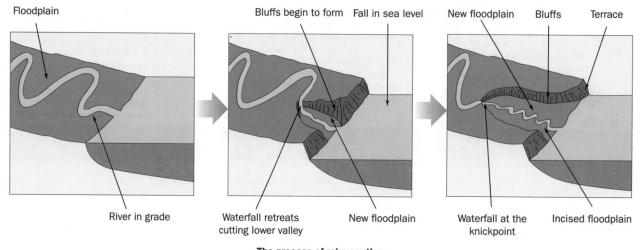

The process of rejuvenation

> Rejuvenation offers explanations for many of the more unusual river landforms. Understand its contribution for both AS and A2 examinations.

A river is said to be **rejuvenated** when its base level is lowered. This can be caused by **coastal uplift** (isostatic readjustment – after an ice age or due to seismic activity) or by a **drop in sea level**. When base level is lowered the river has more **potential energy** and is therefore able to cut vertically into its bed. The result of this is that the long profile is extended. A **knickpoint** marks the point where the graded and ungraded profiles meet; there is often a **waterfall** at this point.

PROGRESS CHECK

1. Why does the velocity of a stream increase downstream?
2. What does the Hjulström curve help explain?
3. Why does clay require more energy to be eroded in rivers?
4. What is the difference between competence and capacity?

Answers: 1. More water from the drainage basin, tributaries, greater efficiency. 2. It shows the relationship between velocity and competence. 3. Clay particles stick together. 4. Competence is the maximum size of particle carried. Capacity is the total amount of material a river can carry.

Fluvial processes

AQA	**AS U1**
OCR	**AS U1**
WJEC	**AS UG1**
CCEA	**AS U1**

> You know the processes, now learn the 'features' – especially for the structured papers!

> It is important that you can draw diagrams to support your knowledge of landforms, waterfalls, gorges, etc.

This section examines the effects of river processes on the landscape. The variation in gradient, volume of water and amount of debris in the different stages of the river from source to mouth lead to the development of characteristic landforms in the different river stages.

The upper course

Features of the upper course include:

- **interlocking spurs**, e.g. River Dane, Derbyshire
- **potholes** where pebbles and debris swirl around in joints and hollows on the river bed, gradually drilling a hole in the river bed, e.g. River Taff, Glamorgan
- **waterfalls and rapids** which are common in the upper course, where there is a variety of different strength rocks, steeper gradients and fast water.

interlocking spurs

'v-shape'

Case study: The formation of Hardlip waterfalls – High Force, County Durham

High Force is a waterfall on the River Tees, some 20 metres high. The Tees plunges over hard rock formed by the Great Whin Sill. The lower section of the waterfall is composed of carboniferous limestone. The carboniferous limestone wears away more rapidly, leaving a narrow (700 m) deep gorge in front of it.

> Over the course of time, waterfalls and rapids migrate upstream. This forms a dramatic transverse profile, called a gorge.

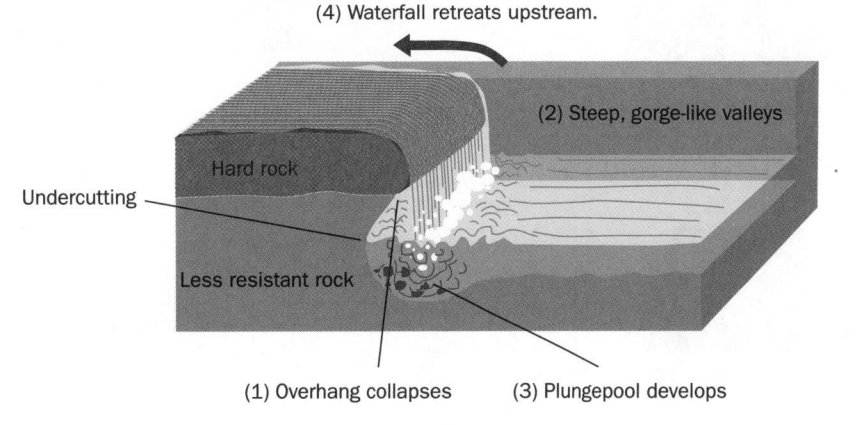

(4) Waterfall retreats upstream.

(2) Steep, gorge-like valleys

Hard rock

Undercutting

Less resistant rock

(1) Overhang collapses   (3) Plungepool develops

Soft rock is undercut (1). This leaves a layer of hard rock which overhangs the layer of soft rock (2). The water flows over the overhang and creates a plunge pool in the soft rock below (3). Eventually the overhang will collapse due to the erosion of the soft rock beneath it. The waterfall then retreats up stream (4). This creates a steep, gorge-like valley.

The middle course

The change in the shape and size of the valley is due to the way water flows through **meanders**: erosion occurs on concave banks, deposition on convex banks. Meanders gradually move downstream.

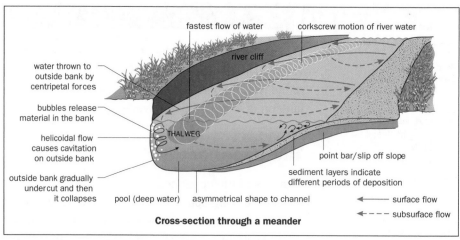

Cross-section through a meander

Meanders are the principal feature of the middle course. It is not clear how they form, but there are a number of misconceptions as to their initiation. Obstructions are unlikely to initiate meanders; it is more likely that the deformation of the river bed holds the key. Meanders may/may not be attempting to release excess energy from the river system, e.g. River Ribble, Lancashire or the River Yare, Norfolk.

> Controversy surrounds the formation of meanders. It makes them a favourite with examiners at both AS and A2.

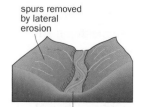

spurs removed by lateral erosion

the valley floor widens as meanders start to wander downstream

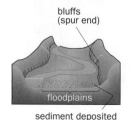

bluffs (spur end)

floodplains

sediment deposited by meandering river

Flood plain formation

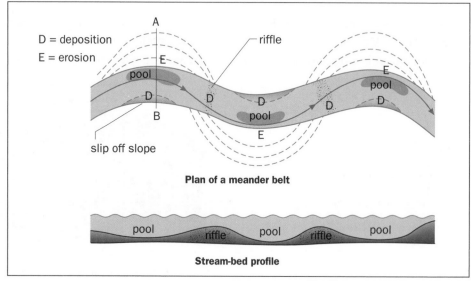

Plan of a meander belt

Stream-bed profile

As the meanders continue to grow in size, the floodplain gets bigger. Bluffs are cut. When the river floods it leaves material (deposits) over a floodplain.

The lower course

Thee are several features of the lower course.

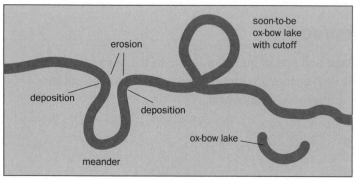

Ox-bow lakes

- As meanders grow in size **ox-bow lakes** can breach their banks. When this happens a cut-off/mort lake or ox-bow forms, e.g. Mort Lake, London or the Trent Valley, Nottinghamshire.
- **Braiding** occurs when a river flows or moves through a series of interlocking channels, rather than through a single thread. They are thought to form because of load *vs* discharge differences, induced

by changes in slope or additions of water from tributaries/floods. They are highly unstable as the river is trying to achieve a more efficient profile (see diagram below).

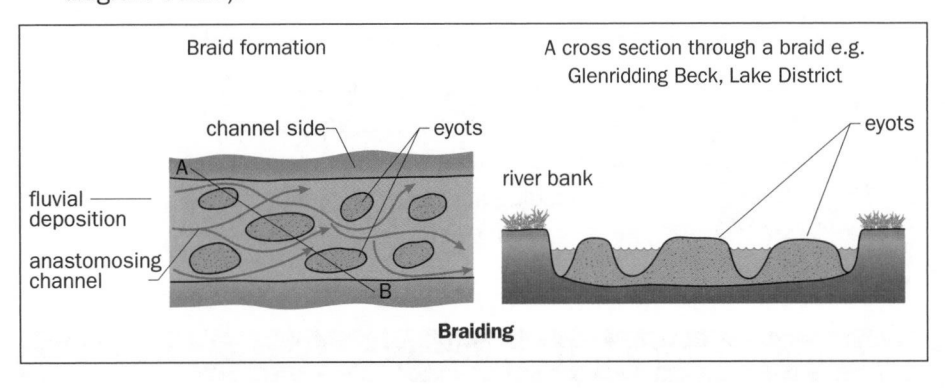

Braiding

Flocculation is where clay particles become attached in a fragile structure called a floc. This heavier floc drops at the flow of the river.

- **Deltas** are the biggest of the world's rivers when they reach the sea, a lake or a lagoon with a massive load of material. This debris is dropped into the calmer water of the receiving areas. The salty water flocculating debris also aids the formation of deltas, as does the shallow angle of the coastal strip. Three main types of delta exist: **arcuate**, e.g. the Mekong Delta, south-east Asia; **birds foot**, e.g. the Mississippi delta, USA; **estuarine**, e.g. the Seine delta, France.

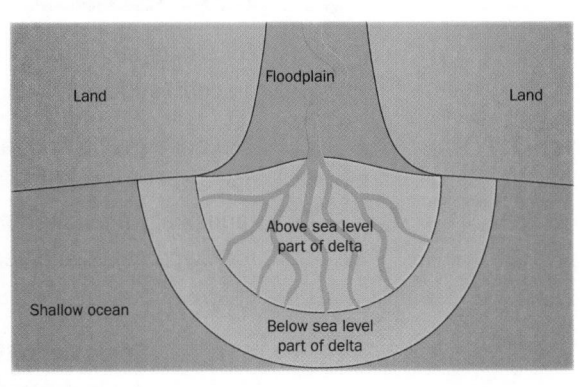

The formation of deltas

The formation of levees

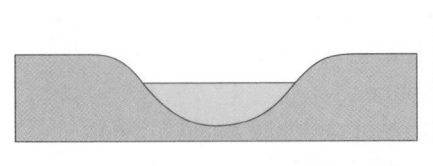

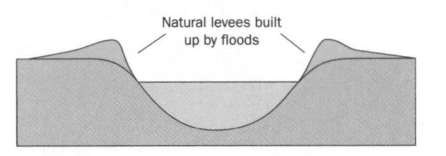

The formation of levees

PROGRESS CHECK

1. Describe and explain how ox-bow lakes are formed.
2. Describe how a flood plain is formed.
3. Describe and explain how levees are formed.

Answers: 1. When the neck of the loop of a meander is broken through. Deposition dams off the loop, leaving the lake. 2. As meanders migrate downstream they cut off interlocking spurs forming a straight sided wide flat valley. 3. Levees are natural features formed after flooding.

4.3 Water management

LEARNING SUMMARY

After studying this section you should be able to understand:

- the effects of variable regimes in rivers
- the importance of hydrographs
- human effects on the river basin
- flooding and flood protection
- managing water as a resource

Variable regimes in rivers

AQA	AS U1
OCR	AS U1
WJEC	AS UG1
CCEA	AS U1/A2 U2

Know in detail the importance of hydrographs for your AS exams.

The regime of a river refers to variations that occur seasonally in discharge. Regimes are affected by climate, e.g. in tropical/equatorial climates there may be a regular or simple cycle exhibited in the regime. In the 'seasonal' temperate climates of Western Europe more complex regimes are common with 'multiple' peaks, affected variably by snow/glacial melt, rainfall and evapotranspiration.

Hydrographs

For single storm/precipitation events the relationship between precipitation and discharge is shown on a **storm hydrograph**.

The appearance and labelling of the hydrograph are shown below. Various factors can affect its shape. Physical variants, such as the nature of the inputs (principally precipitation) and the characteristics of the catchment and shape, rock type, relief and vegetation can cause changes in the storm hydrograph shape.

Some definitions and influencing factors

- Hydrograph size and shape:
  - high rainfall = greater discharge
  - big basin = greater discharge
  - elongated basin = steady discharge
  - large symmetrical basin = flashy discharge
- **Lag-time:** the time interval between the peak of the rainfall event and the maximum discharge. It can be affected by channel steepness and drainage basin shape.
- **Peak-flow:** greatest in the very largest drainage basins. Steep mountainous catchments cause high peaks. Lowland catchments have flatter peaks.
- **Baseflow:** maintains river flow away from flood periods.

How three factors affect hydrograph shape

(a) Density of streams

(b) Basin shape

(c) Steepness of long profile

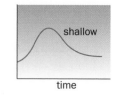

Human activities

Human activities have direct and indirect effects upon the hydrology of a basin.
- The input of water to a river channel by land drainage.
- The storing of water in reservoirs increasing evaporation and regulating discharge.
- Changing patterns of throughput by deforestation decreases interception and evapotranspiration and increases overland flow. Urbanisation creates impervious tarmac surfaces preventing infiltration, increasing overland flow and resulting in shorter lag times and higher discharges. Humans affect outputs by irrigation and water abstraction and induce long-term changes in groundwater supplies.

> Show your knowledge of topical geography here. Use up-to-date examples and collect newspaper articles of actual events.

Floods

AQA	**AS U1**
OCR	**AS U1**
WJEC	**AS UG1**
CCEA	**AS U1/A2 U2**

> Small basins are especially prone to flooding – 80% of lives lost to drowning occur in such floods. Know some case studies.

Populations have always chosen to live near water.
- Soils near rivers and on floodplains are very fertile and can be used for agriculture.
- Rivers are sources of drinking water and provide water for agriculture.
- Rivers enable trade to happen.

People generally avoid settling directly on the floodplain, instead building their settlements on the bluffs or terraces. Good examples are Washington, Paris and Budapest where people are relatively safe from floods. However, as populations grow they need extra land which is situated nearer the rivers.

Causes of flooding

> Remember flooding is a natural process and is simply the response to too much water in the system.

Many floods are directly related to the weather. **Rain** falling at extremely **high rates**, or for an unusually **long period** of time, is a common cause of flooding. Areas that receive a great deal of snow in the winter experience springtime flooding when the **snow** and **ice melt**, such as in the Rhone Valley in southern France. Rainfall and snowmelt can sometimes combine to cause floods.

> You must know all the reasons why rivers flood, including those that are caused by human and physical factors.

Floods also occur as a result of a combination of factors that indirectly involve weather conditions, e.g. low-lying coastal areas that experience **high tides**, storms and high winds from low pressure systems. Furthermore, tsunamis, volcanic heating and rapid melting of a snow pack or even dam failures can cause flooding.

> When excess discharge is present in a river increased erosion takes place. If increases in discharge are too rapid, water moves outside the channel and onto the floodplain.

Without the presence of people on the floodplain floods would not present the risk that they normally do. People living on the floodplain contribute significantly to the problem. Normally, vegetation intercepts significant amounts of precipitation and returns it to the atmosphere before it hits the ground and is absorbed by the earth. However, a number of man-made processes and practices, such as **clear felling** and **animal grazing** hamper these processes. The ground has to absorb more moisture than it would otherwise. When the field capacity is reached, the likelihood of flooding increases. The construction of concrete and stone **buildings** and **tarmaced roads** also contribute to the flooding problem. Rain is absorbed into porous materials – it is not absorbed into man-made material. The result is additional run-off and flooding.

> Water weighs 64 lbs/cubic foot and travels at 6–12 mph. Under certain conditions 2.54 cms of rainfall might have the same energy potential as 54 000 metric tonnes of TNT!

Floods are probably the most dangerous form of natural disaster. More people are killed by floods than all the other categories of natural disaster. When a flood occurs it is virtually impossible to stop. This is because water is heavy and can move with great speed. The case studies covered later in this chapter develop the above causes and many others.

Coping with flooding

The potential of a flood to destroy life and property means that people have had to develop ways to prepare for this destructive type of natural disaster. Dams have been constructed to limit excess water from inundating drier areas. Water from flooding has been directed away from populated areas to planned **flood storage areas**. Such **reservoirs**, or **sumps**, give the soil time to absorb the excess water. The system is used extensively along the Mississippi in the USA. **Excess water** is returned to the river after the flood has passed.

Techniques and examples of soft engineering are given below. Other methods that are used include **hazard zoning** (where planners predict areas that might flood) and by restricting **encroachment** on floodplains (where buildings are not allowed to restrict the flow of flood water or allow it to pond).

As long as people choose to live on low-lying floodplains and flood prone areas engineers can only do so much to protect them from floods.

> The frequency of flooding can be predicted, e.g. 4 m high flood occurs in 1801, 1903 and 1999. The return interval (RI) is about 100 years. This needs to be understood when you write about flooding in essays.

Soft engineering

Technique	How it works	What are the benefits?	What are the disadvantages?
Floodplain zoning/flood hazard mapping and land management (see diagram below)	• Flood zones are geographic areas that the Department for Environment, Food and Rural Affairs (DEFRA) and the Environment Agency (UK) and Federal Emergency Management Agency (FEMA) (USA) has defined according to varying levels of flood risk. • Information generated usually determines the use of the floodplain. • The Environment Agency website allows people to enter a postcode and to get an instant flood risk map drawn.	• Cheap and avoids damage to property. • With no residential areas damage is avoided. • No buildings mean less rapid run-off and less flooding. • More sport and recreation space available. • More agricultural opportunities. • Increased infiltration allowed.	• Of no use where landscapes are already urbanised or developed. • Planning issues are harder to enforce and realise in LEDCs. • Issues where there is a shortage of housing and space.

Flood risk area

1 in 1000 year annual probability flood

1 in 100 year annual probability flood

Defined floods

Little or no flood risk

Low to medium risk

High risk

High risk

Low to medium risk

Little or no flood risk

Floodway

Technique	How it works	What are the benefits?	What are the disadvantages?
Warning systems	**Solution:** Sirens or TV/internet/text/public announcements prepare the population. **Example:** Regular flooding means that few countries use early warning systems for the onset of flooding (e.g. UK and USA); Jamaica has a highly developed warning system developed to cope with 'hurricane' flood events.	• Weather watching and agency warnings are used all over the world and the information they forward is much more accurate. • MEDCs are far better prepared nowadays. • Electronic communication helps with early warnings.	• LEDCs have problems in telecommunications and getting the message out to evacuate quickly enough. • Some floods occur too rapidly, e.g. Boscastle, UK, 2004.
Washland and water meadow	**Solution:** The river is allowed to flood naturally in to water meadows and washland that exists on the river's floodplain. **Example:** Regular flooding of the River Trent (UK) washlands enriched the meadow soils, creating exceptional grazing areas for sheep.	• Traditional and very cost effective. • Habitats for animals provided. • Silt from flooding is used in agriculture. • Low maintenance.	• Period of under-use during flood and post flood periods. • Less farmland available.
Restoration and afforestation	**Solution:** Managed rivers are returned to their natural state with trees planted and old river courses restored. **Example:** One of the world's most impressive schemes is in South Korea; it is attempting to restore the Han (Korea), Nakdong, Geum and Yeongsan Rivers, to provide water security, flood control and ecosystem vitality.	• Depending on the level of works it can be low cost. • Environment improved and habitats restored. • Sustainable. • Maintenance is minimal once the scheme is established.	• Unless properly planned restoration can be an eyesore and alien species that are introduced may sap the system (eucalyptus) or acidify it (coniferous trees). • Local flood risk might be exacerbated.

Hard engineering

Techniques and examples of hard engineering are given below.

Technique	How it works	What are the benefits?	What are the disadvantages?
Flood relief channel/canal/ spillways	**Solution:** These work by channelling excess water away from flooded rivers to other rivers or storage areas.	• Such schemes prevent flooding of properties close to the river, as discharge is reduced.	• They can be expensive to build. • Lots of land is required to 'cut' the channels.

Technique	How it works	What are the benefits?	What are the disadvantages?
	Example: The River Exe, runs through Exeter in Devon; it has three spillways, costing £8 million; the two most recent, the Exwich Spillway and the Trew's Weir Relief Channel, collect water above Exeter and return it below the city.	• The 'stored water' can have a 'multi-purpose' use.	• The increase in discharge into the river below the diverted area can cause downstream flooding. • If the canals also become over loaded with water the extent of problems is spread even wider.
Straightening and channelisation	**Solution:** Meanders are eliminated and the river is widened and/or deepened to allow water to flow through the system more quickly. **Example:** See Kissimmee River example pages 99–100; Ogunpa river channelisation, to relieve flooding around Ibadan, Nigeria; at low flow is used as a domestic waste dump; at high flow this causes flooding; still incomplete despite NGN 10 billion spent out already.	• It moves water away from vulnerable areas very quickly. • Navigation is improved.	• Flooding occurs downstream. • Wild life habitats are lost or changed. • Possible increase in erosion as river speeds increase. • Channelisation is ugly.
Dams	**Solution:** Water held back during floods so discharge is reduced. Water release during low flow conditions and therefore avoids downstream flooding. **Example:** Whitney Point Dam, constructed in 1942 on the River Otselic, USA; built to protect New York and Eastern Pennsylvania; controlling flooding in the Tioughnioga and Chenango Rivers.	• Hydro electric power production. • Multi-purpose, i.e. sailing, fishing and general recreation. • Use for irrigation of farmland. • Habitat formation.	• Hugely expensive to build. • Sedimentation can be an issue. • Reservoirs behind some dams are massive; they swallow up settlements and agricultural land. • Habitat destruction.
Levees and embankments	**Solution:** Banks are 'raised' to contain high flows in the river. The biggest levees are man-made. **Example:** Extensive systems on the Ganges (India) Brahmaputra (Asia) and the Mississippi (USA) rivers; the Mississippi levees stretch 1600 km from Cape Girardeau, Missouri to the Mississippi Delta; New Orleans relies upon levees for protection.	• The banks provide new habitats. • If armoured with concrete erosion is decreased. • Can be used as route ways. • Floodplain can be used.	• If breached they can actually temporarily increase flood levels. • They need constant maintenance if they are to retain their strength and integrity. • They can be extremely ugly.

Technique	How it works	What are the benefits?	What are the disadvantages?
Flood walls	**Solution:** Made of concrete or metal shuttering it is placed vertically at the channel's edges. **Example:** A new flood wall built on the River Ouse (cost £750 000), protects 450 homes at Cawood and Ryther, near Selby.	• Quickly and in some cases cheaply erected. • Used where space is at a premium.	• Never completely water-tight. • Can be unsightly.
Flood retarding storage areas/ sumps; flood detention/atten-uation/ balancing reservoirs	**Solution:** A hollow or prepared reservoir for the temporary detention of water during flooding. **Example:** River Yare, Norfolk.	• Sustainable in terms of the landscape. • Looks natural and attracts wildlife.	• A blot on the landscape in some instances • Large area needed for it.

Case study: Cumbrian flood, UK, 2005 (MEDC)

Background: Carlisle is situated on the flood plain of the River Eden. The catchment covers 2400 km². Its flood history extends back to the seventeenth century. The Little Caldew and River Petteril also add water into the system.

Cause of the flood: High rainfall from warm moist tropical air from the south-west fell in the period 6–8th January, 2005. During this two-day period the equivalent to two months worth of rain fell (200 mm fell in some areas of the catchment!). The rivers Eden and Kent flooded at the same time as the rivers Caldew and Petteril. Carlisle received late notification of the flood and local defences were overwhelmed.

Effects: 27 000 homes were affected and three people died. Cost of the flood is put at £400 million. Parts of Carlisle were 2.5 m deep in water. Two streets, Warwick Road and Willow Hulme Road were worst affected. Hydrologists believe 50% of the water came from surface water and drains and from silted water-courses.

Short-term help: Some people were rescued from rooftops and local schools helped with displaced people, feeding them and providing a bed. There was early help with insurance claims. Some families were housed in caravans. The government offered financial support.

Long-term management of the problem
- Established a flood management plan and 'flood zone' the floodplain.
- Rebuilding of the defences costing £12 million in phase one and £24 million for phase two.
- The new embankments, sluices, flood walls and pumping station will be built to cope with a one in 200 year flood.

Case study: The River Koshi flood, Nepal, 2008 (LEDC)

Background: The city of Purnia is located on the banks of the River Koshi, Nepal. The catchment area covers 59 280km^2 and houses 2.5 million people. The area has a history of annual flooding caused by rapid changes in course, steep gradient and youthfulness (a fluvial term implying speed and erosive ability of the river). The massive silt load and the monsoon rain also contributed to the river flooding. The water of the Ganga and Mahandra also add water into the system.

Cause of the flood: On 18th August heavy monsoon rainfall caused breaches on the River Kosi. Intense rainfall in the catchment exacerbated the flood situation. Flooding is common in September and October, so the area was due to experience three months of flooding. The late de-silting of the river and construction of unstable/high embankments also contributed to the flood.

Effects: The flood in the valley that was 15 km wide. Two million people were affected, 100 000 people were evacuated into 162 camps, run by the government and NGOs, 85 people were killed, though disease in the camps probably caused more deaths, and 100 000 hectares of already marginal farm land was destroyed. Roads and railways were breached and blocked. Some of the catchment was covered in 3 m of water for two weeks. Diseases, such as dysentery, increased. Cost of the flood was estimated at $14.25 million.

Short-term help: 25 000 food parcels were dropped; clean drinking water was provided from over 200 standpipes; life-jackets were provided; plastic sheeting for temporary shelters distributed. The National Defence Response Force (NDRF) and the Army and Air Force were mobilised with boats and helicopters and the Disaster Management Department are convened. UNICEF and UNDP conducted an assessment. Red Cross distributed disaster packs.

Long-term management of the problem

- Strengthen the Koshi's embankments and barrages.
- Discourage floodplain settlement.
- Dam building might control the flood waters.
- Earlier evacuation.
- Prepare seed banks to help reinvigorate farming.

Case study: The Boscastle flood, UK, 2004 (MEDC)

Background: Three rivers converge on Boscastle – Valency, Jordan and Paradise. Boscastle has steep sided valleys, which means the water runs off quickly. It also has a narrow floodplain, thin soil so little storage available; the soil saturates quickly, sandstone base rock with limited permeability, agricultural ground with little interception storage and a drainage basin only 40 km^2.

Cause of the flood: High intensity rainfall. Three million tonnes of rain fell; 185 mm arrived in just five hours, the majority falling in the first two hours from a localised low pressure system.

Effects

- 1000 residents and tourists affected. There was one casualty – a broken thumb!
- 58 properties (including historic buildings) and 84 cars damaged.
- Residents' furniture and possessions were destroyed; a bridge was washed away; the church was filled with 2 metres of flood water; up to 100 people were rescued by Navy and Air Force helicopters.
- The habitat of local wildlife was destroyed; the weight of water eroded gardens and ripped up roads.
- Coastal pollution caused as debris and fuel from cars flowed out to sea.
- Disruption to the village, as a major rebuild project had to be carried out.
- Cost of repairs to the properties was put at £2 million

Longer-term effects

- **Economic losses:** Rural Cornwall is one of the UK's poorest rural counties. Most shops stayed shut for the rest of the season and the bad publicity reduced tourist numbers.
- **Wider regional impact:** Tourism accounts for 30% of Cornwall's GDP. The population doubles during July and August each year, with tourists

spending up to £1 billion throughout the county.

- **House price falls:** People may find the value of their homes permanently reduced, now that Boscastle is associated with a serious flood risk.
- **Mental trauma:** Many residents suffered following the flood with properties often inhabitable for up to 6 months.

What of the future?
- A culvert was built for the River Jordan by the Environment Agency.
- The channels of all the rivers will be deepened and widened, but this could affect wildlife.
- More trees could be planted to increase interception and evapotranspiration.
- A dam could be built to hold back flood-waters.

Managing water as a resource

Environmental solutions to flooding

Since the end of the nineteenth century we have 'trained' rivers, mostly to control flooding. Environmental solutions to flooding has seen the most promising of reforms over the last decade. Many environmental solutions are now used to control floods and flooding events.

- Pools, riffles and meanders are now reconstructed, as they are seen as the best, most stable alternative to artificial straightening.
- The re-introduction of vegetation is seen as important, as it replicates and promotes the bank stability of the natural stream channel.
- Stream maintenance, which is small scale and in harmony with the watery environment, is seen as important.
- Bio-technical methods copy and reproduce the natural symmetry/asymmetry of the stream channel.

Restoration returns degraded rivers back to their original state.

Case study: River restoration

River Cole, UK (MEDC small scale)
This river runs north-east of Swindon in Wiltshire. Over 900 years of adaptation has meant that the river and its ecology had suffered.

Habitat appearance was the reinstatement of the physical features of the river, flood storage and retention was improved and sustainable flooded meadow land was developed. Re-establishing reed beds, the backwaters and habitats of the old river course and for the restoration project re-shaping and engineering the river bed, were also important considerations.

The benefits of the scheme have been very quick to appear.
- Flooding is better managed, with low lying meadows collecting silt from the river.
- The river flows faster now, meaning that low flow issues are avoided.
- Water quality has improved as wildlife, plants and trees have been planted.

- National targets for improvements in biodiversity have been achieved.

Kissimmee River, USA (MEDC regional scale)
Pre-1940: The Kissimmee river, Florida meanders the 100 miles between Lake Kissimmee (15 m above SL) to Lake Okeechobee (5 m above SL) on a two mile wide floodplain. Between 40–50% of the time there is overbank flooding. The floodplain was biologically, and ecologically diverse and dependent upon frequent inundation. The river had a thriving bass fishery and wildlife population. Human settlement on the floodplain was sparse.

Post-1940: Human settlement of the floodplain increased; ranching and farming became the dominant land use. However, flooding and inundation from hurricane events led to the State of Florida requesting the Army Corps of Engineers to design a flood-control scheme for the state. In the end this took some 12 years to plan and authorise.

Between 1962 and 1971 the Kissimmee River was channelised, impounded and regulated. The meandering river was turned into a 56 mile, 30 ft deep, 300 ft wide canal (called C38). This all had a massive effect on the ecology of the Kissimmee River. Up to 31 000 acres of wetland ecology was lost and encroachment of vegetation in the low flowing Kissimmee River led to increased biological oxygen demands. Wading birds declined to be replaced by species from the plains. Sport fish, such as bass, were replaced by low oxygen tolerant species. Spawning sites were lost.

The Kissimmee River revitalisation and restoration project, initiated in 1992 under the Water Resources Development Act, had the goal of restoring ecological integrity whilst ensuring hydrological control for the system. Phase one of the project was completed in 2007. Wading birds returned and oxygen levels were up in the river. This was despite a state wide drought, lasting two years.

The final water regulation scheme will be implemented by 2013. Some $650 000 million of work was shared between the Corps of Engineers and the South Florida Management District.

Lake Bam, West Africa (LEDC regional scale)
Lake Bam, is a naturally occurring lake in Burkina Faso. It dries out in the dry season and floods the nearby towns and fields during the wet season. The lake suffers badly from sedimentation and growing populations. Global climate change is probably affecting it too. The lake is part of the Nakamb/ Volta river system. The catchment for the lake is 2600 km^2. The mean annual rainfall is 600 mm and the dry season (with temperatures of up to 40°C) evaporates about 1500 mm. The volume of the lake is 20–34 million m^3 in the wet season and by the end of the dry season 0.9 million m^3.

The local population of 60 000 people live on the banks of the lake. The largest source of income comes from irrigated crops, whilst fishing goes on in the lake. There were several main recommendations.

- The construction of sediment traps in feed rivers and in the lake.
- Terraces, stone dykes and afforestation.
- Raising the level of the lakes outlet (adding a massive 40% to the lakes volume).
- A control system for the lakes outlet.
- Raising the level of the fields around the lake to limit flooding.
- Excavation of internal reservoirs to increase the volume of the lake.
- Support for fishermen, improvements in fish habitats and stocking of the lake.
- Livestock management, tree protection and sewerage management.
- The overall costs of the improvements are estimated at £44 million.

Kissimmee area

Response to floods – in summary

Know examples of flood control schemes in the UK. Many are well documented.

Response		Control
Adjustments *in actions on the floodplain*	Abatement *of problems in the catchment*	Protection *along the channel*
<ul><li>None</li><li>Emergency action</li><li>Flood proofing</li><li>Land use regulation</li><li>Financial disincentives</li></ul>	<ul><li>Afforestation</li><li>Change agricultural practices</li><li>Change the vegetation use</li><li>Effect of urban areas</li></ul>	<ul><li>Walls and embankments</li><li>Channel improvements</li><li>Diversion schemes</li><li>Reservoirs</li><li>Barrages and flood barriers</li></ul>

KEY POINT

Remember that all flooding questions will at some point ask you for the causes and impacts of flooding. If it is AS the questions will develop these points as the structure of the question unfolds. If it is in an essay you will have to use case studies to develop the theme.

- Remember the main cause of flooding is prolonged or heavy rainfall.
- Remember there are physical factors that increase the risk of flooding: steep slope, circular drainage basins, high drainage density, sparse vegetation and impermeable ground.
- Remember there are human factors that influence flooding too: climate change, urbanisation, deforestation, flood management strategies and poor agriculture practice.
- Remember that flooding affects people, the economy and the environment.

PROGRESS CHECK

1 What human activities increase the risk of flooding?
2 What is river restoration?
3 What is hard engineering?
4 How can studies of river hydrographs aid understanding of river flow?

Answers: 1. Urbanisation, road building and poor agriculture practice. 2. Environmental solutions to river management, where trained rivers are released from constraints. 3. Hard engineering involves major engineering work. 4. Analysis of hydrographs and regime graphs allow hydrologists to plan for floods.

Exam practice question

1 Using case studies and examples discuss the following statement that, 'Human and physical factors have to be studied to properly explain the size and frequency of river flooding events'. **[15]**

The challenges of the coast

The following topics are covered in this chapter:

- ● Processes and landforms
- ● Management and planning

5.1 Processes and landforms

LEARNING SUMMARY

After studying this section you should be able to understand:

- the importance of energy transfer through the coastal system and its effects on erosional, depositional and transportational processes
- landforms of the coastal strip
- the effects of varying sea levels

Processes

AQA	**AS U1**
OCR	**AS U1**
Edexcel	**AS U2**
WJEC	**A2 UG3**
CCEA	**A2 U2**

Human activity is a threat to the coastline. You must show knowledge of this.

The coast is particularly interesting to study because it is constantly changing (by the hour if large storm events are in progress). Coastal geomorphologists are interested both in the mechanics that cause change and the landforms that develop. Additionally, as a big proportion of the world's population lives near the coastline, we have to deal with the threats and problems posed to human habitation. Problems include flooding, rising sea levels, accelerated erosion, the effects of industrial pollution and the effects of tourism. Careful, sustainable management of the coast helps to deal with these problems. The coast is defined in geographical terms below.

You need to know these terms.

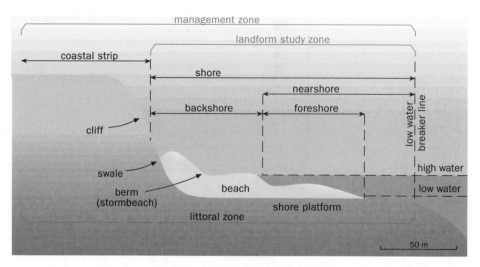

Geographical features of the coast

Coasts can also be classified whether they are an emergent/ submergent, primary/ secondary or high/low energy coast.

Traditionally, the approach adopted by coastal geomorphologists has been to try to classify the coast, according to whether the dominant process is erosional or depositional. Currently, the relationships of mechanisms and processes to landforms are again being explored, the so-called 'process-landform response approach'.

Note: 'sediment transport' is a coverall for erosion + deposition. There is no permanent loss of sediment, it is just moved back and forth.

Inputs of energy →	Processes →	Landform response →	Outputs
Tides, wind and waves	Sediment transport	2d – beach (slope/shape) 3d – landforms (stacks/cliffs etc.)	Of energy (breaking waves) and sediment on the sea floor

Waves, currents and tides

The interaction between waves, tides and coastal currents shape, modify and mould the shoreline. Waves provide the energy for the coastal system, tides spread energy over a larger vertical area of the coast, and currents spread and redistribute energy/sediment along the coastline. All three can act against and work with one another.

Waves

Remember: wave size is determined by the fetch (the distance over open water that the wind has blown), the strength and duration of the blow and by other factors, e.g. offshore gradient.

The force behind the formation and shaping of the coast is the wave. The drag effects of wind across the sea cause undulations on the surface. As these undulations build (because of pressure contrasts on the windward and leeward sides) so water starts to move in an orbital or oscillatory fashion inside them. This movement is related to their height. Energy conversion, from potential to kinetic, occurs continually within these waves. Waves are, then, a means of moving energy through water with only small displacements of water particles in the direction of energy flow.

Waves are described according to their height, velocity, length and period. Questions will expect confident answers that use such terminology.

Waves that break can be either **constructive** or **destructive**.

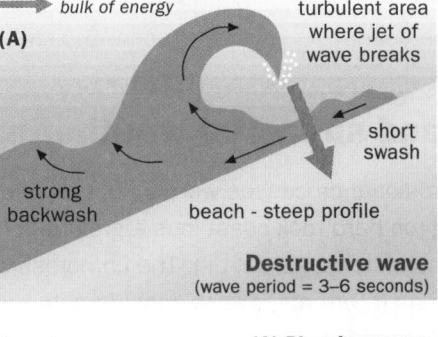
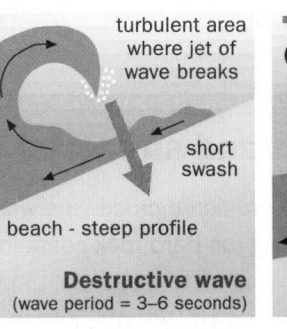
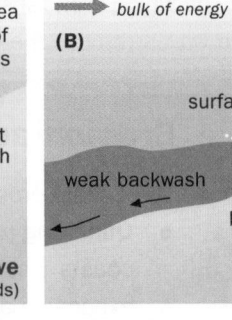

Swash runs up the beach. Backwash runs down the beach.

(A) Plunging waves and (B) spilling waves

Four types of wave have been identified; surging, collapsing, plunging (as in A) and spilling (as in B).

As waves move nearer the coast, the submarine contours/sea bed starts to affect the waves. Locally both friction and drag start to increase. This is known as **wave refraction** and has the effect of varying the available energy along a coastline.

Currents

Waves approaching at 30° move most sediment along a shore!

Normal currents establish a cell circulation in the near-shore zone. A large amount of sediment is moved up the beach by the swash and is

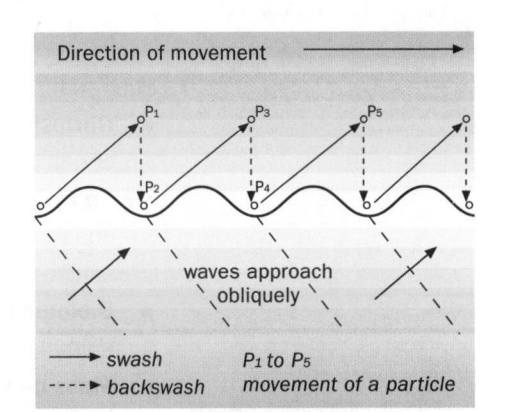

Longshore drift

Most sediment on beaches has a river or offshore origin.

Most sediment moves west to east in the south of the UK. Along East Anglia's coasts it moves north to south.

This coastline needs protecting: 85 000m² of farmland is at risk, Easington Gas Terminal brings 25% of the UK's gas ashore, Spurn Head is a site of special scientific interest (SSSI) and there are lots of tourists that visit the area.

balanced by the rip of the backwash running back down the beach. The water moving back down the beach forms a **riphead**, a deep (up to 3 m) energy/water dissipation hollow offshore. The effect of the strong, sediment-laden swash is to form a beach cusp, the smallest of the beach depositional features. It is composed of graded/sorted sediments in a horned shape.

For the most part waves approach coasts at a slight angle (less than 10° is normal). These wave-normal/oblique currents, aided by winds and submarine currents, carry sediment up the beach at the same angle as the wave/current, and returns it in 'rips' perpendicular (at right angles) to the beach. The net effect is to move material along a beach; this longshore movement is called **longshore drift (LSD)** – see diagram on page 103.

Tides
Erosional processes are concentrated between high (HTM) and low tide (LTM) marks, in the splash zone above the HTM and the wave base below LTM.

Case study: Holderness, UK – the management of coastal erosion

Holderness is a lowland region of eastern England, part of the East Riding of Yorkshire. The Holderness coast is Europe's fastest eroding coastline. It erodes 2 m/year. The main reason for this is because the bedrock is made up of tills. This material was deposited by glaciers around 12 000 years ago. Erosion of the Holderness cliffs begins when soft till cliffs become saturated with rain water, weakening the cliff which then falls as a block or slurry slide and the waves remove the debris

from the cliff foot. At Flamborough the chalk is more resistant and forms a headland. Mappleton, which lies south of Hornsea, is a small, but important, holiday location. It was protected with two groynes in 1991. The effect of this was to starve the coast of sand further south leading to massive and near uncontrollable cliff erosion. Further south, Kilnsea has been lost completely and at the furthest southerly point lies Spurn Head (a sand and shingle spit of some 5.5 km).

Erosion and subaerial weathering

30 tonnes pressure/m² of 'explosive' effect.

Three main erosional processes work to destroy the coastline.
- **Quarrying** (on hard rock coastlines – granite), or **hydraulic action** (on soft rock coasts – sandstone, glacial till). The compression, as a wave hits a coast, and expansion as it retreats, exerts considerable pressure on the coastline.
- **Abrasion/corrasion** – waves use the pebbles and cobbles to corrode the cliff base, undercutting it rapidly. Weaknesses in rocks and joints, are differentially exploited by this form of erosion.

Can be heard working on pebbly beaches!

- **Attrition** – reduces the size and rounds off individual clasts (pebbles) as they bash into one another.

Subaerial processes include:
- **Corrosion and solution** – the chemical dissolving of rock in acid-saline conditions.
- **Sea weathering** – slaking caused by the alternate wetting and drying of the coast, along with salt crystallisation are most prevalent, but hydration and oxidation also play a part.
- **Biological weathering and bio-erosion** – the effects of plant roots and animals on the coast also have a profound effect.

Remember that most features on beaches are temporary, or transitory, and the tides, weather and man all affect them.

- **Mass-movement** – rock falls (on hard rock coasts, e.g. north-west Scotland) and mudslides, slumps and slides (on soft unconsolidated rock, e.g. the tills of the Norfolk coastline) can deliver more material to the coastal sediment system than other marine erosional processes.

Landforms of the coastal strip

AQA	AS U1
OCR	AS U1
Edexcel	AS U2
WJEC	A2 UG3
CCEA	A2 U2

Cliffs

Note: different combinations of height, orientation, steepness, determine cliff height. A relationship between rock type, erosion and cliff morphology exists.

Cliff vary in height, orientation and steepness, as well as their lithology and structure. For instance, the cliffs on the soft boulder clay/till of north Norfolk rarely exceed 8 m, whereas the Cumbrian cliffs of St Bees Head reach 25 m.

Waves concentrate energy at the cliff base, forming a **wave-cut notch**, the size of which is determined by the tidal range. As time passes, the overhanging cliff eventually collapses. This material is used as ammunition to accelerate the creation of the wave-cut notch.

Shore platforms

As shorelines, coastal slopes and cliff lines are eroded there is a marked retreat that leaves behind a platform. These platforms have an overall convex shape and an average slope of 0°–3°. Most show a break in slope, marking the LTM. Present-day processes have produced most platforms, e.g. off Bembridge, Isle of Wight.

Note: if asked to explain beach profiles you must use diagrams in your answer. Know your processes!

Wave-cut platform

Sea level changes can either be **eustatic** (a worldwide change caused by tectonic or climate change) or **isostatic** (a local change caused by compression or decompression or through tectonic activity).

Sea level change results in more frequent coast floods, the submergence of low lying islands and changes in the coastline.

Features of **submergent** coasts include:

- rias – submerged river valleys, e.g. Plymouth Sound
- fiords – drowned glacial valleys, e.g. Hardangerfjord, Norway
- dalmatian coastline – where valleys running parallel to the coast are drowned, e.g. the Croatian coast.

Features of **emergent** coasts include:

- raised sea beaches, on the coast and river terraces in the estuary areas, e.g. on the Isle of Arran and Cardigan Bay, Wales.

Other features of coastal erosion

AQA	**AS U1**
OCR	**AS U1**
Edexcel	**AS U2**
WJEC	**A2 UG3**
CCEA	**A2 U2**

For a long period of time **abrasion** was held responsible for the formation of shore platforms. The modern view is that shore platforms have a multiple-process origin. They may be caused by a combination of the processes below.

- **Abrasion** – sand grains moved by waves plane the platform surface. This is particularly effective in the upper shore section.
- **Mechanical wave erosion** – the process of quarrying, through wave hammer, compression and pressure release, picks out and exploits variations in lithology. This process causes cliff recession and roughens the platform surface.
- **Weathering** – wetting and drying can cause hydration, oxidation and salt crystallisation. These processes are slower than the violent processes outlined above.
- **Subaerial processes: solution** – the chemical solution of calcareous rocks (limestone/chalk), e.g. Norfolk's chalk shore platforms, are affected by solution; slight rises or falls, away from the atypical sea temperature, produce rapid chemical stripping of the submerged platform.
- **Tides** – vary the level of process activity.

> **KEY POINT**
>
> **Concordant or longitudinal** – where the 'grain' of the rock is parallel to the coast, e.g. Lulworth Cove.
> **Discordant or transverse** – where the 'grain' runs at right angles to the coast, e.g. Swanage.
> The grain reflects the structure – the alignment of the folds or the different types of rocks. This leads to **differential erosion**.

Arches and stacks

- The sea erodes along a line of weakness (e.g. a fault) in a headland to form a cave.
- Caves formed on opposite sides of headlands join to form an arch.
- An arch will eventually collapse to form a stack.

Geos and blow-holes

Where a fault in cliffs at right angles to a coast is eroded by the sea, a long, narrow inlet known as a geo may form, e.g. Huntsman's Leap, Pembroke. The first stage may well be a cave which connects to the surface by a chimney to form a blow-hole.

Case study: Physical features and processes found along the Dorset coast

Pronounced headlands alternate with wide bays where the softer clays have been more easily eroded. Wave refraction concentrates erosion on the headlands.

Extensive sandy beaches have accumulated in the shallow and sheltered waters at the head of the bays.

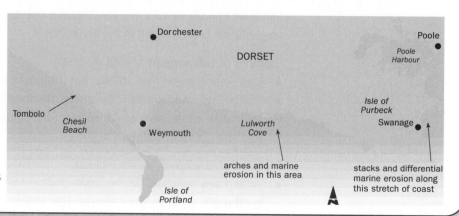

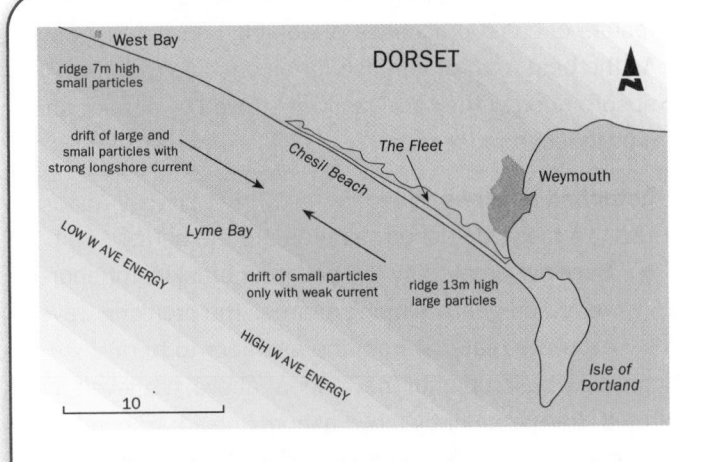

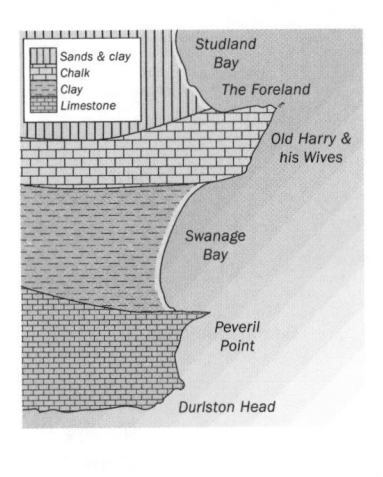

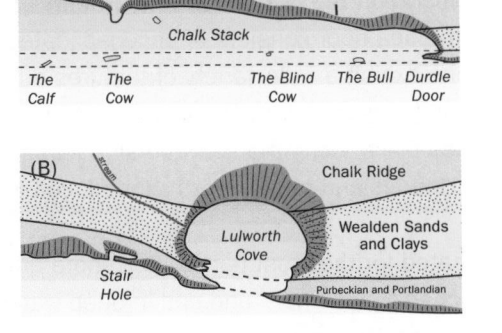

Limestone has almost completely gone, a few isolated rocks remain at low tide. Durdle Door is a fine natural arch.

At Stair Hole the sea has broken through the Limestone (via caves and joints) and is beginning to attack the soft Wealden clays. Lulworth Cove is a near circular bay, extending east to west, its growth impeded by chalk. Its growth was initiated by the stream flowing into the sea through the limestone.

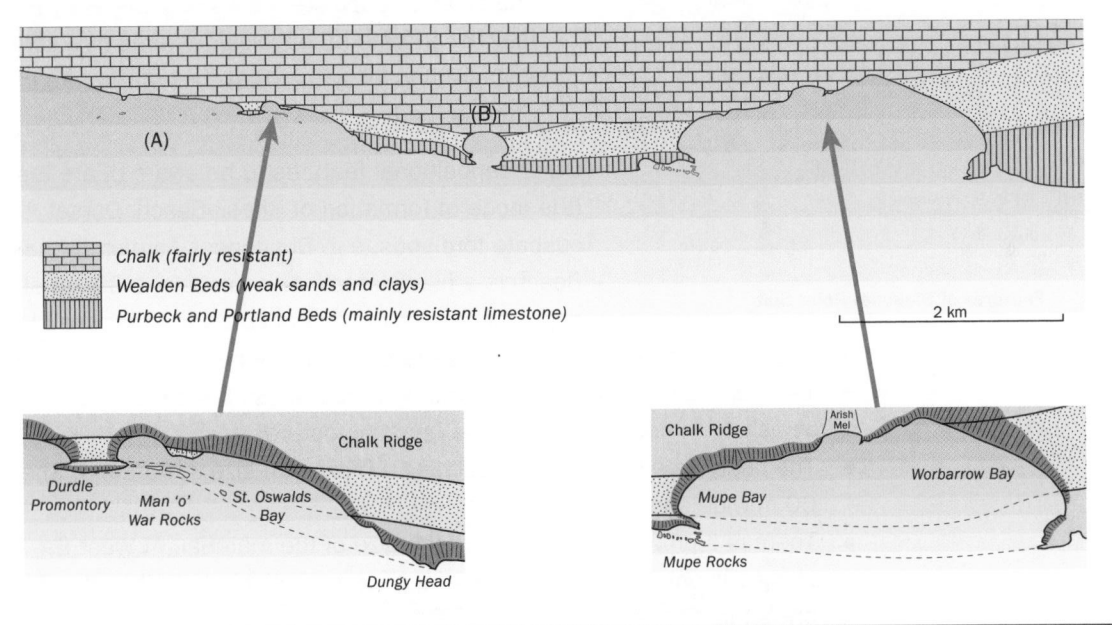

Landforms of coastal deposition

The debris from coastal erosion is moved and deposited by waves and currents. Major depositional landforms are shoreline beaches and detached beaches (spits and bars). Each landform has its own dynamic sediment store/budget, with material also being lost and gained.

Beaches

These are perhaps the most widespread of depositional landforms. They are geomorphologically successful because of the mobility of their loose sand sediment. These complex systems exhibit a range of minor landforms, shown in the following diagram.

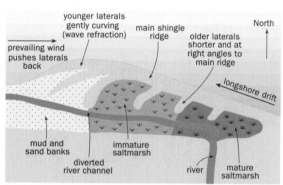

Beach landforms

Most of the landforms shown in the diagram are the results of coast processes re-working beach sediments. As the best form of defence for a coastline (as they absorb energy) they are easily destroyed by natural and man-induced processes.

Detached beaches

The two major detached features are the bar and spit.

- **Bars** are created by the action of breaking offshore waves on gently sloping shores. The breaking waves excavate material from the sea floor to form submarine bars, which slowly build up until they appear above sea level; many quickly become vegetated. Bars frequently move ashore/inland; trapped sea water in the form of a lagoon is filled by sediment and marsh vegetation; sand dunes tend to be the final relic of the bar (e.g. Looe in Cornwall). The best examples of bars in the UK are found off the Norfolk Coast between Hunstanton and Sheringham, where the sea floor consists of unconsolidated glacial material. The shingle formations exhibit a wide variety of features due both to straightforward wave action and longshore drift. The usual form in Norfolk is crescentic; drifting that takes place along the seaward side leads to modification and complication. Scolt Head Island off the North Norfolk coast, is a fully developed offshore bar.

> There is an alternative theory that suggests bars are flooded dune systems – a relic from lower sea level during the last glacial period.

- **Spits** are created for the most part by longshore drift and are attached to the shore and end in deep water. The diagram opposite shows how Blakeney Point Spit may have formed the typical features of a compound spit and it is likely that Blakeney started its life as an offshore bar!

Other depositional features to be aware of are tombolos (the mode of formation of Chesil Beach, Dorset, UK) and cuspate forelands, e.g. Dungeness Foreland. Chesil Beach is a fine example of a tombolo, a 30 km shingle ridge connecting the Isle of Portland to the mainland. From West Bay to Abbotsbury it hugs the coast. Then the elongated lagoon – the Fleet – separates it from the mainland. Whether Chesil Beach is a spit or a bar driven ashore is not clear. It has two unusual features.

- The beach is a simple ridge reaching a maximum size at Portland, where it is 60 m wide and 13 m high.
- There is a progressive grading in the size of the material. At West Bay the shingle is small, about the size of a pea; by Portland it reaches the size of one's fist.

Features of Blakeney Point Spit

5.2 Management and planning

LEARNING SUMMARY	After studying this section, you should be able to understand:
	• that the coastal strip is used extensively by man
	• that the coast has to be protected and preserved

Managing coasts

AQA   **AS U1**
OCR   **AS U1**
Edexcel **AS U2**
WJEC   **A2 UG3**
CCEA   **A2 U2**

Analysis of the last UK population census shows that 16.9 million people live within 10 km of the coast. Buildings, roads and recreation facilities occupy 31% of the coastal frontage of the UK and 40% of the UK's manufacturing industry is situated on, or near, the coast. Economic pressure for further expansion of these facilities is likely in the future. With so much of the coast developed, or being developed, there is a demand for coastal defences to protect the population and their economic well-being. Approximately 2100 km of defences presently defend coastlines. The bulk of coastal defences are along the east and East Anglian coast. Management of coastlines in the UK is the responsibility of **Maritime Local District Councils**, the **Ministry of Agriculture**, **Forestry and Fisheries (MAFF)**, the **Environment Agency** and other groups. All defence schemes need planning permission and consultation before building can begin.

Coastal protection

Our coast is vulnerable – know how it's protected.

Once a coastal issue has been identified, a number of factors have to be taken into consideration before deciding the best management scheme for a coastline.

	Cliff face strategies	Cliff foot strategies	Beach management schemes
Hard engineering	• Cliff pinning. • Cliff modification. • Drainage of cliffs. • Gabion baskets £500–2000/m.	• Sea walls (£1500–2000/m). • Revetments (£300/m). • Rip-rap.	• Breakwaters (£120 000 each). • Groynes (wooden at £6000 each). • Beach pumping. • Reef systems (£ millions).
Problems associated with hard engineering	• Over drying and subsistence.	• Expensive to build and maintain. • Walls cause accelerated erosion of the beach and allow beach levels to fall. • Revetments have a short life cycle.	• Groynes interrupt and reduce LSD as they are 100% efficient at trapping sand. • Groynes are visually intrusive. • Reefs change beach plans and profiles.
Soft engineering	• Revegetation.	• No engineering problems.	• Beach nourishment and replenishment (at a cost of £20/m^3).
Problems associated with soft engineering	• Problems only relate to poor vegetation choice.	• No problems.	• Near-shore dredged material can affect the sediment cell, impeding and disrupting replenishment. • Recharged sediment needs to be of the same calibre to the natural sediment.

Case Study: Sea Palling, Norfolk

With large areas of the eastern Broads being put at risk with an excess of 70 000 m^3 of material being lost from the area every year and up to 6000 h of residential, agricultural and commercial properties being put at risk, it was decided that nine shore-parallel reefs would be built to reduce losses.

Completed in 1997 each reef is 250 m long, 45 m wide and contains thousands of tonnes of imported rock. Since completion the scheme has been beset by problems, most controversially being the formation of embayments along the coast and the extreme effect it is having on the sediment budget in the area.

Planning and funding

AQA	AS U1
OCR	AS U1
Edexcel	AS U2
WJEC	A2 UG3
CCEA	A2 U2

All Councils who have management responsibilities for coastal defences have **Shoreline Management Plans** (SMPs) in place, based on the so-called **Sediment Management Cells** (a way of dividing up the UK and Welsh coastline). These SMPs identify the options available to Coastal Managers and are known as **Management Units**. The options are:

- no action is needed to build or maintain defences
- hold the line – interventional to hold defences where they are at present
- advance the line/change – new measures that move defences seaward
- managed retreat.

After a problematic coastal issue has been identified and action appears sensible, funding is criterion tested. The first stage is to convince MAFF, and other agencies, that the proposed scheme will be feasible in terms of the **engineering**, that it is **economical** and that it is **environmentally** sound. Once past the three E's test, it is criterion scored, based on priority, urgency and economics. Only those schemes that score are even considered for funding. The result of this type of scoring is to favour urban areas over rural areas.

Rip-rap is loose slabs of rock placed at the cliff base to break wave energy.

Case study: Coastal defence schemes used on the Isle of Wight, UK

The SMP for the stretch of coast below Bonchurch and Ventnor and between Monk's Bay and Steephill Cove was to **hold the line**. The following examples of coastal protection schemes, completed over the last fifteen years, are all located along The Undercliff, a coastal zone reputed to form the largest built-up coastal landslide complex in north-west Europe. By protecting this stretch the tourist businesses of Bonchurch and Ventnor have also been assured and property protected.

Use a map of the Isle of Wight to locate these areas.

Monk's Bay: Cliff failures at Monk's Bay had been frequent (because of high wave energy, high rainfall and coastal storms). The scheme involved the construction of an **offshore breakwater**, six rock groynes, **beach nourishment** using 17 000 cubic metres of sand and gravel and **rock revetment** to reinforce the existing sea wall. Almost 25 000 tonnes of Swedish granite was off-loaded onto the coast by barge. Re-working the cliff profile and the installation of **land drains** stopped the mass movement in this area. The scheme cost £1.5 million. There has been a significant loss of sand from the scheme since the project finished.

Wheeler's Bay: Property at Wheeler's Bay had become unsellable because of old sea walls collapsing into the sea. Completed at a cost of £1.7 million, the scheme saw revetments built, coastal cliffs were regraded, land drainage and 16 000 tonnes of Norwegian granite dumped as **rip-rap** at the cliff-slope foot.

Western Cliffs: Below western Ventnor blocks of chalk brought down by landslides provides a natural protection from marine erosion for the town. This was further strengthened with Mendip limestone revetment and a series of **groynes** at a cost of £1.2 million.

Coastal concerns

AQA	**AS U1**
OCR	**AS U1**
Edexcel	**AS U2**
WJEC	**A2 UG3**
CCEA	**A2 U2**

Human impact on the oceans

Increased awareness of the environment has led to added pressure to safeguard our natural coastline and has contributed to land use conflict. The pollution of surface water and sub-surface water by chemicals released through industrial activities, either directly into rivers or into the air, and by fertilisers used in agriculture, has led to pollution of coastal water. Examples include:

- Nitrogen phosphates in sewage and fertilisers which lead to algal blooms, e.g. in coastal areas in Italy.
- Mercury, cadmium, hydrocarbons and dichlorodiphenyltrichloroethane (DDT) from industrial activity and farming, which lead to food chain disruption and human genetic problems, e.g. discharges from the River Rhine into the North Sea.
- Oil from exploration and transportation accidents, and from 'tank' cleaning, causes food chain disruption and faunal losses, and ecosystem crashes, e.g. the Exxon Valdez (1989), Sea Empress (1996), Braer (1993) and Jessica (2001) disasters. Such disasters dropped from 8/year to 3/year in the period 1990–2009.
- Chemical and radioactive through waste disposal, e.g. into the Beaufort Channel in the Irish Sea.

Nobody denies damage is being done, disagreement centres on the severity of the damage.

> Florida's coastline is classified as fragile and threatened. After reading this section you should be able to write about the threats posed.

Case study: Florida, USA – a stressed coastline

Issue	Solution
The population of Florida is expected to reach 30 million by 2030. New settlements being built up to the coastline and dunes cause problems after storm events, as there is no room to build buffers against storm surges.	Laws need updating relating to the Coastal Construction Control Line (CCCL), Florida's coastal management programme. The buffer between settlements and the sea have to be widened to prevent damage to homes and beaches. Reduce the provision of insurance for coastal buildings – this will limit the desire to be on the coast.
Florida is a biodiversity hotspot. The Florida Keys is the world's third biggest coral reef and is threatened by pollution from agriculture and sewage.	Better standards for water quality are required. Storm-water regulations need to be put in place, limiting the amount of waste released into the sea. Improve industrial compliance and pollution legislation. Enforce cruise ship 'dumping' regulations.
Off-shore oil spills threaten Florida's multi-billion dollar tourist industry.	Ban or control drilling in the eastern Gulf of Mexico. Drop subsidisation of drilling operations.
Florida's marine environment needs help to prevent fish and mammals dying out.	End over-fishing. Add environmentalists and academics to voting positions on State legislature. Support the dwindling stock of rare fish and mammals.
During the last century the sea level has risen by 1 m and this affects coral growth, tidal mud flats, mangroves and salt marsh.	Reduce pollution and emissions and encourage renewable energy.

Case study: The Deepwater Horizon oil spill, Gulf of Mexico

It is absolutely vital that you use case studies at both AS and A-Level.

On 20th April 2010 on the Deepwater Horizon oil rig, 64 km off the coast of Louisiana Gulf of Mexico, a massive explosion initiated the release of 90–180 million gallons of crude oil, causing a massive environmental disaster in the Gulf of Mexico. The well was eventually capped and controlled in August of 2010.

Case study: The importance of Thailand's coastline

Thailand's once abundant and healthy marine and coastal resources are under significant pressures. Nearly a quarter of Thailand's population (13 million people) live in the 22 coastal provinces (not including Bangkok). Tourism and manufacturing have increased along the Thai coast. Furthermore, industrial/port development and urban development has led to increased demands for freshwater and the generation of massive amounts of industrial and urban waste. Excessive groundwater extraction can cause sea water intrusion and land subsidence, which has increased coastal erosion. High biological oxygen demand (BOD) is an indicator that water pollution is high off Thailand's coast.

Coastal areas are important tourism destinations for the Thai population, as well as international travellers. Tourism revenues are very important for Thailand, but with the tourists come inevitable environmental damage. Specific pressures from tourism on coastal resources are shown below.

Tourism-related activities and events

- Land clearing for construction.
- Changes in freshwater run-off and sedimentation from construction and development.
- Placement of buildings and other structures on the beach or in coastal waters.
- Increased waste generation, sewage and wastewater disposal.
- Increased freshwater demand.
- Over-fishing to supply restaurants.
- Walking and collection of souvenirs, e.g. on reefs.
- Harbour maintenance and boat anchoring.
- Sand mining for beaches and construction.

Environmental impacts

- Damage or loss of wetlands, mangroves and other coastal habitats.
- Increased salinity levels impact mangroves.
- Increased sedimentation rates degrade mangroves, sea-grass beds and coral reefs.
- Changes in sedimentation patterns increase erosion and elevate risks during natural disasters.
- Pollution of near-shore waters.
- Water shortages and increased groundwater usage, possibly resulting in land subsidence and increased erosion.
- Unsuitable fishing practices.
- Physical damage to reefs and removal of organisms beyond sustainable limits.
- Destruction of submerged and fringing vegetation; damage to coral reefs from anchors.
- Increased erosion in other areas.

Case study: The threat to Zanzibar's coastal system

The ad hoc development of tourism has caused social and environmental problems for Zanzibar. Kiwenga was the first tourist development (centred on three villages Kiwenga Kairo, Kiumba Uremo and Gulioni).

In 2005 almost 125 000 tourists spent $117 million in Zanzibar (20% of its GDP). Tourism has brought

Location of Zanzibar

electricity, water and telecommunications to the area, helped by 60% of all the foreign investment that comes into the country.

Tourism, however, is threatening the fragile coastal ecosystem in a number of ways. Zanzibar will need to manage its future tourist development carefully to ensure that coast damage is not irrecoverable.

Pros	Cons
Economical Community is benefitting from better means of communications, security facilities, availability of socio-economic facilities, economic infrastructure and improved social welfare environment. This enhances employment and trade opportunities and may raise living standard of locals and improve the life of individuals in the long-term.	*Economical* Economic opportunities are attractive to migrants which increase pressure on resources and services; local employment is taken by migrants; unskilled locals engaged in unskilled activities, which may lead to social unrest in the long-term. Increased pressure to existing resources, due to the increase of consumers, may result in conflicts between resource consumers and increase numbers of poor.
Social Maintaining of historical sites, high-class structures, e.g. hotels, villas and houses; good infrastructure increases land value and upgrades village status.	*Social* Low purchasing power of locals, loss of land rights and displacements; interaction between local and migrant communities leads to land conflict and family disputes and social disorder.
Environmental Establishment of gardens, environmental groups and development of land use planning and guidelines; environmental policies and guidelines lead to secure environment.	*Environmental* High competition between the community and hoteliers in terms of resources and facilities utilisation leads to depletion and degradation of resources.

Coral reefs and atolls

Coral reefs and **atolls** are widespread between latitudes 30°North and 30°South in the western parts of the Pacific, Indian and Atlantic Oceans. They are either **fringing**, **barrier** or **atollic** in origin. The coral reefs and atolls come from a biological 'source' – the remains of polyps, algae, foraminifera, molluscs and other shell-like organisms contributing to their formation. Ideal sea conditions in which reefs form are where the salinity of the water is in the order of 27 to 38 ppm (parts per million), with a mean sea temperature of 18°C and with an adequate circulation of sea water.

Changing sea temperature distributions (attributed to El Niño) and global warming (causing sea level changes) are affecting the continued development and growth of coral reefs. Destruction by human activity (tourism, over-fishing and wholesale destruction for export) and the crown-of-thorns sea star infestation has brought destruction to vast areas. Pollution and sediment disruption also clouds water and slows/stops the growth of coral.

Case study: Coastal environments under threat – Great Barrier Reef, Australia

Australia's coral reefs, including the Great Barrier Reef, require clean, clear, unpolluted, neutral (not acidic) and warm water with temperatures between 20 and 32°C. They are important because they form a natural breakwater, are ecologically important, are a tourist hotspot and provide safe protected bathing. Australia's coral reefs are under threat from:

- tourists and divers damaging the reef
- tour boats' propellers churning sand and choking the coral
- dynamite fishing and pollution from coastal resorts
- acid rain run-off bleaches the coral.

Australia

Great Barrier Reef

PROGRESS CHECK

1 How does hard and soft engineering on coast differ?
2 Why are cost benefit analyses undertaken?

Answers: 1. Cost and their effect on the environment are the major differences. 2. Cost benefit analyses allows decisions to be made about the coast when changes are contemplated (decisions about cost and effect on the coastline).

Global warming

Global warming is the term given to increased temperatures on the Earth's surface, resulting from carbon dioxide and other gases trapping the incoming solar radiation. Global warming results in the world's oceans increasing their volume (so-called **thermal expansion**) or a **eustatic** change. The predictions for sea levels as a result of global warming for the next 50 to 100 years are alarming – the capital of Norfolk could easily be renamed Norwich-on-sea! Combined with an increase in storminess, increased extreme tidal events and storm surges, lowland coastal areas will need enhanced protection from the sea.

It has been suggested that levels of carbon dioxide might be reduced by pumping iron sulphate into the sea. Plankton would thrive in this iron-rich environment and the increased numbers of plankton might absorb the excess carbon dioxide.

Sea surges (storm surges)

Sea surges are an increasingly frequent phenomenon. They result from a number of concurrent, but freak conditions, e.g. North Sea surges result from a combination of:

- high tides
- strong northerly winds (influenced by the presence of low pressure)
- high pressure to the west of the North Sea and low pressure to the east
- the bottleneck that is the southern North Sea and its lowland coast exacerbating the problem.

Events in 1993, 1995 and 1999 caused flooding and damage to large parts of the Broads and coast of Norfolk, as well as parts of coastal western Europe.

Coastal flooding

The physical causes of coastal flooding are:

- low pressure storm surge conditions
- water forced against a coast by strong onshore winds
- tidal currents forced into a bottleneck
- submarine landslides/volcanoes/earthquakes.

The effects of coastal flooding are exacerbated by human activity through:

- poor management of river or coastal systems
- building on lowland coastal areas
- coastal reclamation.

Integrated Coastal Zone Management (ICZM)

AQA	AS U1
OCR	AS U1
Edexcel	AS U2
WJEC	A2 UG3
CCEA	A2 U2

To ensure the coast is managed effectively account is taken of all the different groups that may influence what happens at the coast. This enables planners to establish an **Integrated Coastal Zone Management Plan (ICZM)**. It identifies all the groups, their impact and possible solutions for the coast.

Case study: Strategies to manage Australia's coast usage

Increasing pressures on Australia's coasts	ICZM solutions to reduce coastal problems
More people are living at the coast in Australia – 84% at present – as Australia's ageing population retire to the coast.	Have planning controls on the size and nature of new communities. Encourage planners to design 'sustainable communities' which produce less pollution.
Tourists to Australia increasingly want to be at the coast – to surf, sail and scuba dive. Eco-tourism at the Great Barrier Reef is a growth area.	Educate people about the sensitive eco-systems and encourage people to be responsible in the way they act.
Australian farms are using increasing amounts of chemical fertiliser and pesticides to increase production – especially the wine industry. The chemicals get washed off into rivers.	Check the pollution levels of rivers. If there are high levels of chemicals, trace them back to where they are coming from. Have fines for those causing the pollution.
Trawling for fish is reducing fish stocks and scoops up other species at the same time leading to fewer fish species in the sea off Australia's coast.	Scientists to measure the number of fish stocks. Only permit trawling far from coral reefs. Encourage fishing by line rather than by net.
Fish farms and shell fish farms (aqua-culture) are growth areas. Thousands of fish are 'grown' in enclosed cages at the coast and then 'harvested'.	Measure the amount of marine pollution. Only give licences for activities in areas where they won't be in tourist areas or may affect the growth of coral.
Australia is the main source of many minerals for China and Japan, such as copper, iron-ore, tin and uranium. This is increasing the amount of ships going to and from Australian ports.	Increase the number of marine pilots to navigate safe routes. Make sure the shipping routes don't go through sensitive marine environments.
Global warming is leading to higher sea temperatures and rising sea levels. It is also contributing to coral bleaching where the sea is more acidic.	Plan for the growth of marsh plants and mangrove forests along the shore to 'soak up' (absorb) the rising water. Speak at international conferences about the negative effects of global climate change.

Sample questions and model answers

1. Study the diagram below.
   (a) Suggest why coasts are dynamic environments. [3]
   (b) Explain how, despite being dynamic, coastlines can become irregularly shaped. [5]
   (c) How does man manage coastal retreat along the coastline? [7]
   Comment also on any disadvantages and advantages.

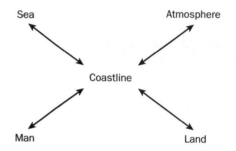

The first sentence rather wastes precious space.

Rather weakly expressed towards the end.

Offers two points (a wave refraction point and a lithology point); neither are taken on fully.

This candidate clearly understands the question, and in the space available selects strategies and shows some insight into cost–benefit analysis.

(a) This is the case because there are so many factors now, which can alter coastal processes and therefore change the coastline dramatically. For instance, through coastal sea defences, man has changed the natural processes of a specific area; only to increase it in another. Through changes in the atmosphere, sea levels can rise and fall and the wind can change the direction of longshore drift.

(b) Wave refraction focuses the waves onto the headlands causing an uneven erosional pattern. Irregularly arranged hard and soft rock can have a similar effect on the coastline. Man's attempt to hold the line can also lead to an irregular pattern to the coastline.

(c) There are many strategies for coastal management. Most often groynes, wooden constructions, mainly in the breaker zone are put on the beach. They build up the beach locally, but sediment deficits downdrift cause the beach to shrink and erosion to increase. Artificial reefs may be constructed around a particularly bad cliff line, but this too can have bad effects on the coastline, causing embayments and build-ups of sand behind reefs. Beach replenishment is another option, when sand is dredged from the sea. Sand is the best energy absorber available. Adding sand can upset the dynamic equilibrium of the coastline. In fact the government has decided that it is best not to use these schemes, it is best to let the sea do its worst and then to compensate landowners. The effect of these schemes is to protect those areas that are economically important and to let the sea take those areas of little worth.

6 Arid and semi-arid environments

The following topics are covered in this chapter:

- **The causes and distribution of deserts**
- **Desert processes and landforms**
- **Desert conditions and animal and plant adaptations**
- **Man in arid lands**

6.1 The causes and distribution of deserts

LEARNING SUMMARY

After studying this section, you should be able to understand:

- the causes of aridity
- the distribution of arid and semi-arid areas around the world

Aridity

AQA	AS U1
OCR	AS U1
Edexcel	A2 U4
WJEC	A2 UG3
CCEA	A2 U2

A straightforward topic, but not normally examined at A2.

Aridity is a lack of water and generally it can be classified by mean annual rainfall. Temperature also has an effect as it can determine evapotranspiration. Surprisingly, many areas with similar rainfall regimes to the UK may be classified as arid due to other factors at work.

The 'rainfall definition' of aridity		
250–500 mm/yr	Semi-arid	Sparse vegetation such as grassland, few trees grow.
25–250 mm/yr	Arid	Plants only appear along river courses.
< 25 mm/yr	Extremely arid	Plant growth only after rainfall.
These classifications cover one-third of the world.		

The causes of aridity

Pressure

Deserts are found in over 60 countries between 15° and 30° North and South. About one-third of the land surface of the world is classified as arid, semi-arid and/or dry.

Cold ocean currents

Cold air above ocean currents ensures that there is little moisture available to cool and form clouds. The coasts of Western, North and South America and Africa display such conditions. Both continents have west coast deserts slightly inland.

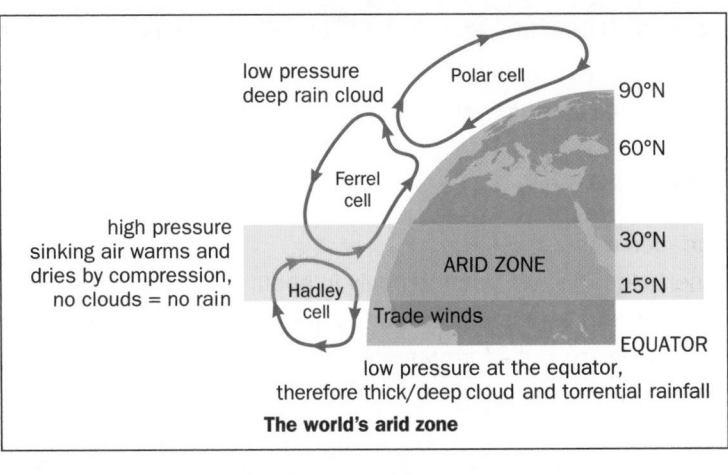

The world's arid zone

low pressure deep rain cloud — Polar cell — 90°N

Ferrel cell — 60°N

high pressure sinking air warms and dries by compression, no clouds = no rain — Hadley cell — 30°N — ARID ZONE — 15°N

Trade winds

EQUATOR

low pressure at the equator, therefore thick/deep cloud and torrential rainfall

Rainshadow and continentality

Air descending from mountainous areas warms and dries by compression, little rainfall forms and aridity is the result. Central areas of continents are dry because the air moving over landmasses does not absorb large amounts of water vapour. During the last ice age, conversion of water to ice resulted in larger continental areas. This extreme continentality is thought to have facilitated the spread of deserts during the ice age.

Worldwide distribution of deserts

AQA	AS U1
OCR	AS U1
WJEC	A2 UG3
CCEA	A2 U2

The majority of the world's most arid areas lie between 15° and 30° North and South.

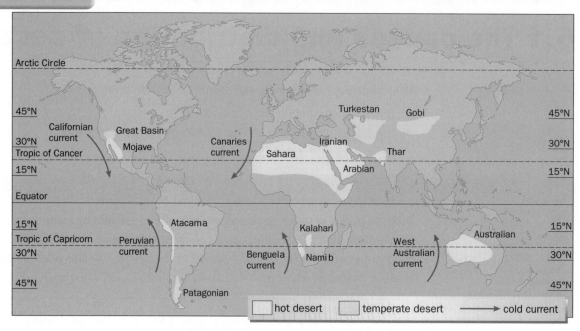

The world's major deserts and associated ocean currents

6.2 Desert processes and landforms

LEARNING SUMMARY

After studying this section, you should be able to understand:

- the processes of weathering in deserts
- that water and wind processes cause differing landforms
- that the origin of some landforms is difficult to ascertain

Processes

AQA	AS U1
OCR	AS U1
WJEC	A2 UG3

With regard to desert landforms, controversy has raged for many years over the part played by water and wind (**aeolian effects**) in forming desert landforms.

> Sand covers 20% of the Earth's surface. Over 50% of this area is deflated desert pavements.

Desert surfaces

A range of desert surfaces exist.

- **Ergs** – sandy deserts/sand seas, common only in about 30% of deserts. Their distribution seems to be climate linked (i.e. less than 150 mm of rain).
- **Pavements, gibber plains or reg** – form as a result of wetting and drying. They are hard, rock covered areas.

Weathering

Desert weathering is a controversial topic. **Chemical weathering** is limited by:

- the lack of water
- low rates of penetration into the rocks
- the amount of capillary action
- the alkaline nature of chemicals taken into rocks creating no aggressive acids.

Other weathering processes include:

- **Salt weathering** – rocks in deserts often contain efflorescent salts which set up stresses in the rock and produce fractures. This process is seen in porous and poorly cohesive rocks.
- **Exfoliation** – granular disintegration and chemical rotting also have an effect.

Aeolian processes

Winds that blow across deserts often produce an effect similar to fluid in motion. The lack of vegetation reduces surface roughness permitting smoother wind/land contact.

Wind erosion

In deserts water and wind processes are equally important. Know how they relate together!

- **Abrasion** – occurs when small particles are hurled by the wind against rock surfaces and occurs slightly above the ground. **Ventifacts**, rocks smoothed by wind abrasion, are common in deserts.
- **Deflation** – wind blows away rock waste and lowers the desert.
- **Attrition** – rock particles rub against each other and wear away.

Aeolian transport

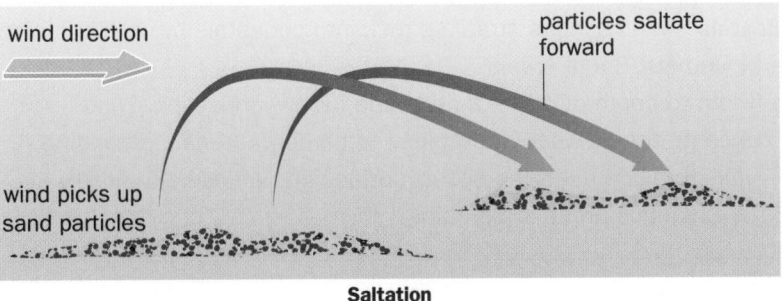

Saltation

- **Saltation** moves small particles in the direction of the wind in a series of short hops and skips. It normally lifts sand-size particles no more than one centimetre above the ground and proceeds at one-half to one-third the speed of the wind. A saltating grain may hit other grains that jump up to continue the saltation. The grain may also hit larger grains that are too heavy to hop, but they slowly creep forward as they are pushed by saltating grains. This is called **surface creep**. Velocity is an important variable – a critical velocity has to be reached before particles will move (see diagram).

Deposition

Three processes have been recognised.

- **Sedimentation** – settling occurs and there is no further effect on other sand particles.
- **Accretion** – occurs when sand grains come to a rest.
- **Encroachment** – the process of continued growth of sand accumulations. Once sand has accumulated it traps more and more sand, ripples turn into dunes and dunes into '**draa**'.

Fluvial processes

Desert lakes are generally ephemeral and are called **playas**. They vary in size from a few metres to several thousand km^2. They are very salty.

Three main types of river are found in desert areas.

- **Exogenous** rivers – sources outside the desert.
- **Endoreic** rivers – these form near the desert.
- **Ephemeral** rivers – these flow for only part of the year.

Drainage systems

The mountain areas of deserts and the lowland deserts have hugely different drainage systems.

Mountainous areas – a high amount of scouring occurs producing very rocky beds and lots of debris/sediments in the upper areas of mountains. As slope decreases, sediments concentrate in lower areas, so there is a rise in the deposition of alluvial fans at changes in slope.

Lowland areas – the nature of the surface over which water flows determines the drainage pattern. Few permanent rivers exist; where they do, they are shallow, sandy, straight and lack sinuosity. In flood conditions they are choked with sediment.

PROGRESS CHECK

1. What is the maximum amount of rainfall that describes a desert?
2. How does the rainshadow effect contribute to aridity?
3. What are the differences between an endoreic and exogenous river?

Answers: 1. 250 mm. 2. When the air descends the leeward side of mountains it warms and dries by compression. 3. Endoreic has a source near the desert and never shows beyond it while exogenous has sources outside the desert.

Landforms

AQA	AS U1
OCR	AS U1

Features produced by wind erosion

- **Rock pedestals** – wind sculpts stratified rock into pedestals by wind abrasion, e.g. Gava Mountains, Saudi Arabia.
- **Yardangs** (width to depth of 4:1) – a ridge and furrow landscape. Wind abrasion concentrates on weak strata; leaving harder material upstanding.
- **Zeugen** – wind abrasion turns the desert surface into a ridge and furrow landscape, e.g. various areas in Bahrain.

> The Sphinx at Giza may be a modified yardang!

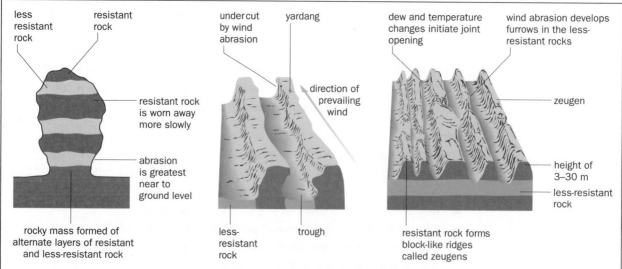

Inselbergs

Wind (and water) attacks the original surface leaving round-topped **inselbergs** (through **exhumation**). The material removed has a deep-seated 'decay' origin and may display extensive 'unloading' (subsurface weathering). There are two major forms: domed inselbergs (**bornhardts**) and boulder inselbergs (**Kopjes, rubbins**), e.g. Matopos, Zimbabwe.

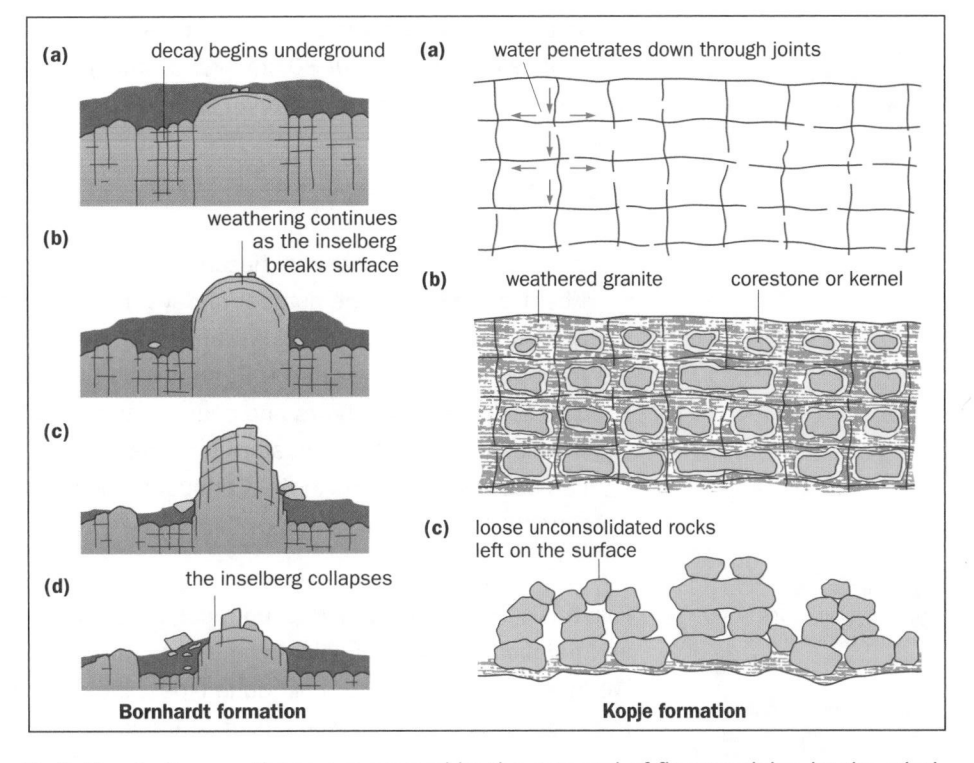

> Good diagrams and supportive labelling is a must if you hope to convey landform understanding.

(a) decay begins underground

(b) weathering continues as the inselberg breaks surface

(c)

(d) the inselberg collapses

Bornhardt formation

(a) water penetrates down through joints

(b) weathered granite   corestone or kernel

(c) loose unconsolidated rocks left on the surface

Kopje formation

Deflation hollows – these are caused by the removal of fine particles by the wind, lowering the surface and creating a hollow. The best known example is the Qattara Depression.

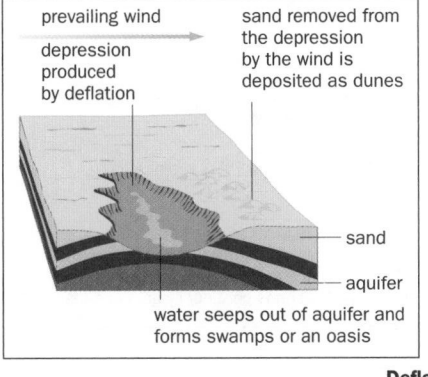

prevailing wind

depression produced by deflation

sand removed from the depression by the wind is deposited as dunes

sand

aquifer

water seeps out of aquifer and forms swamps or an oasis

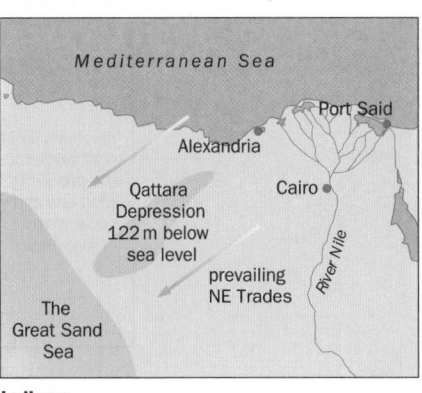

Mediterranean Sea

Port Said

Alexandria

Qattara Depression 122 m below sea level

Cairo

prevailing NE Trades

The Great Sand Sea

River Nile

Deflation hollows

Features produced by wind deposition

Wind-deposited materials occur as sand sheets, ripples and dunes.

- **Sand sheets** – these are flat areas of sand with sand grains that are too large to saltate; 45% of depositional surfaces are of this type, e.g. Selima in South Egypt.
- **Dunes** – the wind eventually blows sand into a network of troughs, crests and ripples that are perpendicular to the wind direction. They are the consequence of saltation.

> Dunes can move fast! Between 1954 barchans in China's Ningxia Province were moving at more than 100 m per year!

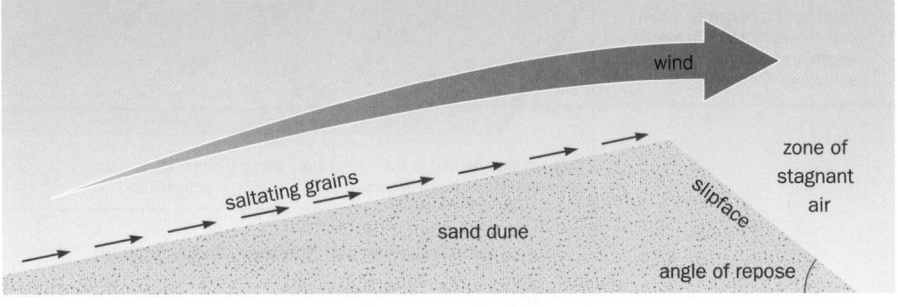

wind

saltating grains

sand dune

slipface

zone of stagnant air

angle of repose

Dunes produced by wind deposition

> Outside of the USA, sand dunes make up 25% of arid landforms.

Accumulations of sand build into mounds and ridges and they become a dune when the slip face is about 30 cm high. Dunes grow as sand particles move up the gentle upwind slope by saltation and creep. They fall onto the slipface inducing movement.

KEY POINT

Sandstorms are a seasonal hazard in north-east Africa and are called Khamasin (fifty) for the number of days on which they occur. They strike around April with the onset of warm conditions. Hot rising air lifts dust up to 4500 m above the desert. Returning as brown rain in winds of up to 110 kph it closes airports and causes many accidents. On average 20 people die due to the sandstorms per year. When the sand moves west it can destroy coral reefs in the Caribbean and has been linked to hurricane formation.

> Star dunes are pyramidal in shape. They are the tallest dune.

> Parabolic dunes have 'arms' pointing upwind.

Five basic dune shapes have been recognised: **crescentic, linear, star, domes** and **parabolic**. Ralph Bagnold, an engineer, working in Egypt prior to the Second World War, recognised the main dune types, the **crescent or barchan dune** and the **linear or seif dune systems** (*seif* is Arabic for sword).

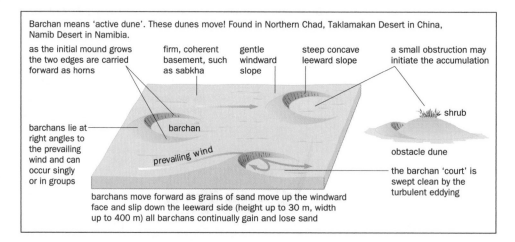

Barchan means 'active dune'. These dunes move! Found in Northern Chad, Taklamakan Desert in China, Namib Desert in Namibia.

- as the initial mound grows the two edges are carried forward as horns
- firm, coherent basement, such as sabkha
- gentle windward slope
- steep concave leeward slope
- a small obstruction may initiate the accumulation
- shrub
- barchans lie at right angles to the prevailing wind and can occur singly or in groups
- barchan
- prevailing wind
- obstacle dune
- the barchan 'court' is swept clean by the turbulent eddying

barchans move forward as grains of sand move up the windward face and slip down the leeward side (height up to 30 m, width up to 400 m) all barchans continually gain and lose sand

As to formation these dunes are either:
- the result of obstacles getting in the way
- an erosional phenomenon
- as products of a vegetated landscape or,
- as products of complex wind regimes, secondary wind flow patterns (and large amounts of sand).

Seifs have a height of 100m and are up to 190 km long! length > width and they are regularly spaced.

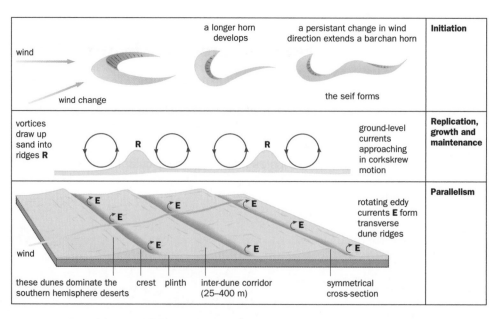

	Initiation
wind — a longer horn develops — a persistant change in wind direction extends a barchan horn	
wind change — the seif forms	

	Replication, growth and maintenance
vortices draw up sand into ridges **R** — ground-level currents approaching in corkskrew motion	

	Parallelism
wind — **E** crest plinth inter-dune corridor (25–400 m) symmetrical cross-section — rotating eddy currents **E** form transverse dune ridges	
these dunes dominate the southern hemisphere deserts	

Linear dunes (self, traverse, dunes or draa)

Features produced by water in deserts

More people drown in deserts than die of thirst!

Rain does fall occasionally in deserts and desert storms are often violent. A record 44 mm of rain once fell within 3 hours in the Sahara. Large Saharan storms may deliver up to 1 mm/minute. Normally dry stream channels, called **arroyos** or **wadis**, can quickly fill after rain and **flash floods** make these channels dangerous.

However, the evolution of arid landforms is often affected by events that occurred long ago. Past climatological conditions, reflected in many desert landforms, began to develop during pluvial periods several thousand years ago. Channels which were once part of a perennial drainage system now receive the run-off from torrential storms, unhindered by vegetation, leaving deep alluvial debris over the wadi floor, though wind deflation may later remove much fine surface material. Surface channels on old shield areas often lead into inland drainage depressions. Many of these lake depressions have been filled by materials washed forward from the foot of the enclosing mountains, where streams entering the basin have deposited their load and created alluvial fans. Some fans have coalesced to form a mass of material known as a **bajada**. Some of the material from the bajada washes forward over the pediment levelling the landscape with unconsolidated deposits.

Water can also dissect out deep wadi systems. Stream-dissected scarps overlook the older plains, where detached **outliers** form **mesas**, **buttes** and **pinnacles**.

In the USA 32% of arid landforms relate to the action of water in deserts.

KEY POINT

- **Sabkhas** – an occasionally flooded desert area with an extensive thickness of evaporates (salts), usually coastal in nature.
- **Duricrust** – hardened areas of minerals that cover deserts. They are impermeable and thought to be due to weathering.
- **Desert varnish** – iron and magnesium oxides and silica coat rocks. Due to evaporation.

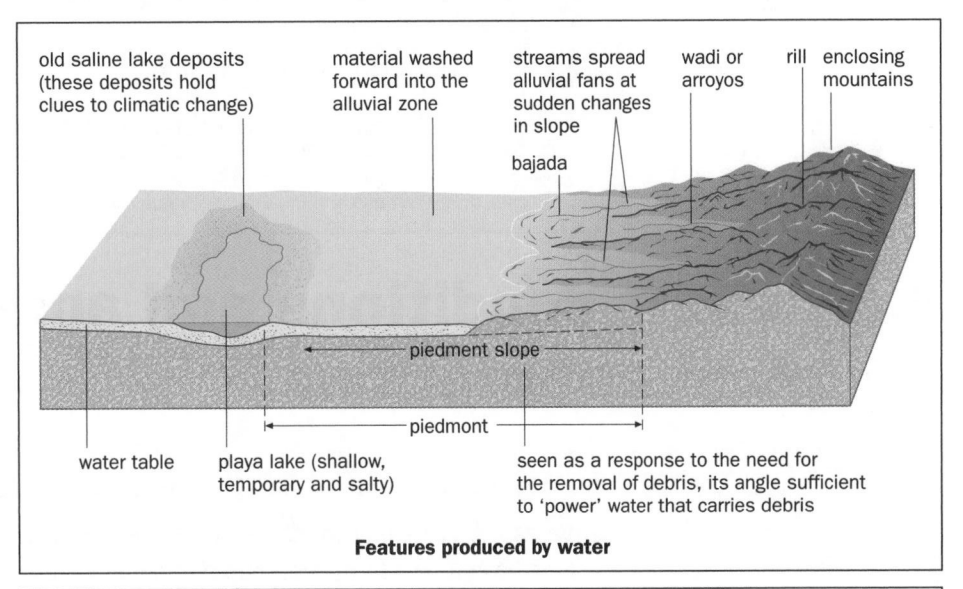

old saline lake deposits (these deposits hold clues to climatic change)

material washed forward into the alluvial zone

streams spread alluvial fans at sudden changes in slope

bajada

wadi or arroyos

rill

enclosing mountains

piedment slope

piedmont

water table

playa lake (shallow, temporary and salty)

seen as a response to the need for the removal of debris, its angle sufficient to 'power' water that carries debris

Features produced by water

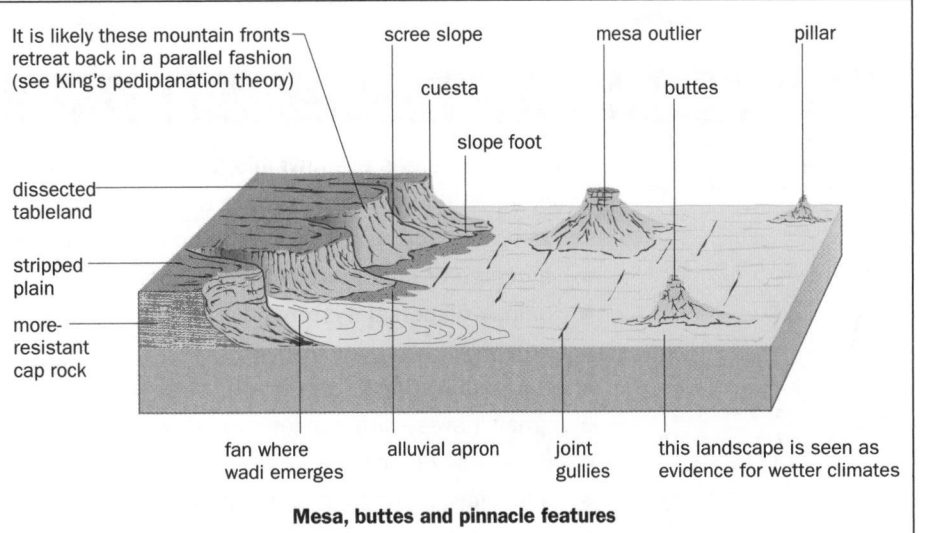

It is likely these mountain fronts retreat back in a parallel fashion (see King's pediplanation theory)

scree slope

mesa outlier

pillar

cuesta

buttes

slope foot

dissected tableland

stripped plain

more-resistant cap rock

fan where wadi emerges

alluvial apron

joint gullies

this landscape is seen as evidence for wetter climates

Mesa, buttes and pinnacle features

Case study: Lake Bonneville, USA – a playa within a desert

Ground speed records are commonly established on Bonneville speedway. The space shuttle lands on Rogers Lake Playa, at Edwards Air Force Base, California.

Lake Bonneville is the relic of a large lake that existed during the last ice age (12 000 years ago), though it probably has an origin from 50 000 years ago. It was 52 000 km², almost 300 m deep and was 1525 m above sea level. Climatic change caused the lake to fall below the lowest outlet. When the water evaporated it left an arid depression composed of billions of tonnes of salt and other minerals. The remnants of Lake Bonneville include Utah's Great Salt Lake, Utah Lake and Sevier Lake.

Equifinality

Few landforms have proven origins. Many landforms can come about through different processes and quite different conditions, e.g. pediments, dune formation inselbergs and deflation hollows may have many origins.

PROGRESS CHECK

1. What are the differences between bornhardts and kopjes?
2. Outline the reasons for the differing formation of seif and barchans dunes.
3. What are wadis? How do they form?

Answers: 1. Principally the way in which they emerge on the surface and the way in which they form. 2. Seif dunes form when the wind blows from a certain direction for some time. Barchans dunes are the result of obstacles and complex wind regimes. 3. Dry stream channels, they are cut during rainy periods when water flows in the valley.

6.3 Desert conditions and animal and plant adaptations

LEARNING SUMMARY

After studying this section, you should be able to understand:
- plants have to adapt to survive in the desert environment
- soils have a unique appearance and composition in desert areas

Adaptations of plants

AQA   **AS U1**
OCR   **AS U1**

Plants in arid areas are **physiologically specialised**, adapted in form and structure. Within the desert there are niche locations for plants even though the soil is susceptible, skeletal, saline and immature. These include:

- a degree of **ephemeralism**, remaining dormant in the soil as fruits or seeds
- unique **dispersal systems**, i.e. barbs and bristles
- **xerophytic**, drought resistant
- **root adaptations**, tap roots or heavy lateral branching
- small leaves, with sunken or restricted stomata or pale, reflective, leaves
- hairs, spines or thick waxy-walled leaves
- **succulents**, plants capable of storing water
- plants with a small surface-to-volume ratio

The lack of vegetation in deserts means that adaptations are necessary. A favourite topic for examiners to pick on!

- **cell sap variations** allow varying amounts of water to escape
- ability to tolerate desiccation
- **halophytic**, salt tolerant, e.g. Creosote plant.

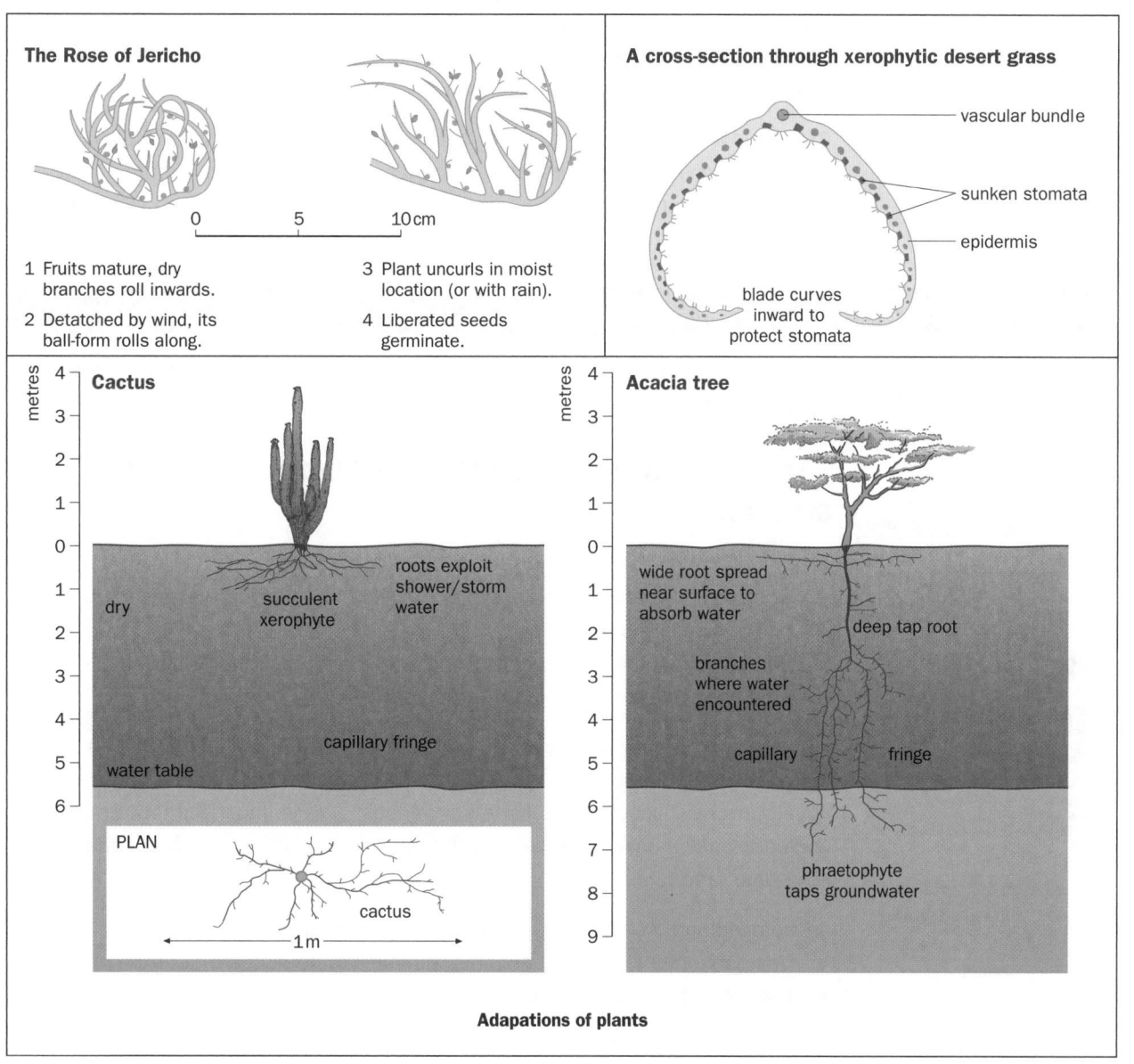

The Rose of Jericho

0   5   10 cm

1 Fruits mature, dry branches roll inwards.

2 Detatched by wind, its ball-form rolls along.

3 Plant uncurls in moist location (or with rain).

4 Liberated seeds germinate.

A cross-section through xerophytic desert grass

vascular bundle

sunken stomata

epidermis

blade curves inward to protect stomata

Cactus

metres

roots exploit shower/storm water

dry

succulent xerophyte

capillary fringe

water table

PLAN

cactus

1 m

Acacia tree

metres

wide root spread near surface to absorb water

deep tap root

branches where water encountered

capillary   fringe

phraetophyte taps groundwater

Adapations of plants

Deserts typically have a plant cover that is sparse, but in places such as the Sonoran Desert in the American south-west it can have the complex desert vegetation. The giant saguaro cacti provide nests for desert birds and serve as the trees of the desert. Saguaros grow slowly, but may live 200 years. When fully grown, saguaros are 15 m tall and weigh as much as 10 tonnes. They dot the Sonoran and reinforce the general impression of deserts as cacti-rich land.

Although cacti are often thought of as characteristic desert plants, other types of plants have adapted well to the arid environment. They include the pea family and sunflower family.

When 9 years old, they are about 15 cm high. After about 75 years, the cacti are tall and develop their first branches.

Soils

| AQA | AS U1 |
| OCR | AS U1 |

Soils that form in arid climates are predominantly mineral soils with low organic content. The repeated accumulation of water in some soils causes distinct salt layers to form. Calcium carbonate precipitated from solution may cement sand and gravel into hard layers called '**calcrete**'.

Caliche is a reddish-brown to white layer found in many desert soils. Caliche commonly occurs as nodules, or as coatings on mineral grains, formed by the complicated interaction between water and carbon dioxide released by plant roots or by decaying organic material.

Animal adaptations

| AQA | AS U1 |
| OCR | AS U1 |

Too hot and animal protein breaks down and animals die; too cold and animals' body function slows down; too little water and animals die.

They have thick skin and hair to resist the effects of the sun.

Camels metabolise fat to release water.

They are some distance from the desert surface so remain cool.

Camel

Thermoregulation (control of temperature) in animals

Thermoregulation can be either **behavioural** or **physiological**.
- Behaviour – posture/orientation, e.g. a lizard on a hot rock to warm up or positioning into the wind to cool down.
- Physiological – altering or controlling, e.g. metabolism of fat by camels' metabolic rate to control body temperature.

In this respect torpor can lower the body temperature (hibernation) and **estivation** (occasional summer sleeps) happens during the heat of the summer/day which, usually involves sleep during these hot periods. Size is important too. Small bodied animals heat-up and cool down faster (this relates to a larger surface area to volume ratio).

Advantages and disadvantages of being large include:
- heat is gained slowly and animals will survive the high temperature of the day
- they don't have to eat all of the time as no heat is lost.

Advantages and disadvantages of being small include:
- heat is gained and lost quickly; heat is lost almost as soon as it is gained
- food has to be eaten continually to replace lost energy.

Ears with blood vessels near to the surface release plenty of heat.

Rabbit

Water regulation in animals

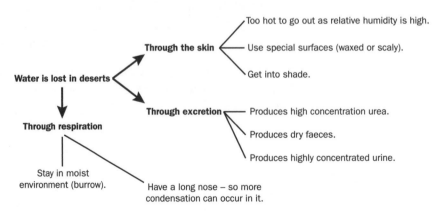

Water is lost in deserts

Through the skin
- Too hot to go out as relative humidity is high.
- Use special surfaces (waxed or scaly).
- Get into shade.

Through excretion
- Produces high concentration urea.
- Produces dry faeces.
- Produces highly concentrated urine.

Through respiration
- Stay in moist environment (burrow).
- Have a long nose – so more condensation can occur in it.

Water is essential for life – it is required by animals to exchange gases for instance.

6.4 Man in arid lands

LEARNING SUMMARY	After studying this section, you should be able to understand:
	• that deserts if managed correctly can be very productive
	• the importance, and effects, of irrigation in deserts

Settlement patterns

AQA	AS U1
OCR	AS U1
Edexcel	A2 U4
WJEC	A2 UG3
CCEA	A2 U2

The effects of people on the arid and semi-arid lands may be seen at three overlapping cultural/technological levels.

- Small groups of people, like the bushmen of the Kalahari and the Australian aborigines, live a **semi-nomadic food-gathering** and hunting life, and have adapted to their environment with remarkable efficiency. Over the years, the hunter-gatherers have had little lasting impact on their environment.

- By contrast, pastoral nomads seasonally cultivate selected areas. **Pastoral nomads** and **shifting cultivators** have greatly affected the natural flora and fauna. They respond to physical conditions through their mobility.

- Many millions live a settled life in a relatively moist environment within arid zones near oases, on a flood plain or delta, or in an area irrigated by water brought from afar. **Irrigation**, with its associated land-use and settlement, has changed entire ecosystems, and ecological repercussions have been felt far beyond the irrigated areas.

Irrigation: a chronology

AQA	AS U1
OCR	AS U1
Edexcel	A2 U4
WJEC	A2 UG3
CCEA	A2 U2

Benefits and problems caused by irrigation

Benefits

Traditionally, people who settled at oases used simple technology to raise and distribute water locally through aqueducts and underground channels or qanats. Evaporation losses from this system are minimal, though some water is lost by infiltration.

Other methods were to use shallow, gravity-fed channels, blocked by mud packing as required and included; cisterns; underground caverns; depressions dug to catch run-off and retain water moving by gravity through the sub-soil; the creation of lakes during the rains by blocking stream beds with boulders and earth dams.

Today electric or diesel pumps raise groundwater from greater depths. Massive dams create deep lakes stretching back hundreds of kilometres. Metal pipelines of decreasing diameter carry water from reservoir to the field and release it through nozzles, or tiny drip feeds, to the plants.

Problems

Problems, such as water-logging and saline accumulation, are common. Water not used by crops, lost by evapotranspiration, or drained away, accumulates as rising groundwater. In desert conditions soils rapidly acquire salt from the evaporation of dilute saline irrigation water. Water acquires more salt as it slowly moves up through the ground. The land eventually becomes too saline for crops to tolerate.

Case study: Pakistan: an 'irrigated' country

In Pakistan, the whole hydrological system of a huge river, the Indus, has been transformed, as engineering systems have evolved progressively to control and distribute its waters.

Most of the flow in the Indus system is from catchments in the Himalayas and its foothills. Monsoon breaks in June and rivers reach peak flood-levels in the foothills in July–August, causing a flood wave to pass downriver. From September the river levels fall, then snowmelt brings another rise in March. Irrigation has brought some problems for Pakistan.
- Dams and diversions have caused floodplains to be deprived of alluvium and its nutrients which leads to needing expensive fertiliser.
- There is increased danger from infections such as bilharzia.
- Where year-round cultivation replaces cropping with a dry fallow period, crop pests may thrive on perennial food sources.
- There are also many instances of plant pests and diseases being carried along water channels.

It is important to read the tropical section where details on desertification and desertification of the savanna can be found.

Despite these problems through the nineteenth and twentieth centuries, irrigation projects have continued apace. Inundation canals were built to carry water from cuts in the bank to land parallel with the river. Gated barrages were constructed to raise the water level and allow accumulations of alluvium to pass downstream. Water could now be diverted throughout the year. Large canals were built to transfer water from the western rivers to supplement those further east. More recently, high dams in the mountains have provided storage and hydroelectricity. Perennial canals now serve most of the cropped land. Despite the aridity, over half of Pakistan's workforce are engaged in agriculture.

Case study: Mineral mining at Evango, the Namib desert, Namibia

Many desert areas hold massive reserves of minerals. Uranium was discovered in the Namib Desert in 1928, but it was not exploited until 1970. Recently Rio Tinto, the owners of the company exploiting the resources, has determined there are sufficient resources left to extend the life of the operation to 2016. All of the resources identified exist in the hyper-arid Namibian desert, but within easy reach for the 'export processing zone' of Walvis Bay.

However, before Rio Tinto started phase two of the area's development, extracting 4500 tonnes/year, the Namibian Government, the major decision-maker, insisted that all future development must be carried out sustainably in terms of socio-economic and environmental impacts. Initial research suggested that advantages included:

- jobs and employment
- more housing
- more disposable income being spent locally
- more schools and medical facilities would be built
- tax revenue for the country would increase.

While negative effects include:
- farming would be affected
- there would be irreversible destruction of habitats and ecosystems
- increases in fall-out pollution
- seepage of pollutants into water-courses
- increased road traffic
- increased water and power demands
- archaeologically important areas could be lost
- noise pollution would increase
- waste disposal will be an issue
- increased incidence of radiation health issues.

Case study: Israel's use of desert areas

Israel's farmers are well-known for their innovative use of desert areas. Drip irrigation was invented in the 1920s west of Beersheva. This system has now largely been replaced in Israel by so-called 'spoon-feeding' where water and minerals are delivered to plants more precisely. Israel's newly created fish-farms rear barramundi (an Australian predatory fish) in brackish water. This water is used to water olive trees and to provide for the potato, tomato and pomegranate industries. In Israel's greenhouses tomatoes are produced at the rate of 20 tons/acre/year with most going for export to Europe.

Dubbed 'Negev 2015', the Negev and Galilee Development Ministry intends to invest $3.5 billion further developing the desert areas in Israel.

Case study: Has the Sahara leapt the Mediterranean?

The Iberian Peninsula is under threat from 'sahelisation' (desertification). It has been assessed that 37% of southern Spain is at risk from desertification. Unless urgent attention is given to the problem 17 tonnes/ha/year of topsoil will continue be lost in the south of Spain through soil erosion.

The problem in the south is aggravated by human mismanagement, drought, fire, overgrazing and poor agricultural techniques and aquifer depletion (because of tourism and settlement growth along the coastal strip and agricultural demands for water). To satisfy the farmer and the tourist 10 000 illegal wells have been dug in the Malaga area alone and 510 000 nationally!

The Spanish instituted a €90 million scheme to plant 45 million trees, on 61 000 ha in the mountains in the period 2009 to 2010 and a ten year plan to aid farmers.

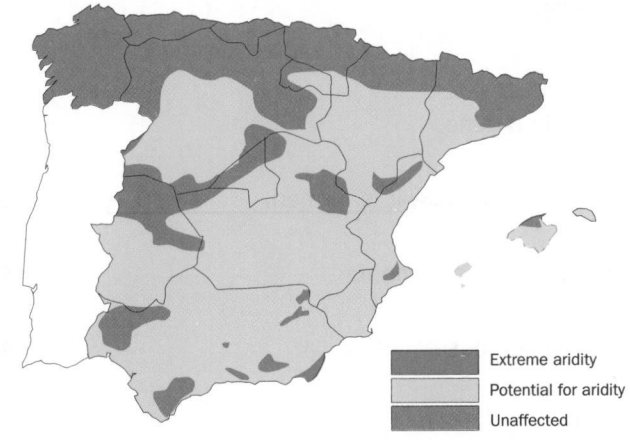

Extreme aridity
Potential for aridity
Unaffected

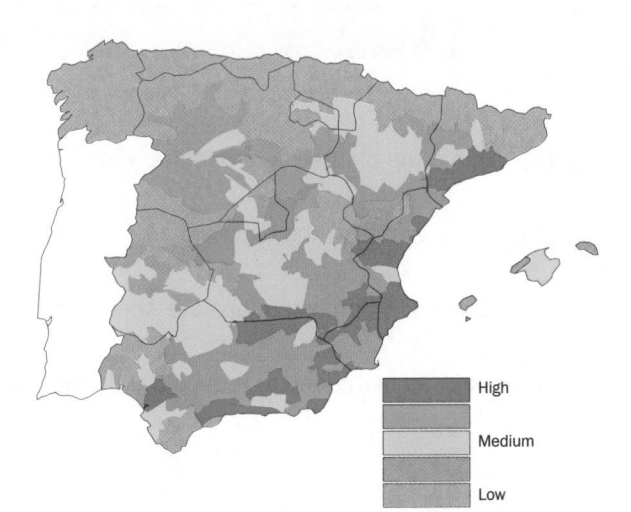

High
Medium
Low

PROGRESS CHECK

1. Outline the ways by which some desert plants have adapted to their environment.
2. Some desert cities carry out xerophytic planting – what is this?
3. Using an example outline the benefits of mineral exploitation in LEDCs.
4. Outline the issues that have led to 'salinisation' in Spain.

Answers: 1. By being xerophytic; by adaptations of leaves into spines (reducing water loss); root and leaf adaptations. 2. Where appropriate desert plants are used in planting, this saves water use in such areas. 3. Jobs, more housing, the country gains more tax revenue. 4. Over use of water by farmers producing goods for export and increases in settlement and tourism in areas of the country requiring water; also a long period of year on year droughts.

Exam practice questions

Structured question

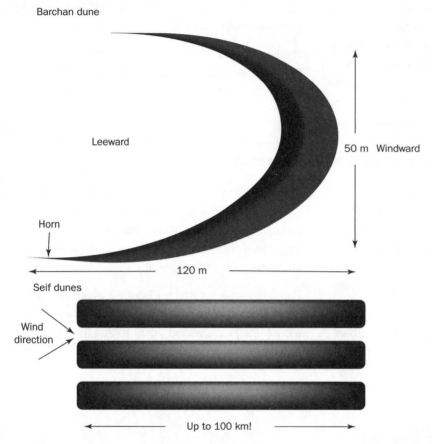

Barchan dune

Leeward

50 m Windward

Horn

120 m

Seif dunes

Wind direction

Up to 100 km!

1 **(a)** Explain the plan section of either the seif or the barchan dune. **[3]**

(b) Draw a profile of the barchan dune and add labels to explain how it is formed. **[6]**

(c) Explain why barchan dunes are seen as a threat to human settlement. **[4]**

(d) What has man done to reduce the threat that such dunes pose? **[4]**

Essay question

2 Describe and explain how the impact of irrigation influences population, settlement and economic development in desert environments. **[25]**

Cold environments

The following topics are covered in this chapter:

- Glacial activity
- The periglacial realm
- Issues in cold environments

7.1 Glacial activity

LEARNING SUMMARY

After studying this section you should be able to understand:

- that present day glaciated areas occupy a smaller area than during the Pleistocene era
- how extremes of accumulation and melting affect the glacial system
- that glaciers can be classified and movement assessed and measured
- that glacial processes modify landscape, landforms and drainage systems
- that glacial processes both threaten and advantage human occupation

Glaciology

AQA	**AS U1**
OCR	**AS U1**
Edexcel	**A2 U4**
WJEC	**A2 UG3**

Approximately 20 000 years ago, glaciers and ice sheets covered 32% of the land surface. Today 10% is covered by ice and 7% of the ocean surface is coated by pack and sea-ice at the maximum winter extent. An additional 22% of the land is underlain by continuous or discontinuous zones of permanently frozen ground. The ice age is not over, it has only diminished in its overall extent and intensity.

At maximum glaciation few areas of the Earth were unaffected by ice activity and action. The last worldwide glaciation occurred during the **Pleistocene** epoch which comprised some twenty extreme cold periods (or major advances of ice) interspersed by periods of milder weather, **interglacials**.

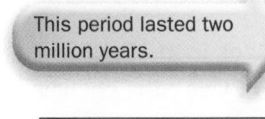

This period lasted two million years.

dispersal centres

glacier ice

ground ice

sea ice

Maximum Pleisocene ice extent

REGION	Present 000km^2	Pleistocene maximu m 000km^2
Antarctica	12 588	13 200
Laurentia	153	13 790
Scandinavia	4	6670
Asia	115	3370
Rockies	77	2500
Greenland	1803	2160
Others – N. hemisphere	131	4070
Others – S. hemisphere	27	1020
TOTAL	14 898	46 780

Extent of glacial masses

Ice age Britain

Most of the UK would not look like it does without the effects of glacial activity.

No area of the UK escaped the effects of the most recent ice age: the landscape we see today was created by ice advances over the land mass. Three main advances have been recognised.

- The **Anglian advance**, 420 000 to 380 000 years ago. This extended the furthest south, leaving major deposits in East Anglia.
- The **Woolstonian advance**, 170 000 to 130 000 years ago. This originated in the Highlands of Scotland and affected most of Northern Britain, largely wiping out the effects of the Anglian advance.
- The **Devensian advance** began about 115 000 years ago with a maximum advance happening about 30 000 to 18 000 years ago. Depositional and meltwater features in Shropshire, Cheshire and Yorkshire mark the limit of the advance. Ice left the land 14 000 to 10 300 years ago with the final brief advance from the Loch Lomond area.

In summary, large areas of the UK have been heavily affected by glacial activity, but three points should be emphasised.

- Erosional landforms are likely to have been influenced by multiple and successive glacial advances.
- Depositional features are the work of the most recent glacial activity though successive depositional/glacial events may have an effect, e.g. the enlargement and deepening of East Anglia's 150 m of depositional material.
- Not all locations in the UK have been covered continually in ice, e.g. the Cheviots.

C V	cirque and valley glaciers
	Loch Lomond advance
	Devensian
	older glaciation
	unglaciated

0   200km

Limits of the main glacial advances

Causes of past glacial periods and ice advances

Glacial periods are cyclic.

Ice ages are the result of a number of factors, such as: changes in the Earth's orbit; changes in the tilt of the Earth's axis (the **Milankovitch Cycle**); the changing nature of the oceans' currents; reduced carbon dioxide in the atmosphere; the changing distributions of land and sea areas and the relative change in the levels of the land and sea. These factors affect the Earth's **albedo** (the reflectivity of the Earth, sea and atmosphere), cooling occurs and an ice age ensues.

- **Glacial systems** – a system is the balance between inputs, storage and outputs. Glacial systems attempt to maintain dynamic equilibrium.
- **Glacial regimes** – the glacial regime describes the gain and loss of snow and ice in a glacier (or ice sheet). The regime determines the size of the glacier and whether it is advancing or retreating.

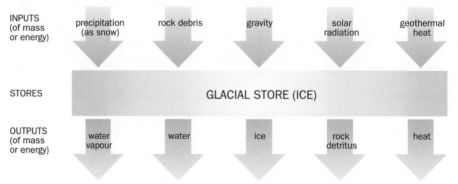

The glacial system

Types of cold environment: glacial at high altitudes and high latitudes; periglacial at high altitudes and latitudes and in continental interiors; alpine at high altitudes; polar found around the Poles.

- **Accumulation** – the gaining of snow (gaining of mass), is the input into the glacial system. It occurs high on the glacier or ice sheet.
- **Ablation** – loss of snow/ice (loss of mass), is the output from the glacial system. It occurs near the margins or snout of the ice mass.
- **Positive regimes** – supply > loss by ablation. The ice thickens and advances.
- **Negative regimes** – supply < ablation. The ice thins and retreats.
- **Balanced regimes** – supply = loss. A steady state, the ice mass remains constant.
- The **glacial balance** or **budget** describes the annual fluctuation in ice mass position. Glaciers advance or grow in winter and retreat in summer, a pattern which can be shown diagrammatically/graphically.

Converting snow to ice can be expressed as a flow diagram.

Snow falls $\Rightarrow$ snow accumulates (**alimentation**) $\Rightarrow$ compression occurs $\Rightarrow$ **firn** or **névé** ('half ice') $\Rightarrow$ pelletted ice with air spaces forms $\Rightarrow$ the following season more snow falls $\Rightarrow$ accumulates $\Rightarrow$ compression occurs $\Rightarrow$ flakes melt $\Rightarrow$ water refreezes between previous seasons' pellets $\Rightarrow$ white névé (density 0.06 gm/cm^3) gradually turns to blue glacial ice (density 0.9 gm/cm^3) as more air is driven out by compaction, re-crystallisation and re-freezing $\Rightarrow$ the ice has reached some 30 m thick after several seasons of accumulation and compaction.

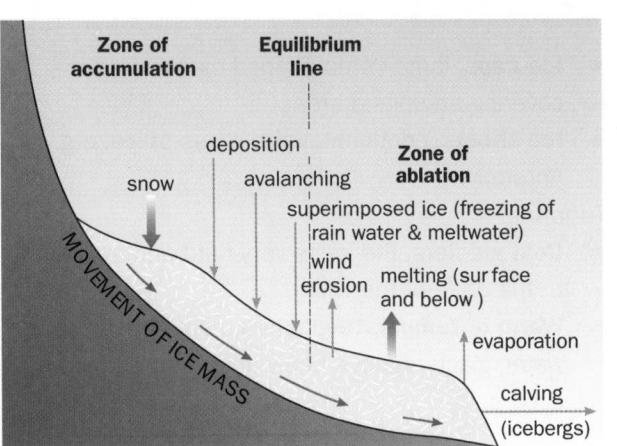

Factors affecting mass balance

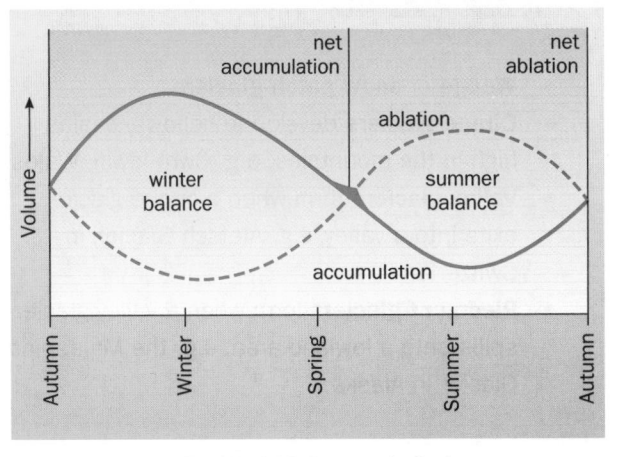

The glacial balance or budget

This helps with lubrication plus subsequent movement of ice masses.

Between the interlocking grains of snow and ice there is an inter-granular layer of chlorides and other salts. These can lower the freezing point, thus ensuring some liquid remains in the system.

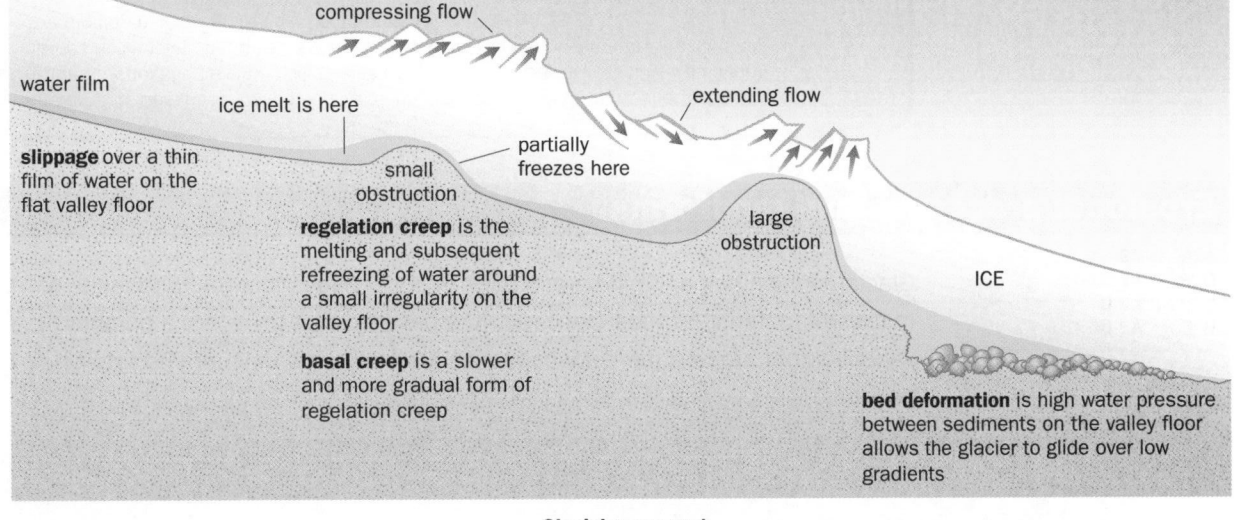

Glacial movement

Ice on the move

Ice movement can be difficult to understand, as ice can behave as a 'plastic liquid'. Glacier movement is controlled by the gradient of the 'rock floor' and the thickness of the ice (which controls the pressure melting point and temperatures within the ice). Smaller, thinner valley glaciers tend to move more rapidly than the massive ice sheets.

Actual types of movement are detailed below.

- **Basal slippage**, where a thin film of melt water allows a glacier to slide over the bedrock.
- **Internal deformation** where gradients are high. The glacier changes shape under its own weight, with ice particles moving relative to one another or in layers – a form of laminar flow.
- **Compressive flow** when compaction causes ice deformation on shallow slopes.
- **Extending flow** when thinning of the ice causes ice deformation on steep slopes.

> Friction holds ice back on the valley sides and trough floor.

> The classification of physical processes is a favourite at A2 Level. Know how glaciers and ice sheets can be classified.

Glacial classification

> **KEY POINT**
>
> **Size**
> - **Wedge** or **snow patch glaciers**.
> - **Cirque glaciers** develop in hollows/basins high in the mountains, e.g. Cwm Idwal, Wales.
> - **Valley glaciers** form when a cirque glacier exits into a valley, e.g. Aletsch Glacier, in Switzerland.
> - **Piedmont glaciers** form when a valley glacier spills onto a lowland area, e.g. the Malaspina Glacier in Alaska.
> - **Ice caps**, large dome shaped mass of ice covering an upland area.
> - **Ice sheets**, continental size mass of ice, e.g. Antarctica.
>
> **Thermal**
> - **Cold glaciers**, the ice is very hard and frozen to the base of the valley.
> - **Warm or temperate glaciers**, heat generates water, glaciers move freely and lots of erosion occurs.

> **PROGRESS CHECK**
>
> 1. What is a glacial regime? What is a negative regime?
> 2. What does the equilibrium line divide?
> 3. What is ablation?
> 4. What is pressure melting point?
>
> Answers: 1. Describes losses and gains on glaciers and determine whether it grow or retreats. A negative regime occurs when ice thins and retreats (supply < ablation). 2. It divides accumulation from ablation. 3. Loss of ice and snow from the glacial system. 4. The melting point of glacial ice under pressure.

Erosional processes and landforms

AQA	AS U1
OCR	AS U1
Edexcel	A2 U4
WJEC	A2 UG3

Glacial erosion has an immense impact, in some cases shaving thousands of metres of material off the landscape. It is believed that there are a number of variables that prepare the landscape and determine the amount and rate of erosion.

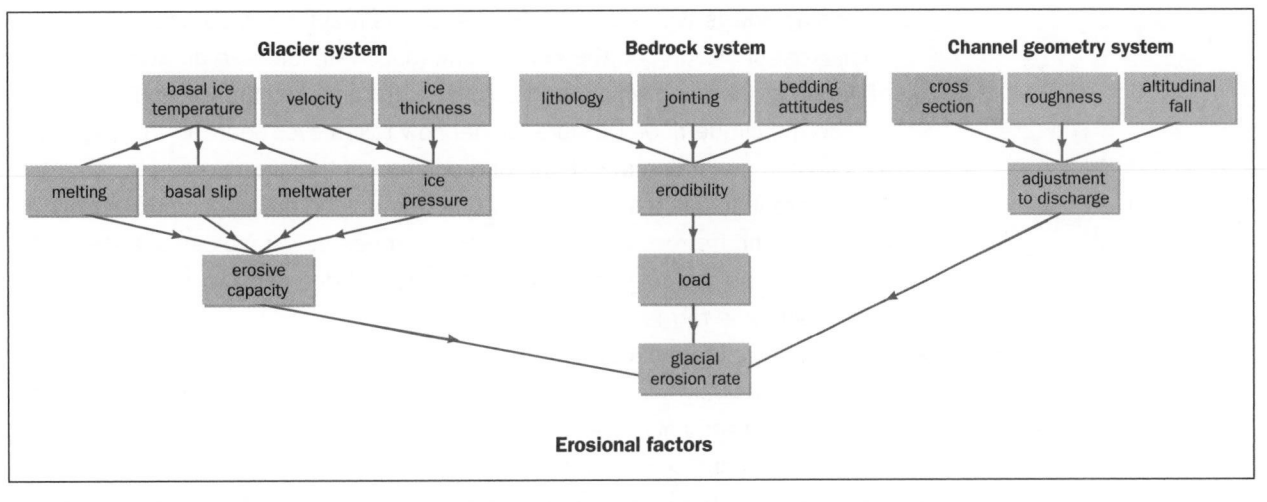

Erosional factors

	Indirect processes		Direct processes	
Surface rotting	**Pressure release**	**Plucking (quarrying)**	**Crushing & fracturing**	**Abrasion**
Frost action, weathering, mass movements, wind and water action in the vicinity of the ice mass.	Glacial advances and retreats constantly change the overburden pressures. Lots of material is made available for subsequent advances.	Water frozen into cracks and crevasses causes rock to be ripped out as the glacier moves downstream. Plucking occurs mostly on the bed and sides of the glacial trough.	Due to the sheer weight of ice.	Debris carried by the glacier scrapes and scratches the glacial trough wearing it away.

Erosional landforms

Effects vary from one area to another depending on relief. The impacts of the Pleistocene era can be seen in the present day landscapes of Northern Europe and North America, particularly in areas of high and varied relief. Landforms produced by glacial erosion include **cirques**.

- Cirques are large, rounded, armchair-shaped basins consisting of steep, frost-shattered backwalls, a smooth overdeepened floor and a moraine-covered threshold, or lip. They vary in size from the huge Antarctic cirques to the smaller British examples, such as Cwm Idwal in North Wales which has a backwall height of 420 m and an excavated area of 1.38 km². In general their length to height ratio is between 2.8:1 to 3.2:1 and shows a preferred orientation, onto shaded north-east facing slopes.

- **Nivation** is thought to be one of the main initiators of cirques. Snow accumulates in small, scattered hollows and, as the depth of snow increases, the hollows are gradually enlarged by frost shattering, meltwater erosion and perhaps chemical weathering too.

- Loaded with large amounts of debris and driven by the steep gradient and large inputs of snow, the cirque glacier starts to move by **rotational slip**.

Formation of a cirque

Extending flow is typical of the upper sections resulting in abrasion of the floor. Frost shattering on the **bergshrund** (backwall) leads to an accumulation of debris at the base which then becomes entrained by the ice and is used to abrade the cirque floor. Compressional flow takes over near the threshold or rock lip and maintains the shape and position of the cirque.

- As the cirque increases in size over time the upland area becomes heavily dissected and fretted. When two or more cirques develop close together a narrow steep sided ridge or **arête** (e.g. Striding Edge and Swirral Edge in the Lake District) is formed.
- When three or more cirques form close together a **pyramidal peak** or **horn** forms (e.g. Helvellyn in the Lake District, the Matterhorn in Switzerland). Both Helvellyn and Striding Edge were formed by the headwall retreat of Red Tarn and Nethermost Cove cirques.
- After glaciation the cirque often contains a small lake or tarn held back by the lip or threshold.

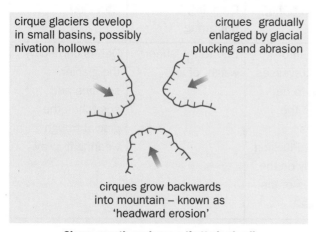

Cirque growth produces a 'fretted upland'

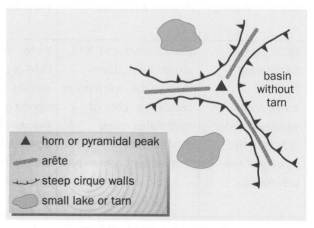

Post-glacial 'dissected upland'

Glacial troughs

> Be able to explain much of this in diagrams. At A2 a diagram can save a hundred words!

After leaving the mountains glacial activity is often concentrated into pre-existing river valleys. **Glacial troughs** tend to be steep-sided and U-shaped. Thick ice in the valleys moves rapidly (because of pressure melting) aided by sub-glacial streams flowing under **hydrostatic pressure**. Down-cutting is intense, i.e. up to 600 m of erosion occurred in the valleys around Windermere. The diagrams below show the characteristics of the glacial trough.

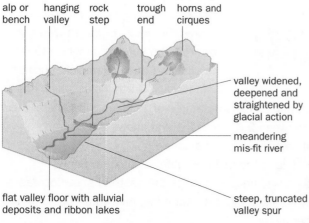

Block view of trough

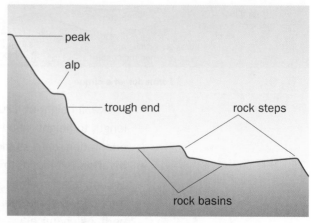

Long profile of trough, post-glaciation

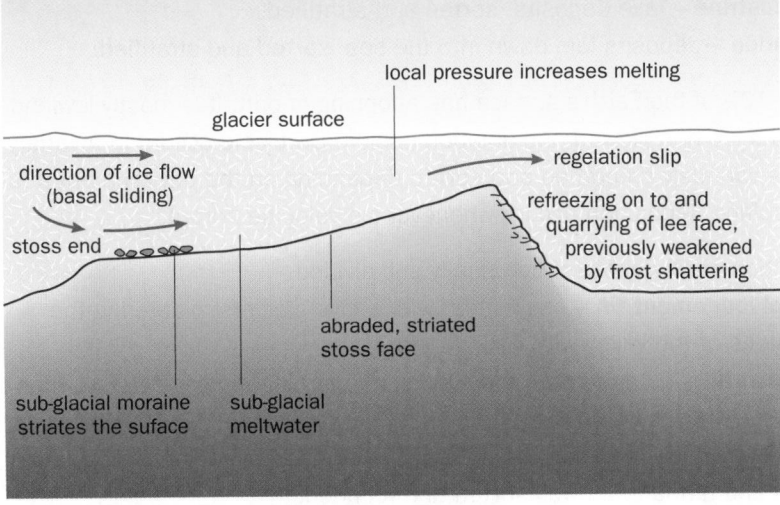

Roche mounonée formation

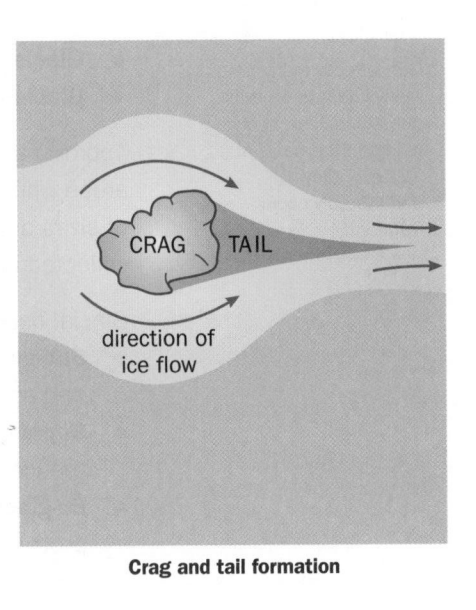

Crag and tail formation

> Rock lips and steps are thought to form because of changes in rock type.

- The head of the trough is marked by a sudden change of slope called a **trough end**. The long profile of the valley displays a series of rock steps and intervening basins that often contain ribbon lakes.
- The erosive effects of the ice ensure that tributary valleys are cut and subsequent drainage has to descend to the over-deepened valley in the form of **hanging valleys**.

Other features of glacial erosion include **Roche Moutonée**, e.g. Grange in the Borrowdale Valley of the Lake District and **crag and tail** formations when resistant rock obstructs ice flow, and material is deposited on its leeward, e.g. Edinburgh Castle rock.

Glacial diffluence

Glacial diffluence occurs when the normal flow of a valley glacier is blocked. The glacier increases its depth and thickness and eventually flows across the lowest point available forming a **col** or **gap**, e.g. the Freshie, Geldie and Dee system in Scotland.

PROGRESS CHECK

1. What is plucking?
2. What is rotational slip in cirque glaciers?
3. What is the name of the ridge formed between adjacent cirques?
4. How do hanging valleys form?

Answers: 1. Ice freezes to valley side then moves and big chucks of rock are ripped out. 2. Ice moves in rotary fashion within the cirque, the "armchair" shape of the cirque encourages this. 3. Aretes. Cols and gaps. 4. They are the truncated remnants of former tributary valleys.

Glacial deposition and resulting landforms

AQA	**AS U1**
OCR	**AS U1**
Edexcel	**A2 U4**
WJEC	**A2 UG3**

> East Anglian drift ranges in depth from 143 m to 175 m.

Drift

The term **drift** refers to all types of deposits left after glacial ice has retreated. Drift is laid down in four distinct environments.

- **Glacial** – dropped directly from the ice mass, unsorted and unstratified.
- **Fluvio-glacial** – laid down by meltwater streams, sorted and stratified.

Drift affects many areas. Many variants on drift questions have been asked in the past. At A2 Level they will undoubtedly occur frequently.

- **Glacio-lacustrine** – lake deposits, sorted and stratified.
- **Glacio-marine** – deposits laid down into the sea, sorted and stratified.

Approximately 10% of the Earth's surface has a topping of drift. It is mostly lowland areas where the ice has slowed and ablation is high. Drift yield occurs in a variety of sizes and is variously smoothed and sorted depending on the environment it is collected in (upland deposition occurs whenever ice velocities drop).

Glacial deposition or till deposits are usually sub-divided.
- **Sub-glacial lodgement till** – associated with active glaciers, plastering the bed and sides of the valley with rock debris.
- **Supra-glacial till** – dumped onto the glacier, as moraine, finding its way, when the ice melts, onto the land surface.
- **En-glacial material** – material within the ice is dumped by melt-out or ablation processes and tends to be well sorted and rounded.

Landforms

The landforms that result from glacial deposition vary enormously according to the environment of deposition – they are literally dumped or moulded by the ice.

Lodgement landforms include:
- **Till sheets** – a veneer of boulders, pebbles and clay, producing low monotonous plateau landscapes, e.g. the Norfolk landscape, where the 90 m Cromer Ridge is composed almost entirely of lodgement till and is probably the most distinctive feature of the drift in East Anglia.
- **Moraines** – material plucked, abraded and weathered from the valley and mountainside. It can be of end/terminal/push or recessional varieties. (N.B. medial, lateral and en-glacial moraines all contribute to this material.)
- **Fluted moraines** – form behind obstructions in/on the valley floor.
- **Drumlins** – rounded mounds formed parallel to ice flow, with broad upstream ends and a tapered downstream end. The streamlined elliptical shape causes minimum resistance to ice flow, with the broad end being the zone of maximum ice pressure. They are found in swarms or drumlin fields. They are between 150–1000 m long and about 100–500 m wide. Swarms form the so-called '**basket of eggs topography**' e.g. in the Afon Morwyn valley in Wales and the Eden valley in Cumbria. Drumlins may be formed in two ways: (a) by **deposition** – material dropped out of the glacier during melting is moulded into its stream-lined form by the retreating ice; (b) through **dilatancy** – stresses in till found near the base of moving ice masses deforms deposits during ice transport, causing the till to settle and compact, and the ice then flows around the mass of till, streamlining its shape.
- **Erratics** or **perched blocks** – rocks carried by moving ice are eventually dropped in areas where the dominant rock type is totally different. They are useful to geographers as they indicate the origins of ice and directions of movement, e.g. the Norber Block, North Yorkshire.

Meltwater or fluvio-glacial depositional landforms

Landforms produced by meltwater often have distinctive characteristics. They are smooth, well sorted and the water ensures the effects of glaciation are carried way beyond the ice front. The source of the material may be en-glacial, supra-glacial or sub-glacial. The water that carries the material appears from sub-glacial channels, sourced from sub-glacial melting (during movement) and general ablation. The excessive load of the meltwater streams and the hydrostatic

Silurian grit

Carboniferous limestone

The Norber Perched Block

pressure, created by constriction, leads to choking of the water and channels that issue from the ice. Meltwater features include:

- **Eskers** are ice contact deposits formed in meltwater streams in the very unstable sub-glacial environment of the glacier, e.g. Punkaharju, Finland.
- **Kames** form in the ice margins, deposited by heavily laden streams running into the glacier, e.g. Haweswater Trough in Cumbria. Sometimes these form terraces.
- Retreating ice often contains blocks of ice lodged in meltwater deposits. When this ice melts the local area collapses to form a **kettlehole**.
- **Sandar** is probably the most widespread meltwater deposit. This zone of deposition extends out well beyond the ice. The meltwater results in size sorting (known as grading) with the bigger particles being left near the ice front. There is a strong seasonal aspect to the production of this particular deposit. The huge load carried by the meltwater causes braiding to occur, e.g. the rivers of northern Iceland. Huge expanses of sandar have been spread over northern Europe forming sandy heathland, e.g. in north Norfolk at Kelling Heath.

> The Blakeney Esker is the best developed kame in the UK at up to 50 m high and 3.5 km long. In the same area kames can be recognised in the Glaven Valley.

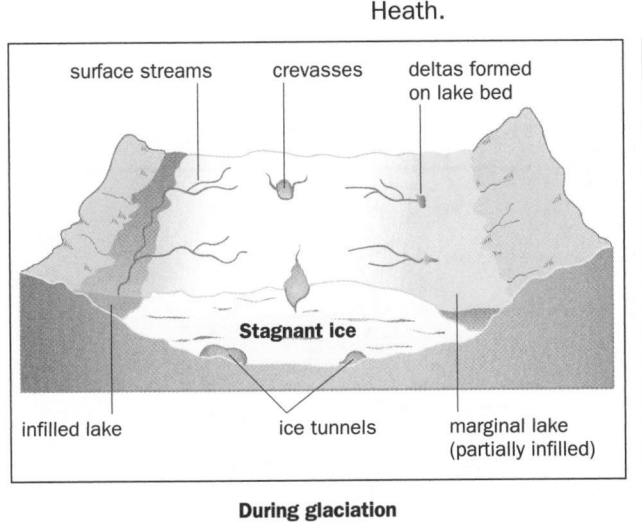

During glaciation | After glaciation

Effects on the landscape and land use

AQA **AS U1**
OCR **AS U1**
Edexcel **A2 U4**
WJEC **A2 UG3**

Glacial diversions and drainage modifications

The Avon and Thames are examples of rivers diverted from their original courses by glacial activity. Many areas of lowland glaciated Britain were affected by meltwater from lakes blocked or impeded by lowland ice flows. Examples include Lake Lapworth in Shropshire, Lake Harrison in Northamptonshire and Lake Pickering in Yorkshire. The **varves** (laminated clay deposits) deposited in these pro-glacial lakes have been variously settled and used for agriculture.

> Understand that past events affect present landscapes.

KEY POINT

In glacially eroded valleys:

- They offer great scope for **HEP production** with their steep, deep valleys for storage and rock lips providing sites for dam construction.
- Tourism: the dramatic scenery and existence of snowfields make glaciated areas favourites for walkers and skiers.
- Communications: glaciated valleys offer natural routeways through high mountain landscapes and lower areas, e.g. the Mohawk Gap that leads to New York.
- Settlement and industry: these hug the glacial trough floor.

continued

- Fjorded coasts: traditionally fjords have provided important centres for farming, fishing, industry and communications.

In areas affected by glacial deposition:

- Rich soils have formed in East Anglia and the Northern Lowlands of Europe. It has enabled a whole range of activities to be undertaken from silviculture (trees), pastoral and arable production. The best agricultural areas are those that have been deep ploughed, fertilised and irrigated.
- Some glacial fringe lands beyond the ice-covered areas were covered by thick layers of loess – fine sediment carried by out-blowing winds.
- The lakeland plateau of Finland, though of limited agricultural use, is a great tourist attraction.
- Glacial 'decline' led to the formation of peat bogs (in Central Ireland), a source of fuel.
- Outwash sands and gravels have been extensively quarried for the building industry, e.g. the deposits of the Thames Basin supply the London construction market.

Case study: North Yorkshire Moors, UK

The valley of the River Esk was blocked by the Scottish ice. Lake Eskdale overflowed through Newtondale Gorge (a typical spillway – flat floored and steep sided) into Lake Pickering. This in turn drained through the Kirkham Abbey Gorge. The present River Derwent rises within a few kilometres of the coast near Scarborough, cuts through Forge Valley (a lateral overflow), turns inland, flows through Kirkham Abbey Gorge (a direct overflow) and eventually reaches the sea 120 km from its source.

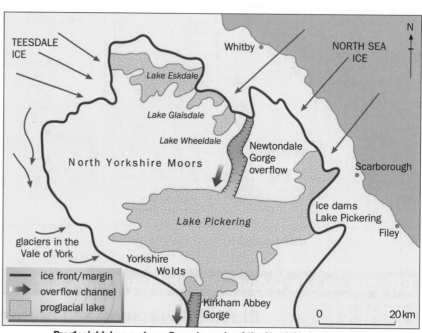

Proglacial lakes and overflow channels of the North Yorkshire Moors

Case study: Gravel/sand quarries

Some gravel/sand quarries have been utilised by local communities for leisure and recreation, others returned to nature.

Dorney Lake, Berkshire, UK: Gravel extraction near the village of Dorney in Berkshire has created the 2000 m Dorney Lake, the new £15m Eton College Rowing Lake. Gravel from the excavation is to be sold to off-set costs. The gravel will be used for a variety of projects in the Thames Basin. Dorney will be used in the 2012 Olympics.

The Cotswold Water Park, Norfolk UK: This is the UK's largest water park, 50% larger than the Norfolk Broads. It was created through gravel extraction in 132 pits.

Felmersham Gravel Pits, UK: Gravel was quarried to build runways at local aerodromes during the Second World War. Groundwater has filled the 52 acres of pits created, to form an important nature reserve near the River Great Ouse.

Whitlingham Country Park, Norwich, UK: Created by the owners the Colman family, gravel and sand diggings are slowly creating new Broads, recreational lakes and a 2000 m sailing lake for the people of Norwich and East Anglia. Aggregates have been used for projects like the southern by-pass around Norwich.

A present day glacial hazard: the avalanche

Case study: Motrac avalanche, February 1999

Avalanches are more likely when:
· slopes are steeper than 30°
· a lot of new snow falls over a short period
· winds are present lead to drifts
· old snow melts and refreezes, encouraging new snow to slide off.
At Montroc, heavy snow fell on Monday and Tuesday, but melting and refreezing of old snow was thought not to be responsible.

1 Tuesday 9th February 2.35 p.m. (1.35 gmt): a tidal wave of snow 150 m wide and 6 m high crashes down into the valley at 95 kph and buries much of the village of Montroc.

2 The force of the avalanche is so great that it sweeps through Montroc and travels 40 m uphill to smash into the village of Le Tour. The avalanche carries some of the chalets as far as 400 m (a quarter of a mile).

3 Snow storms on Tuesday night prevent rescues by helicopter during the first vital hours. The snow is packed so tight that only mechanical diggers can hack their way through to the chalets.

4 18 chalets destroyed and 12 people are killed.

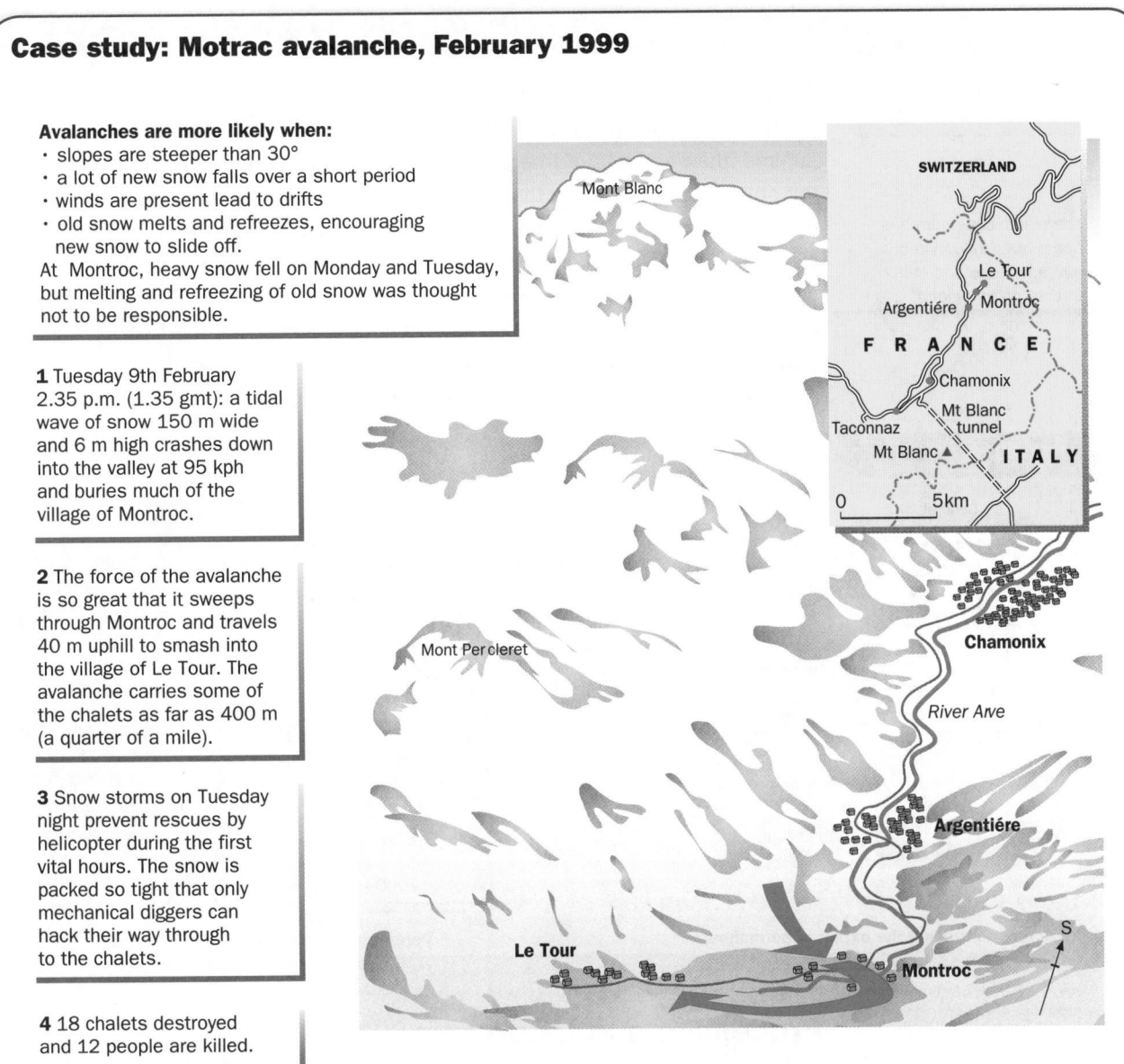

7.2 The periglacial realm

LEARNING SUMMARY

After studying this section, you should be able to understand:

- that areas presently affected by periglacial activity are dependent upon both present and past climate
- that the active/mobile upper layer of the permafrost has an influence not only on the landforms that develop, but also on the use made of these areas today by man

Characteristics of the periglacial realm

AQA **AS U1**
Edexcel **A2 U4**
WJEC **A2 UG3**

> The periglacial realm has had a huge influence on our landscape and still affects polar regions.

Periglacial environments are found on the edges of ice-masses and have permafrost conditions virtually all year around. Other characteristics include a summer thaw and lots of freeze-thaw activity. In the frozen permafrost areas the top soil is known as the **active layer** (this is a highly mobile zone and one which experiences mass movement, as it thaws out in the summer and re-freezes in the winter). Some unfrozen ground **talik** may exist within the permanently frozen sub-soil. Present day northern hemisphere permafrost is distributed as in the figure below. The cyclic nature of the climate (**tundra climate**) and the freezing and thawing of the ground in these areas has led to a peculiar set of geomorphological processes operating and landscape features forming.

KEY POINT

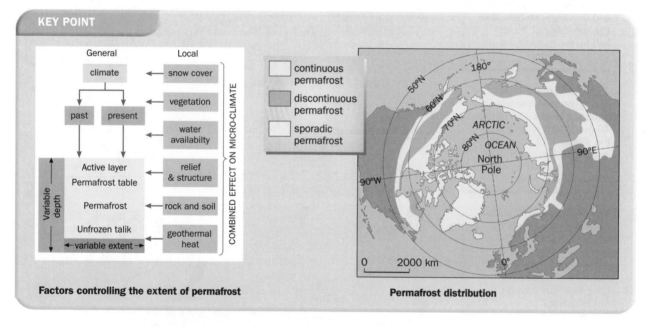

Factors controlling the extent of permafrost

Permafrost distribution

Processes and landforms

Processes creating new deposits

- **Frost shattering** or **nivation** (the mechanical splitting of rocks by ice) is most active where there is a maximum number of temperature oscillations above and below freezing point. The result is thick layers of frost-shattered angular fragments that cover huge areas. Those with the largest chunks are known as **blockfields**.

> Periglacial A2 questions always look to relate man and his environment.

- **Talus** or **scree slopes** may develop as a consequence of this frost shattering activity. Once formed this loose material can become very mobile and is devoid of vegetation. The finer of the deposits constitute a mobile layer on the shallowest of slopes (<2°) of the permafrost/periglacial realm.

- In the warm summer, lubricated thaw mass movements called **solifluction/ gelifluction** move rubble around the landscape in **lobes** and **trails**.
- Wind-blown deposits known as **loess** also blanket such areas (**limon** in France and **brickearth** in Britain), a consequence of the strong out-blowing winds that are known to have dominated in such areas during the last ice age.

Structural features

> Stone polygons vary in size from 0.5 m to 15 m in diameter. Stone wedges are up to 4 m deep.

- In the flat badly drained periglacial areas there is no continuous vegetation and lateral sorting of the active layer takes place to give a highly distinctive ground pattern of **stone polygons**. These are caused by contraction and expansion, a consequence of frost heaving.
- On steeper slopes the rings can become elliptical or elongated into stone **stripes**.
- Where cracks exist on the surface they fill with 'thaw' water, wind-blown sand and stones. An **ice wedge** forms that grows wider and deeper as each winter freeze and summer thaw cycle progresses. This process can lead to a very domed and hummocky landscape, as the ground expands against the ice wedges, distorting surface deposits.
- **Pingos** are mounds or cones covered with deeply fissured drift layers drawn into a mound by an intrusion of ice. They grow very slightly during their life cycle by drawing water to the ice mass from the summer supply of groundwater. These features have been well studied in the Mackenzie Delta in Canada. In Britain the scars of pingos can be seen in the landscape of Breckland, Norfolk.

> These features are similar to limestone 'karst' landscapes (c.f. dolines 'uvalas' poljes).

- **Thermokarst** features are collapsed features in the periglacial zone, formed when permafrost thaws rapidly. Pits, swallow holes, caverns, caves and ravines may coalesce to form valleys in what was previously a fairly flat landscape.

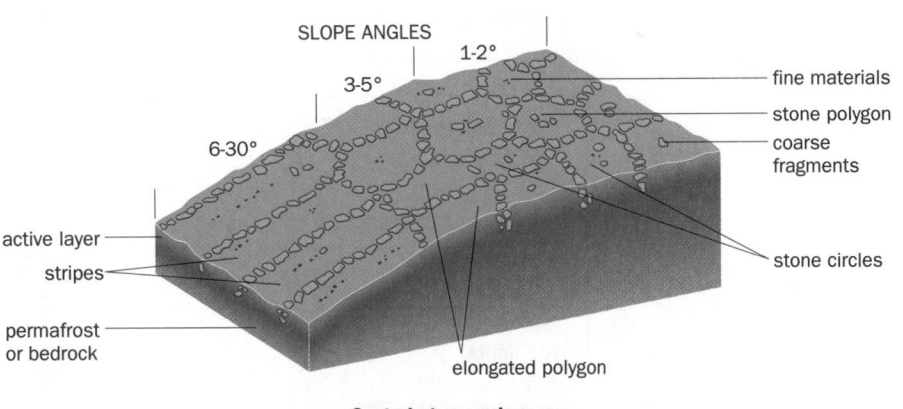

Sorted stone polygons

PROGRESS CHECK

1. What is the difference between glacial and fluvio-glacial deposits?
2. How can you tell the difference between glacial and fluvio-glacial deposits?
3. What are eskers?
4. Outline the way people use glacially eroded valleys.

Answers: 1. Glacial deposits are literally dumped onto the landscape and fluvio deposits are spreads by water. 2. Fluvio deposits are well sorted, smooth and orientated. 3. These form in sub-glacial river, they have the appearance of a river of deposits, appearing as the ice melts. 4. For HEP, tourism, communications, settlement and industry.

Landform modification

- **Angular free faces** are common and **tor-like forms** are found where the joint structures intersect at right angles to the ground surface.
- **Altiplanation terraces** form into a hillside by the combination of frost shattering, nivation and solifluction.

Rock lips are replaced by **protalus ramparts** in nivation hollows.

- Small **nivation hollows** (similar to the cirques of highland glaciated areas) form on the sheltered north-facing slopes of the permafrost/periglacial area. The snow that occupies these hollows both deepens and extends them.
- On the south-facing warmer slopes disintegration processes and mass movement (**solifluction**) leads to a decline in slope angle. This leads to the formation of **asymmetrical valleys**.
- There may be evidence of river activity in the periglacial landscape e.g. in the UK, erosion is seen in the form of **dry valleys** and deposition in the **coombe deposits** left in the bottom of the dry valley.

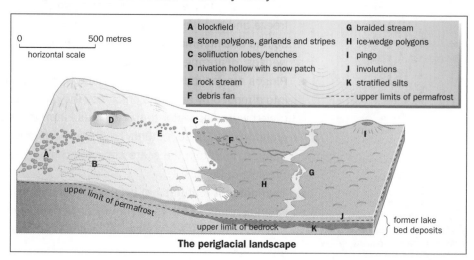

The periglacial landscape

In the UK the Pleistocene landscape of periglaciation is seen in much of Southern England, see figure below.

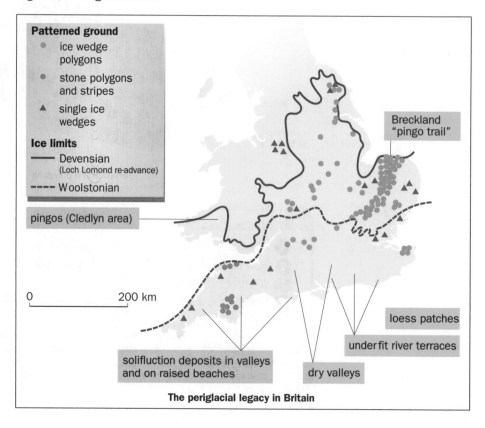

The periglacial legacy in Britain

PROGRESS CHECK

① What is the difference between periglaciation and permafrost?
② How do asymmetrical valleys form?
③ Describe how ice wedges form?

Answers: 1. Periglacial environments are found around the edges of glacial areas/permafrost is the "permanently" frozen soil. 2. South facing slopes in periglacial areas collapse leading to a decline in slope angle. 3. Cracks on the periglacial surface fill with thawed water and stones re-freeze and grow.

7.3 Issues in cold environments

LEARNING SUMMARY

After studying this section you should be able to understand:

● that the cold areas of the world contain many hidden resources
● that all the cold environments that exist are fragile in terms of their use and exploitation by man
● that cold environments have to be managed

Futures for cold environments

AQA   **AS U1**
Edexcel **A2 U4**
WJEC   **A2 UG3**

Cold environments contain valuable resources that attract development, such as fishing, mineral exploration or tourism. They are also fragile environments, principally because of the harsh climate with a short growing season.

● The cold environments can be also easily damaged.
● Fishing can deplete stocks and damage the sea floor.
● Power production disrupts fishery migrations or stored water can hold too much heat so changes microclimates.
● Tourism damages biodiversity and pollutes the environment.
● Oil pollution from sea spills, such as the Exxon Valdez in 1989.
● Mining pollutes waterways and the sea.
● Local populations are affected by the 'loss' of their environment, such as unemployment or change of lifestyle.

All these issues can be managed.

Case study: The Nenet of the Yamal Peninsula, Russia

The Yamal Peninsula is a region in Russia with a deep frozen landscape and temperatures lower than −55°C. The nomadic Yamal-Nenets herd some 500 000 reindeer in this tundra environment. Under the tundra lies 90% of Russia's gas reserves and 12% of its oil reserves. Traditionally, the Nenet reindeer herds have had their winter pastures in a small area south of the Arctic Circle and migrate north as the thaw takes hold. The Nenet manage their land sustainably.

Since oil was discovered in the late 1970s under the tundra, and drilled in the 1980s, contact with the outside world has increased, as has the infrastructure. The infrastructure changes have had a massive effect on the region's permafrost and the Nenet way of life.

The gas and oil programmes are crucial to Russia, but pressure on the tundra pastures may be a problem for the Nenets. The Russians have in fact chosen not to recognise tribal rights, and legislation does not require impact assessments to be made of industrial exploitation sites of this type. The Nenets are low down the priority list.

Since the 1980s Gasprom Oil-12 has drilled in the Bovenkovskoye deposit (the biggest natural gas source in Russia). To enable the full development of the oil and gas fields a new 572 km railway was completed in 2010, as well as a gas pipeline and bridge constructions. These have led to more tundra pasture destruction, water pollution and fish deaths.

As many of the old tribal lands have been destroyed many Nenets have opted for the new market economy that oil and gas has developed.

Case study: Change in the 'Little Alpine Tibet', Livigno, Italy

Before Livigno's economic transformation, people in the village lived in isolation from the outside world and were content in a near self-sufficient agricultural economy. Furthermore, they were completely blocked off from the rest of Europe by deep snow between December and May.

What happened in Livigno's valley is one of the Alps' greatest landscape transformations. The Swiss built a reservoir/HEP system 10 km from Livigno. As compensation the Swiss built 25 km of snow-shedded and tunnelled roads linking Livigno with the Swiss road system. The village turned into a ski mecca. The Italian government's decision to make it a tax-free resort further boosted its appeal.

The multiplier effect is where one factor affects another – usually positively, so building up Livigno has had an effect throughout Italy!

Unemployment is now near to zero as tourism is so labour intensive. The multiplier effect has spread the benefits through the whole of this part of Italy. The biggest loser because of Livigno's development has been the environment.
- The thin alpine meadows and slopes have been stripped by skiers.
- The snow-making machines have damaged the micro-climate.
- There has been loss of species.
- Water is in high demand, with supply reaching its limits.
- The effective disposal of waste and sewerage is an issue.
- Traffic congestion and air pollution has increased.
- Agricultural skills have been lost.

Case study: Are the glaciers growing or going?

Both the Guyot and Yahtse glaciers in Ice Bay in Alaska seem to have advanced during recent summers, when in fact they have the highest thinning rates of any glaciers in Alaska. The Hubbard glacier in Yakutat does seem to be advancing, but this tidewater glacier also goes through typical advance and retreat cycles. Of 2000 glaciers in Alaska 99% are thinning and retreating.

Warmer temperatures are reducing the numbers of glaciers in Montana's Glacier Park. There are now just 25 named glaciers over the 25-acre threshold.

France's largest ice cap, the Cook ice cap, some 12 000 km from the French Alps on the Kerguelen Islands, in the Southern Indian Ocean is melting. In 1963 it covered 500 km^2, by 2003 it covered 403 kms^2. In 40 years 20% had melted.

If the glaciers go what are the consequences?
Even with small glaciers the issues should not be underestimated.
- Fishing and white water rafting would be restricted.
- Unique plants and animals are being lost.
- Exposed rock, land and water retain heat – so when glaciers melt they create a feedback loop that accelerates overall warming.
- Sub-alpine trees now grow at altitudes 50 m higher than they did in 1980.
- Warm winters have allowed the insects to infest half a million hectares of pine.

In the Tajikstan Pamir Mountains warmer winters (increases of 1°C to 2°C since 1940) mean rapid glacier melt. Up to 20% of the country's 8000 glaciers have retreated or disappeared. The biggest glacier, the Fedchenko Glacier, is melting at a rate of 16–20 m per year. The consequence of all of this in Tajikstan are as follows

- Drought – rivers running dry and a lack of drinking water. Three quarters of Tajikstan's agriculture depends upon glacial melt-water for irrigation.
- Crop yields are dropping.
- Tajikstan's hopes of producing HEP are diminishing.

- There has been a 20–25% decrease in natural plant species/metre.
- Tajikstan feeds 50% of Central Asia's rivers. Dwindling water is bringing tensions to the area – disagreements are now common between the neighbouring countries of Kyrgyzstan and Uzbekistan.

Case study: Antarctica

Antarctica is covered in an ice sheet nearly 5 km thick. It contains 90% of the world's ice and 70% of its freshwater! It is classified as a cold desert with precipitation receipts of only 150 mm.

Antarctica has a profound effect on the energy budget of the southern hemisphere. It is a zone of net cooling (it receives less solar energy than it loses by infra-red cooling), i.e. heat is transferred from the mid-latitudes by weather systems. The atmospheric temperature of Antarctica is expected to rise between 0.5°C and 2°C by 2020 which could actually lead to an increase in ice mass on Antarctica, due to an increase in precipitation (snowfall). If, in the longer-term, melting occurs the sea level might rise by 1 metre within 500 years or 4 metres after 5000 years.

The ecosystems of Antarctica are of great interest, as they have few species and, most importantly, there are no alien species. At sea the krill accumulations lead to abundant wildlife accumulations. Fishing, as a whole, is tightly controlled using precautionary principles and sustainable single species and ecosystem approaches. The first visitors to Antarctica went in 1958 – 40 000 per year are now common and this is set to increase. There are inevitable concerns about the environment, such as disturbance of wildlife, littering and environmental degradation.

Environmental issues and management

The protection of the Antarctic environment has been paramount. The Antarctic Treaty, of 1959, established Antarctica as a region where scientific effort was coordinated and where military and nuclear activity was to be banned and where international disputes over minerals was to be highly regulated. The Madrid Protocol of 1998 gave further protection to Antarctica, ensuring protection for the environment and its ecosystems and maintaining the continent as an area for scientific research.

Exam practice questions

1 Study the photograph of part of the Torngat Mountains in Labrador, Canada. The cirque shown still possesses a small perennial glacier, though in recent years it has retreated.

(a) Name the processes at work at A and B on the photograph. How do they differ from one another? **[4]**

(b) Name the landforms at A and B, and explain how they were formed. **[4]**

(c) The table below shows the orientation of 100 cirque basins in the same Torngat area.

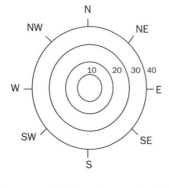

Orientation	No. of cirques
W	5
SW	4
S	4
SE	20
E	18
N	22
NE	27

   (i) On the diagram plot the cirque orientations. **[7]**

   (ii) Describe the pattern of orientation. **[4]**

   (iii) What climate factors have affected this pattern of orientation? **[4]**

(d) What do you understand by the terms glacial flow and glacial retreat? **[5]**

(e) 'The effects of periglacial processes, though perhaps less dramatic than glacial processes, still produce features that are recognisable in today's British landscape.' Describe some of these features. **[10]**

Essay

2 Using examples, discuss the relationship between human activity and the glacial environment. **[25]**

8 Settlement issues

The following topics are covered in this chapter:

- Settlement distribution
- Settlement morphology
- Processes and problems of urbanisation
- Process and problems in rural areas
- Rebranding urban areas
- Rebranding rural areas

8.1 Settlement distribution

LEARNING SUMMARY

After studying this section you should be able to understand:

- settlement models and patterns
- models of settlement distribution

Settlement models and patterns

AQA	**A2 U3**
OCR	**AS U2**
Edexcel	**AS U2**
WJEC	**AS UG2**
CCEA	**AS U2**

Quotations give colour to answers in exams.

Settlements are usually defined as 'a place in which people live and where they may be involved in various activities'.

Early settlement

Settlements first began approximately 10 000 years ago. Before settling in an area, early farmers would have studied a range of factors relating to the **site** (the area upon which a settlement is built) and the **situation** (the relative location of a settlement to other features). These would have included the availability of:

You could be asked to annotate a map with these features.

- clean water
- freedom from flooding
- level land for building and farming
- timber for fuel and building
- good soils for cultivating and grazing
- trade and commercial possibilities.

Today, however, socio-economic factors are a good deal more important.

Urban and rural settlement

What is 'rural' and what is 'urban'? A range of criteria have been used to distinguish between rural and urban settlements in the UK.

Many of the entries in the following table will apply equally well to other more economically developed countries (MEDCs) and less economically developed countries (LEDCs), though variations between countries can be marked. The problems of settlement classification and the dynamism of the settlements themselves are best exemplified through the study of a particular place. Peterborough, the example in the case study on pages 150–1, like many other small English cities, continues to change in both its size and the functions it performs. The rapid speed of these changes affects not only the city itself, but the hamlets and villages it subsumes as it grows. In the Norfolk case study it charts how sudden and permanent changes in a small community can have far reaching socio-economic affects. In Peterborough and Brancaster, Norfolk you see in microcosm all the dilemmas of settled life and how it compounds classification.

Rural and urban settlements – differences and similarities			
Rural	**Urban**		
Agriculture, professional and managerial.	• Commerce, services and manufacturing. • Lack of investment.	**Employment**	Economic
Wealthy land owners, isolated rural poor. Generally wealthier than urban areas.	• Generally poorer than rural areas. • Isolated very rich people.	**Wealth**	Economic
Infrastructure poorly developed. Small number of services/ functions.	• A wide range of functions able to operate in a well-developed infrastructure. Many transport options.	**Function**	Economic
High proportion young people (aged 5–24) and over 65s. White predominates. Low density population. Socially homogenous and inward looking.	• Aged 24-54 predominate. • High proportion of ethnic minorities. • High density population. • Biggest percentage of population born outside of the urban area. • Socially heterogeneous and outward looking. • Poverty leads to poor health.	**Population**	Social
Large, private, estates; dispersed, isolated.	• Small, high-density, local authority, flats and residences in re-developed areas. • Housing is generally of poorer quality. • Length of residence tends to be short.	**Housing**	Environmental

This comparison is a common question at AS Level.

These stereotypes rarely illuminate, but confuse issues for geographers! Beware...

Examiners frequently complain about lack of detailed case study knowledge. It is important to know the detail in the table previously and in the Peterborough and Brancaster information below.

Case study: Different areas in Peterborough and surrounds have different characteristics

Chesterton. Is a small **rural farming** village about 5 miles (6kms) west of Peterborough City centre. Conveniently located adjacent to the A605 and the Great North Road. Housing is of **very large semi and detached housing** and isolated farms. It provides a small base for Peterborough's **commuting population**. Other than the church **no real service provision. Low ethnic mixing**.

Longthorpe. Small **residential "village"** within the bounds of Peterborough. Two miles (3 kms) from the city centre. **Range of shops** (schools, pubs, nursery, shops, bus service, garages etc.). **Low ethnic mixing**. At the **rural-urban** interface.

Central Ward, Peterborough. High density terraced housing (dating from 1841, many houses built to house engineers working for the Great North Railway) and some modern houses mixed with **local shops**, take-aways, mosques etc. and **close to the main shopping** area in Peterborough. Endure many of the typical **inner-city problems**, dowdy environment, street crime, run-down housing, unemployment. **Mixed ethnic background**, though predominantly Pakistani. In 2004 200 Pakistani youths clashed with Afghanis, Iraqi's and Police around the Central Ward/Gladstone Street; street battles saw cars and house set alight and windows smashed.

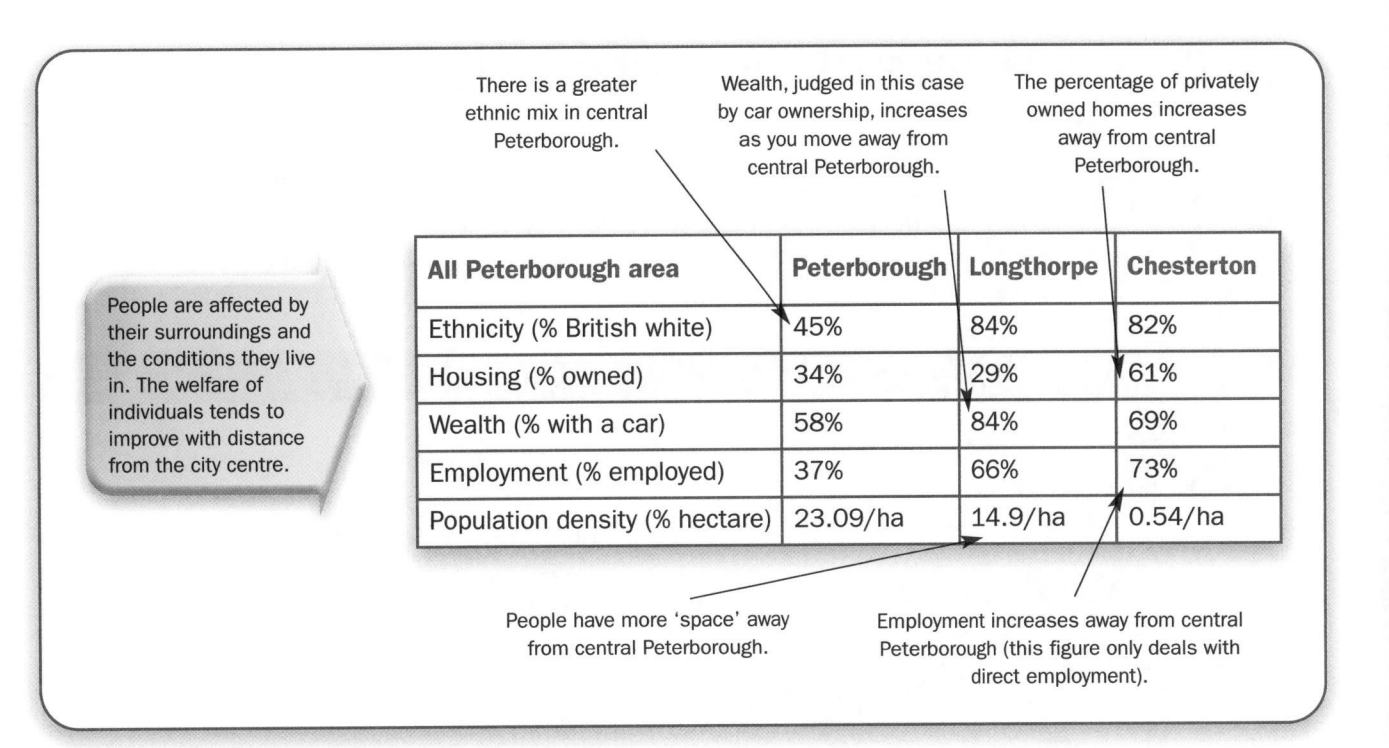

There is a greater ethnic mix in central Peterborough.

Wealth, judged in this case by car ownership, increases as you move away from central Peterborough.

The percentage of privately owned homes increases away from central Peterborough.

People are affected by their surroundings and the conditions they live in. The welfare of individuals tends to improve with distance from the city centre.

All Peterborough area	Peterborough	Longthorpe	Chesterton
Ethnicity (% British white)	45%	84%	82%
Housing (% owned)	34%	29%	61%
Wealth (% with a car)	58%	84%	69%
Employment (% employed)	37%	66%	73%
Population density (% hectare)	23.09/ha	14.9/ha	0.54/ha

People have more 'space' away from central Peterborough.

Employment increases away from central Peterborough (this figure only deals with direct employment).

Case study: Second homes in Norfolk

The 2001 UK census states that 8% (nearly 5000) of houses in North Norfolk are second homes. In some parts of the region the figures are higher, e.g. 66% in Burnham Market and 50% in Brancaster. The average house price in North Norfolk has doubled to £189 000 since 2001 and most houses are now worth 12 times the average salary.

The trend for second homes in North Norfolk is having a socio-economic effect on the area with houses now in short supply and prices therefore being driven up. Local residents describe communities, such as Brancaster, as ghost towns in the winter months. The decrease in full-time residents in Brancaster has led to the closure of local shops.

Settlement hierarchies

AQA	**A2 U3**
OCR	**AS U2**
Edexcel	**AS U2**
WJEC	**AS UG2**
CCEA	**AS U2**

Settlements can be classified according to size of population. There are no hard and fast rules, but typical sizes might be:

Capital	5 000 000 or more
City	1 000 000
Towns	10 000–100 000
Villages	500–2000
Hamlets	11–100
Isolated dwellings	1–10

Within a settlement, a range of activities take place that provide for the people who live there, e.g. it may have a hospital. This is a **function** (or service) the settlement performs.

Large settlements, i.e. cities, house many thousands of people and can support specialised high-order goods and services. Smaller settlements support fewer people and lower-order functions.

8.2 Settlement morphology

LEARNING SUMMARY

After studying this section you should be able to understand:
- the terminology and processes linking rural and urban developments
- the range of MEDC and LEDC models that help explain urban land use

Rural patterns

AQA	**A2 U3**
OCR	**AS U2**
WJEC	**AS UG2**
CCEA	**AS U2**

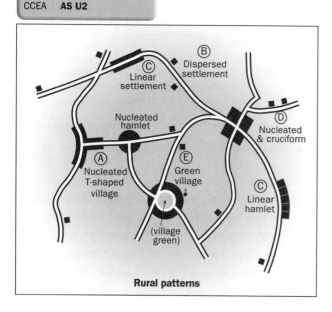

Rural patterns

(A) **Nucleated** – houses and buildings are clustered. Traditional in the UK because of the enclosure system, defence and the need for water, e.g. Urchfont, Wiltshire.

(B) **Dispersed** – farms and buildings widely scattered. Common in sparsely populated areas, e.g. Oby, Norfolk.

(C) **Linear** – spread along a trade or transport route, e.g. Sutton Row, Norfolk.

(D) **Cruciform** – settlement occurs at the intersection of roads, e.g. Pottern, Wiltshire.

(E) **Green Village** – a cluster of dwellings around a village green, e.g. Scorton, North Yorkshire.

Urban models in MEDCs

All these models are simplifications of reality. Cities are dynamic and unique.

Urban land use models attempt to explain the development and zonation of cities. The idea of 'bid rent theory' – that the value of land decreases away from the most central areas – underlies many of these models.

KEY POINT

Burgess' concentric model (1920)

1. CBD
2. factory zone/zone in transition
3. zone of working men' homes
4. residential zone
5. commuter zone

- City grows because of immigration and natural increase.
- Social class increases with distance from the CBD.
- CBD is dominated by commercial activity.
- Population density peaks in the high-density, low-cost housing zone.

Hoyt's sector model (1939)

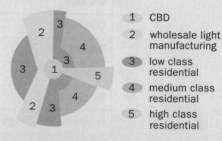

1. CBD
2. wholesale light manufacturing
3. low class residential
4. medium class residential
5. high class residential

- Hoyt emphasised the role of transport related to sector development.
- Hoyt thought that certain activities deterred others.
- Better housing is away from industry.

Harris and Ullman's multiple nuclei model (1945)

1. CBD
2. zone in transition/light manufacturing
3. low class residential
4. medium class residential
5. high class residential
6. heavy manufacturing
7. outlying business district
8. residential suburb
9. industrial suburb
10. commuter zone

- Emphasised that cities don't have a single centre.
- Cities grow and envelope other centres.
- New industrial sites arise.

Mann's model land use in the UK city (1960)

1. city centre
2. transition zone
3. zone of small terraced houses
4. post-1918 residential areas with post-1945 development mainly on periphery
5. commuting distance villages

- East–west split.
- East = working class and industry.
- West = wealthy people's homes.

Models of urban land use in LEDCs – Latin America

You need to know LEDC models as well as MEDC models.

Many cities in Latin America have the look of haphazard growth and a jumble of different activities and buildings without a clear structure. Large areas may be dedicated to residential land use without any commercial centres, industries may be located in central urban areas and airports may be within sprawling cities as expansion exceeds capacity to build infrastructures. High-class houses with all of the luxuries of excellent services may sit next to shanty towns with a glaring lack of even the most basic amenities.

Theories of urban growth

When looking at the models of urban growth in LEDCs you should always bear in mind the following points.
- The rate of **urbanisation** and in-migration.
- Lack of **urban planning** or **finances**.
- The driving force of expansion is **residential growth**, with administrative, commercial and industrial coming a poor second.

Griffin-Ford Latin American concentric ring model

The diagram opposite is a concentric model similar to Burgess', with the oldest area in the centre – the old Colonial City. The growth of the cities occurred since the 1950s and was largely unplanned, and as most migrants could not afford to buy homes they engaged in self-build housing, e.g. *callampas* in Santiago de Chile, or *favelas* in Rio de Janeiro, usually achieved by illegal occupation of land on the urban periphery. Exceptions occurred when there were sites available in the city centre such as steep hills, e.g. *morros* in Rio de Janeiro, riverbanks/flood plains, e.g. *Rio Bogota* in Bogota or on building sites. Slowly shanty dwellings improved, with more permanent materials, the provision of roads and utilities, and became part of the city. Once this happened new areas of temporary accommodation developed further out from the

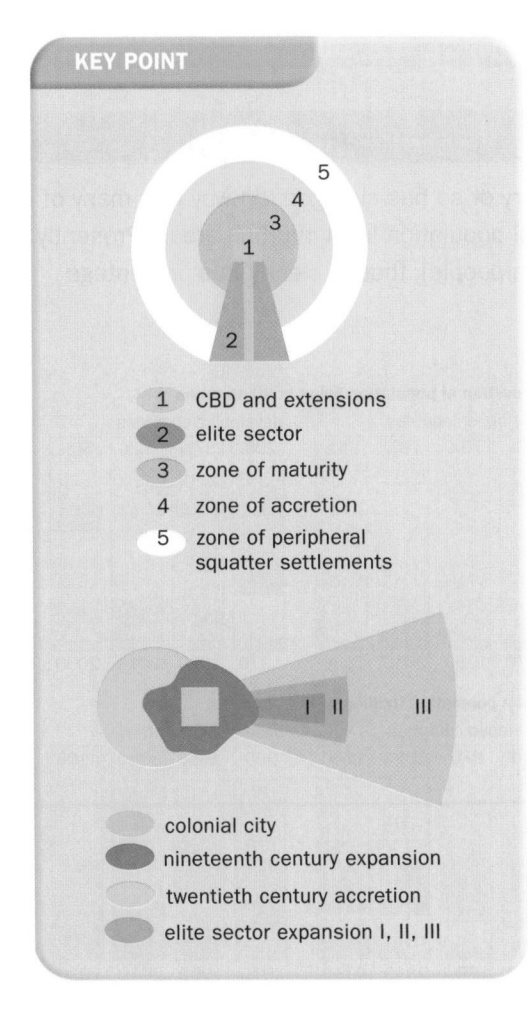

1. CBD and extensions
2. elite sector
3. zone of maturity
4. zone of accretion
5. zone of peripheral squatter settlements

colonial city
nineteenth century expansion
twentieth century accretion
elite sector expansion I, II, III

centre and the process began again. Notice that there is little allowance for industry, recreation or commercial land uses.

Sectoral model of urban growth

This is a different kind of model, which relies on less rapid growth and strong physical restraints such as mountain topography, common to cities of the Andes along the Pacific Coast, e.g. Caracas, Venezuela which is sited in a narrow east-west valley with steep slopes prohibiting development.

The growth is linear rather than concentric. The outer expansion is not due to shanty town growth, but to movement of the élite, occupying large areas at a low density. The poor serve to fill up rather than form the growth on the edge of the city.

8.3 Processes and problems of urbanisation

LEARNING SUMMARY	After studying this section you should be able to understand: ● the causes and consequences of nineteenth century growth on urban areas in MEDCs ● how rapid urbanisation has affected, more recently, the LEDCs and continues to affect the MEDCs ● the rise of the mega-city, primacy and the rank-size rule ● the changes that have occurred in the inner-city areas of MEDCs and LEDCs ● that urban growth and re-urbanisation creates many problems which need to be managed

Urban development and urbanisation

AQA	A2 U3
OCR	AS U2
Edexcel	AS U2
WJEC	AS UG2
CCEA	AS U2/A2 U1

Urban development over the last century or so has changed the way that many of us live. In 1900, about 2% of the global population lived in urban areas. Presently it is in the order of 50% (nearly 3 billion people), though clearly this percentage varies across the globe.

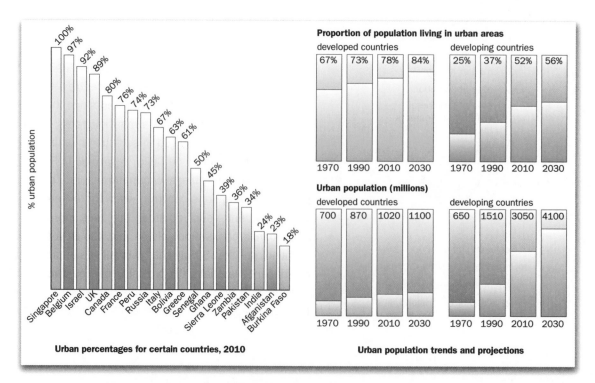

Urban percentages for certain countries, 2010

Urban population trends and projections

It is suggested that in the past agricultural surpluses caused urban growth and development. Today movement to the city is seen as a change in location, and lifestyles, as people are absorbed into a global society and economy.

The urban world is today changing in three different, but connected, ways.

- Through **urban growth** (due to population increases). World urban growth correlates with world population growth. Most of this growth is occurring in the LEDCs; for the most part MEDCs have static urban growth. In those LEDCs that are experiencing rapid urban growth they invariably still have a high degree of rural dwellers.

- Through **spreading urbanism** (the social and behavioural characteristics of those living an urban existence). This is usually obvious through shared activities and tastes. Urbanism or urban 'culture' has changed completely the views of many in the urban areas of Europe and the USA. The urban areas of most African and Asian cities have yet to taste this 'culture'.

- Through **urbanisation** (increases in the proportions of those living in towns and cities). This involves a shift in populations from rural to urban locations. Whether in LEDCs or MEDCs, urbanisation is seen as a cyclic process with populations moving from rural to industrial economies. There seems to be a balance of about 3 : 1 urban rural at which growth flattens out.

Urbanisation seems to have run its course in MEDCs with people returning to the countryside. Currently urbanisation is greatest in Asia and Africa. It is generally agreed that urban growth and urbanisation over the last half century have been driven by the **burgeoning global economy**. The cities being the command and control centres of national, regional and, in some cases, global management of finance.

In Asia the **global economy** has almost single-handedly caused **urban development**. Transnational/multinational companies concentrate their **factory production** in countries and locations (i.e. **near ports** for export reasons) where there is a **cheap and available labour pool** and where there may well be a growing home market for products. Such industrial provision makes the city look even more attractive than usual for the rural poor!

Urbanisation and urban growth in Britain

At the beginning of the eighteenth century the UK was predominantly a **rural farming** economy. The largest city was London. Other cities and towns, such as Bristol and Norwich, acted as market centres for agricultural products. Then notable changes in the agricultural landscape occurred, brought about by **agricultural intensification** leading to greater food production; driving many off the land, out of rural areas and into the cities in search of work.

The percentage living in urban areas in 1801 was 35% and by 1901 it was 78%. The nineteenth century was a period of **rapid industrialisation**. Industry was located in the towns and cities and the increased job opportunities and higher wages attracted thousands to the city areas of the UK. Rapid rural to urban migration occurred: the young, innovative and energetic were attracted to the blossoming cities.

Cities in Victorian Britain became noisy, dirty, congested and dangerous places, packed with factories and plagued by disease. People who could afford to, moved to the suburbs (**suburbanisation**) away from the filth and squalor. Aided by the

> It's important you know why urbanisation has occurred. The reasons are similar in both MEDCs and LEDCs.

> By 1901 there were 33 cities with a population over 100 000. In 1801 there had been only one, London!

development of the tram and railway networks the larger towns and cities of the UK began to spread beyond their original boundaries.

Urban areas have developed in the UK through:

- **isolation** – for effective government in the towns and boroughs of the UK
- **interaction** – roads and rail bringing trade and people closer together
- the **growth of conurbations**, following industrialisation
- the **growth of an axial belt**, stretching between Merseyside, through the Midlands to London and the south-east
- the developing **metropolis**, with the continued growth of London and the south-east.

Urbanisation in Latin America

AQA	A2 U3
OCR	A2 U2
CCEA	AS U2

In recent years the urban population of Latin America has risen **exponentially**. At one time, Latin America was nearly totally rural in make-up, now over 50% live in urban areas. The size, speed and scale of urbanisation has been massive.

City areas are not new to Latin America. Many date back to the time of the Aztecs and Incas, though most were initially established as ports during the colonial rules of European countries for export, administration and commercial purposes. Most of the biggest urban areas are on Latin America's coast – on the whole, few people live in the interior.

> The rapid change and growth of cities in LEDCs make this a popular exam question.

The biggest period of urban growth has been over the last 60 years, with growth rates at 2–5%. At a conservative estimate, Mexico City sees at least 15 000 new arrivals per week.

In most Latin American countries one city still dominates, though the single **dominant primate city** is becoming a thing of the past. Many of the middle rank cities are taking on the roles of government, both political and financial, as they continue to grow. Some of the reasons for the rapid growth of the Latin American city are the same as that experienced in Europe, though for this area it has been compressed into a much smaller time frame (those reasons being **migration and natural increase**, at a time when the population was increasing).

KEY POINT

The speed of urban growth and the growth of mega-cities have been the focus of attention for many geographers. What is now becoming obvious is that the rate of urban growth has actually declined in most regions. What is true is that absolute populations, especially in Asia and Africa, have grown and that some mega-cities do dominate in some countries. In addition only about 5% of urban dwellers live in mega-cities and many of these are actually losing more people than they are gaining.

> Dhaka is susceptible to flooding as it sits on low-lying ground.

In the next decade Dhaka and Lagos are expected to grow the fastest of all the mega-cities, but the rate is only expected to be 3%.

Case study: Fastest growth in the world – Dhaka, Bangladesh

Dhaka is the capital of Bangladesh and sits on the banks of the River Bunganga. It is the ninth most densely populated area in the world with a population of 13 million. It suffers many problems because of pollution (dumped, untreated waste mostly), congestion and a general lack of services. Only 25% have the facility to pipe sewage away from their homes and only two-thirds of homes have piped water. Dhaka's dilemma is whether to continue to grow or to limit growth.

> EPZ's do away with trade barriers/tariffs and bureaucracy to encourage new business and investment.

For growth: as the commercial heart of Bangladesh, Dhaka's middle-class population dominates and drives the economy. Two **export processing zones** (EPZs), with over 400 industries, act as magnets for workers and have both sparked and perpetuated the building boom.

To limit growth: a rather half-hearted attempt has been made to slow Dhaka's growth through encouraging the growth of satellite towns well beyond the city boundary.

Whatever route is chosen it seems likely that Dhaka will continue to be the fastest growing city in the world well into this century.

The rise of the mega-cities

AQA	**A2 U3**
OCR	**A2 U3**
Edexcel	**AS U2**
WJEC	**AS UG2**
CCEA	**AS U2**

A **mega-city** is a city with a population exceeding 10 000 000. In 1950 there were just two of these – London and New York. Today there are at least 25 mega-cities and 19 of them are in LEDCs. Factors influencing mega-city growth, include:

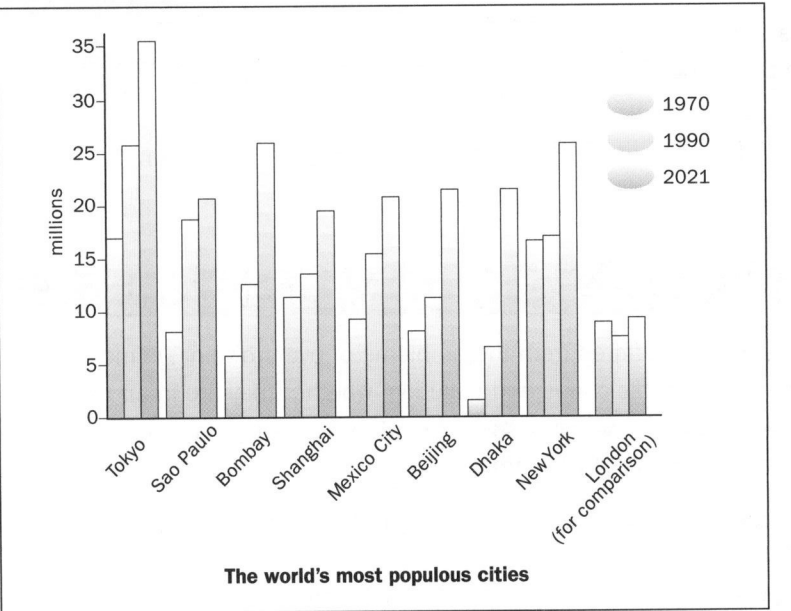

The world's most populous cities

- rural to urban migration
- former ports continue to develop as trading sites
- industrialisation related to international production continues in huge (often transnational, foreign owned) sites; industrial technology transfer also attracts people
- decreasing mortality rates due to better medical provision resulting in natural population increase
- national development policies encourage or force people into urban areas.

Most mega-cities have slower growth now than the **millionaire cities** (those with populations between 1 and 8 million). What the challenge is now is how to manage these sprawling settlements. It is a problem both for MEDCs and LEDCs.

Case study: Primacy and the rank–size rule

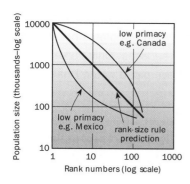

The **rank-size rule**, first drawn up in 1941 by Zipf, discovered an inverse relationship between the size and rank of a given settlement, i.e. the population of the second city in a country will be about half that of the biggest. This suggests that in any country there will be a few large places and many small ones. This means that when we investigate a country it will have high primacy if it exceeds the rank-size rule prediction and low primacy if it is less than the prediction.

Generally rank-size patterns appeared in countries that are smaller than average, have only recently urbanised or have simple economic and social structures. Primacy is linked to development. The rank-size rule does not explain settlement distributions, but does help to establish relationships between the size and importance of towns and cities.

KEY POINT

The problems of mega-cities include:

- they become magnets for immigration
- squatter settlements spontaneously appear
- hospitals and universities are located in the wealthier urban areas; rural areas have little or no provision
- informal employment grows, but contributes little to the national economy
- poor air quality and a lack of clean water
- health problems associated with the above, e.g. asthma and cholera
- habitat loss, crime and violence, drug-related problems
- planning and administration are not co-ordinated.

The advantages of mega-cities include:

- concentration of industry and finance ensures economies of scale
- general education and services are better in the big urban areas than they are in rural areas
- migrants deal with the housing problem in LEDCs.

The problems far outweigh the advantages.

Urban problems

AQA	A2 U3
OCR	AS U2
Edexcel	AS U2
WJEC	AS UG2
CCEA	AS U2

Big cities have always suffered from the problems of overcrowding, crime, psychological stress, traffic chaos and pollution, whether they are in an LEDC or MEDC. The two places below are vastly different, but growing cities, and both experience the same sorts of problems.

Norwich's urban problems
1. Urban sprawl
2. Water supply
3. Traffic
4. Homelessness
5. Heritage deprivation
6. Under-serviced
7. Poverty/homelessness/prostitution
8. Industrial economic instability

Bogota's urban problems
1. Traffic
2. Pollution
3. Water and electricity
4. Political corruption
5. Crime and safety
6. Earthquakes
7. Lack of open space
8. Industry, housing and urban sprawl

Shanty towns

There are in excess of 250 million people living in the shanty towns of the LEDCs. The main reason shanty towns exist is that the authorities just cannot keep pace with the influx of migrants to the city. Unable to live in permanent housing, people are forced to live in spontaneous settlements that they have built themselves, that use cheap waste materials, lack services (electricity, running water and rubbish collection) and are often sited in dangerous and vulnerable positions. They are **overcrowded** and harbour criminals.

These conditions result in **health problems**, mostly related to dirty water, but this is compounded by a poor diet. The shanty dwellers are the urban poor, the old rural poor and nothing changes for them. The young and better-educated are the migrants that 'make' it in the growing LEDC city. There are a range of problems that the authorities have to attend to if they are to avoid shanty towns spreading uncontrollably and blighting the city.

Case study: Badwita, Sri Lanka

Badwita is the biggest of 20 settlements in which 5000 people live in dirty and insanitary conditions. The area is situated on the edge of Columbia's metropolitan area. Homes are constructed out of cheap materials and the vast majority have no electricity, water or toilet facilities.

Badwita was chosen for resettlement and rebuilding by the government and NGOs. This meant resettlement onto a plot of land of 50 m² (with clean water supply, drainage, refuse collection, etc.). The project began in 1997 and took about ten years. But, already, there are problems with drainage and sanitation and in the future the area will prove difficult to maintain with water contamination by nearby industrial developments and the influx of more migrants.

Problems of shanty towns include:
- visual **ugliness** – they are an eyesore
- high incidence of **disease** and health problems
- they are a fire hazard
- most shanty towns are on **vulnerable and hazardous sites**, i.e. steep slopes, land that is regularly flooded or likely to collapse when earthquakes strike
- organised street **crime** and the drug trade use them as a base for trading
- house prices in nearby areas are severely depressed.

Many solutions have been suggested for the shanty problem, some of them very extreme, e.g. bulldozing the shanty towns. However, this is not the best 'cure' for the problem. Possible solutions include:
- making **land available** for new arrivals to the city
- making some sort of **tenure available**, e.g. a ten-year lease
- making cheap material and waste **materials available** for building purposes
- introducing advisors into the 'towns' to help with **construction advice**
- encouraging a **sense of community**
- laying the **foundations** for 'self-help' housing and providing basic services, that new arrivals can build on.

KEY POINT

The main causes of urban inequalities
- **Rural to urban migration** – rural population drawn to the city.
- **When house prices increase** this causes gulfs to develop between the haves and have-nots.
- When there is a **decrease in urban employment**.
- When there is an **increase in urban employment**. In the UK houses built to accommodate factory workers during the industrial revolution were inadequate for the families that filled them in terms of space and facilities, leading to poor health and social inequality.

KEY POINT

How to deal with urban inequalities
- Programmes of urban improvement – Bangladesh has pledged to improve the lives of 100 million slum dwellers by 2020 through their Urban Partnerships for Poverty Reduction (UPPRs).
- Transport inequalities, e.g. in South Africa this has been partly addressed with cheap minibus taxis that bring the suburban poor into central city locations. In the UK the rise of the park-and-ride scheme has really helped with the reduction of environmental damage from traffic pollution in city areas.
- Initiatives to encourage residents to help themselves. Involvement in parish councils and local affairs, resident associations and neighbourhood watch are the methods used in the UK.

The suburbs in the UK

> How change affects the suburbs (and inner-city) is a favourite in exams.

By the end of the First World War London's urban area extended some 11 km from the centre, enabled by the growth of the railway. On the whole the suburbs were built around the villages that surrounded the city. The advent of the car speeded up the process, as did the relocation of industry to the suburbs. The imposition of **green belt** status has slowed the process, but London is effectively surrounded by a suburban doughnut stretching over 80 km from the centre in places. Similar processes happen around all cities.

Characteristics of suburbs
- Linear in nature.
- Low density, low storey housing with gardens and owner-occupied.
- Middle income families.
- Cul-de-sac, crescent and avenue layouts.
- Near modes of mass transport.

Advantages of suburbs
- Cheaper land prices.
- Modern houses with amenities.
- Pleasant, clean, pollution-free environment.
- Better schools?
- Parks and gardens.
- Less crime.

Disadvantages of suburbs
- Increasing cost of housing.
- Lack of things to do in predominantly residential areas.
- Young people prefer the city, which changes the population structure.
- Difficulty of travel to the city: congestion and cost.

Counter-urbanisation

> Counter-urbanisation is the movement away from the city and into smaller, more community-based towns.

During the 1970s and early 1980s a noticeable process of counter-urbanisation began in Europe and in the USA. There are several reasons for this.
- Retirees moving to a more pleasant environment.
- New towns growing outside the city, attracting workers.
- Decentralisation, due to high city rents.
- City dwellers moving away from the congestion, stress and pollution.

Urbanisation is now seen to pass through five important stages or phases.

1. People migrate from the countryside to the city.
2. The pace of migration accelerates, suburbs grow.
3. Inner cities lose their populations.
4. The whole city region loses population due to counter-urbanisation.
5. The population stabilises and people begin to return to the city – **reurbanisation**.

Problems and change in the UK city

The **Central Business District (CBD)** is generally at the heart of the city and is the focus of transport systems. The great accessibility of the CBD means that land is very expensive and limited in availability. CBDs display many common characteristics.

Large cities often have recognisable land use areas, i.e. financial areas, department store areas and other specialist areas.

Concentration of shops	Concentration of offices	Little manufacturing industry	Growth of functional zones	Multi-storey development	Low residential population
Large department stores, such as Marks and Spencer are found at the heart of CBD. They attract large numbers of people from a wide area. Other specialist shops, such as book shops and jewellers, are also concentrated in the CBD.	Regional and head offices of large companies concentrate in the CBD. They are attracted by the accessibility of the city centre. Well-known companies like a well-known location for their head offices.	The CDB is not a suitable location for most manufacturing industries. However, a few specialised industries, such as newspaper and magazine publishers, do locate in the CBD. They need to be near to other CBD services and to have access to road and rail transport for distribution.	Similar activities tend to concentrate in certain parts of the CBD. It is usually possible to find areas given over almost entirely to entertainment, banks and financial services, educational facilities and shops.	The CBD has to grow upwards as well as outwards because of high land values. The most expensive sites have the tallest buildings. In a multi-storey block different activities may often occupy different floors.	There is little housing in the CBD because of the high land values. However, a few people live in luxury flats and apartments.

Main features of central business districts

Core–Frame model

The CBD changes constantly to keep pace with changing society.

The **Core-Frame Model** attempts to reflect some of this dynamism. There are strong links between the core and frame, with areas and businesses constantly being assimilated and discarded from the core and frame. Elements of redevelopment, decentralisation, pedestrianisation, conservation and gentrification further complicate the pattern.

The changing pattern of urban functions

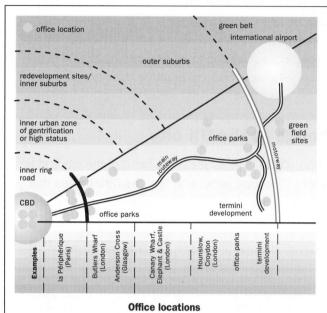

Office locations

Spiralling town and city centre rents have forced offices out of the centre and into a range of cheaper and more advantageous sites around the city.

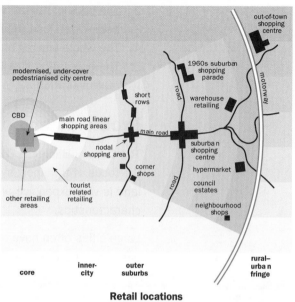

Retail locations

In the days before the car, shopping needs were satisfied locally on the street corner or on the high street. Increasingly location reflects a community's population, wealth and the infrastructure of the area.

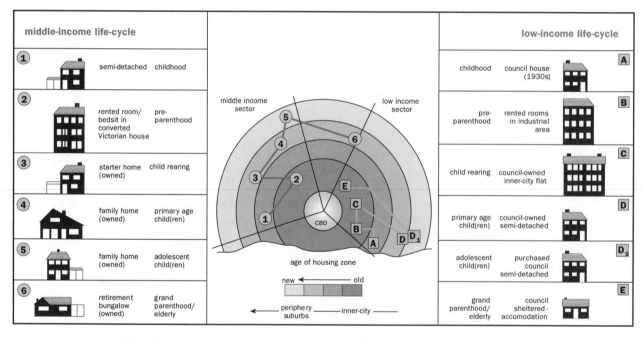

The cycle of residential land-use in the city

Beyond the CBD

The problems for the population that live in the inner-city areas of the UK came to a crisis point in the early 1980s with a number of serious riots, urban unrest, increased crime and racial attacks, e.g. the Toxteth riots, Liverpool in 1981. This happened, despite half a century of major government policy initiatives directed at inner-city problems, starting with the 1947 **Comprehensive Development Areas Policy** (which spawned the **New and Expanded Towns Act**), the **General Improvement and Urban Programme** policies of the 1960s, community programmes and development projects of the early 1970s, the **Urban Areas Act** (1978), through to **Action for Cities** and the formation of **Urban Development Corporations**, **City Action Teams** and the inner-city initiatives in the late 1980s. It was not until the late 1980s that the problems caused by lack of decent housing, jobs and recreation facilities began to be dealt with. In the 1990s the **inner-city regeneration policies** and **development plans** were important tools for planners and since 2000 the **inner-city development partnership** has brought change to the city.

The range of problems that had to be dealt with is below.

Problems in the inner-city

KEY POINT

Environmental problems
- pollution
- vandalism
- dereliction
- lack of open space
- decaying housing
- poorly built tower blocks
- traffic congestion
- social and educational services and recreation are poorly provided for
- overcrowding

Economic problems
- unemployment
- lack of skilled workers
- poverty and low incomes
- poor access
- declining industry
- lack of space for new industry
- high land values

Social problems
- crime problems
- falling birth rates
- concentrations of very old and young
- endemic illness
- lots of single parents
- political activism
- high concentrations of ethnic groups
- dysfunctional families

Despite all the changes that are now taking place, many people say that too little is being done to overcome the problems that continue to plague inner-city areas. That said, several high-profile projects have moved the debate on, e.g. London's Docklands Project and the Cardiff Bay Development Project.

Managing cities

AQA	**A2 U3**
OCR	**AS U2**
Edexcel	**AS U2**
WJEC	**AS UG2**
CCEA	**AS U2**

The subtleties of the problem in the LEDCs are in many ways different to those experienced in the MEDCs.

As urbanisation progresses many problems have to be tackled. Without proper planning and management, problems can get out of hand. In the LEDCs governments and aid institutions have favoured big **capital projects**. But the real needs in the developing–world cities may be more subtle: the creation of an environment in which jobs are available and small businesses can flourish; where land rights are recognised; where public transport and housing needs can be met at a reasonable cost; where child care, schooling and health services are available; where law and order is seen to be fairly applied. In MEDCs overcrowding, loss of agricultural land, urban sprawl and congestion, dereliction of inner cities and pollution are seen as more important and this has spawned a raft of uniquely 'Western' solutions.

The green belt

The green belt is land that is protected from industrial, transport and housing development and is reserved for farming, forestry and wildlife areas. There is a clear and obvious benefit from having and retaining green belt land, but pressures for development, especially in the south-east of England, will test governmental resolve to the full. Already the M25 has cut a swathe through the green belt around London and much more housing will have to be built on green belt land to avoid the spoiling of the landscape beyond the zone of protection. It is estimated that the Channel Tunnel/Channel rail link and the planned development of Stansted Airport will create a need for 100 000s of homes on the green belt to house those moving into the area. Furthermore, without some development, such areas risk being spoilt by fly tipping, dereliction and general neglect.

New towns

The **New Towns Act** (1946) was put in place to provide homes for overspill populations from the biggest of the UK's inner cities, and to attract and develop new industrial agglomerations. They were meant to be self-contained communities separate from the cities. Most are near London, though successive legislation has distributed them over the whole of the UK.

The need for sustainable settlements

The large communities that exist within the bounds of the world's cities provide challenges and opportunities for environmentally conscious planners and developers who want to halt the destruction of the very environment in which they live and work. So how can the lowest ecological footprint be achieved and the lowest quantity of pollution released into the environment of the city?

- Issues related to 'field to fork' processing i.e. fresh food must be brought to the city from the areas that surround it. There is very little extra processing and wastage is curbed.
- Energy is generated from **renewable sources** such as solar panels, soil heat source and wind power. By using less power from fossil fuels energy production is guaranteed for the future.
- Cities produce amounts of heat from a range of sources, e.g. air conditioning. Cities planned with **open spaces** and **water features** cool the city.
- Transport that emits zero carbon is the goal for many cities.

Case study: Norwich – a great city and a clean one!

Norfolk County Council introduced the Norwich Low Emission Zone (LEZ) in July 2008. The goal was to ensure buses spending long periods of time in the city emitted low levels of harmful pollutants such as oxides of nitrogen (NO). The initiative, that had European Union (EU) funding, required 100% of buses to meet specified Euro 3 emissions levels for NO from April 2010. This was achieved.

- Green roofs and walls that absorb and return water to the atmosphere and insulate and cool buildings.
- Marks and Spencer in Norwich are installing a living wall, as part of a revamp of their flagship store, clad in plants, such as English ivy, chrysanthemums, spider plants and aloe vera, which can absorb or filter airborne toxins as part of a 'living' or green wall system. The goal of Marks and Spencer is to be carbon neutral by 2012.
- Sustainable drainage systems that retains water and stops contamination of groundwater.
- **Xeroscaping** which is the landscaping of the urban environment that uses much less water to keep the whole planting system running. In some cases the need for water is completely eliminated.

Xerophytic plants use very little water.

Case study: Eco-town Rackheath, UK

A government directive on eco-living in 2007 has led to the old airfield site at Rackheath being named as a suitable site for an eco-settlement. Rackheath is less than five miles from Norwich and has a range of attributes which make it an ideal location for an eco-settlement. Rackheath was one of twelve (of an original 75) to be selected and was rated Grade A for its suitability. The aim is to provide a sustainable community with a firm focus on walking, cycling and public transport to reduce dependency on the car and so promote low carbon living. The homes that are eventually built at Rackheath will be built to strict environmental standards. They will utilise self-sustaining renewable energy sources, efficient waste management and reduction, will cut emissions and ensure a highly efficient water usage.

No start date has been proposed for this scheme. Local opposition (claiming inaccuracies in the planning) has hampered a number of moves to begin work on the site.

Cities of the future

More people live in cities today than ever before.

Over the last century or so the city has emerged as the most popular type of settlement in which to live. The future for most people looks to be an urban one.

Future city possibilities include:

- the **conflict** city – where class and race hostility is rife
- the **international** city – a centre for trade, finance and high technology
- the **neighbourhood** city – emphasising the sense of community
- the **conservation** city – where environmental concern is important, as is conservation of the historic parts of the city
- the **leisure** city – cities for the leisure age.

8.4 Processes and problems in rural areas

LEARNING SUMMARY	After studying this section, you should be able to understand: - the work of Cloke, who foresaw the decline of the village - the problems, decline and issues for rural areas

A look into the future?

OCR	**AS U2**
Edexcel	**AS U2**
WJEC	**AS UG2**
CCEA	**AS U2**

The diagram on page 166 shows Cloke's model (1979) of the structure of the urban-rural continuum, adapted and applied to part of north–west Norfolk away from Norwich. It shows how land-use might change with distance from the city.

In this model, there is no single typical rural settlement, but rather a **spectrum** between declining villages in the deep countryside to suburbanised villages and overspill towns in the urban fringe.

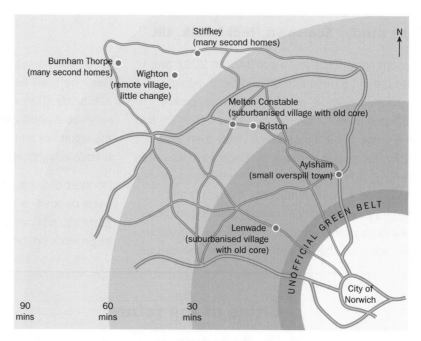

Cloke's model diagram

Rapid and profound change in rural areas

OCR	**AS U2**
Edexcel	**AS U2**
WJEC	**AS UG2**
CCEA	**AS U2**

Rural populations in the UK are growing faster than urban populations and people living longer. Approximately 18% of rural people live in poverty. Local authorities and planners need to be more proactive in meeting rural affordable housing needs, e.g. **fuel poverty**.

All of this is creating stress upon rural settlements. It is worth remembering similar stresses are experienced by the rural dwellers in LEDCs also. Much of the stress does have its origins in structural changes in households. Many rural households have reduced in size, and with population moving in, this makes **demand for houses** very high. Improvements in transport, and leaps forward in technology, mean there has been an increase in **commuting**, working from home and a surge in **second home** ownership. Many rural areas have been subsumed into the suburban landscape.

Causes and effects of rural problems in the UK

The cost of a rural home is generally higher than those in urban areas. But since the early part of the twentieth century the difference between rural and urban has increased significantly e.g. in some areas of the country you'd expect to pay up to 20% more for a house in the country! This has to be compared to income and the ability to afford the housing on offer. In 2007 there were 930 000 households in rural areas with incomes below the official poverty level. **Low incomes** and an inability to pay mortgages (with home costs in excess of eight times higher than the median income in rural areas), or rent a house, it means that young people are leaving the country in their hundreds having no chance of living anywhere near where they want. In some areas of the UK the proportion of 14–15 year olds has dropped by as much as 25%. But why have prices increased? House prices have grown most rapidly where prices have been driven up by incoming commuters, second homers and those who work in new rural businesses.

> The population of rural areas grows by over 100 000 every year. Most are wealthy and the average household income is higher in rural areas than urban. Unemployment tends to be lower, educational achievement is higher and there are more businesses/head of population. But this popularity can mask real deprivation. The socially excluded in rural areas are relatively spread out so tend not to show up in official figures.

> Of the jobs available in rural areas 28% are low paid.

> Many villages have lost their original character, form and function. These are often described as dormitory, commuter or sub-urbanised villages.

> Do remember that all the issues and problems and inequalities in the countryside are interlinked!

| House price increases | Lack of housing for the rural poor | Population decreases taking the children on whom village schools depend and the business which the village shops and pub depend | Decrease in services and increased rural debt | Rich get richer, poor get poorer |

The progression of rural decline

KEY POINT

Changes in the UK villages

Service	Changes
Food shops and post offices	The necessary threshold of population drops and local shops close.
Public transport	Closure of routes that are uneconomic.
Village schools	School closes as there are too few children.
Libraries	Fixed service cut.
Primary health care	Decline in population lead to dental and medical surgeries closing.
Village halls	Underused, vandalised and closed.

> All of these initiatives under the watchful eye of the Campaign for Rural England.

Recognising the **inequality** that exists in UK villages has spurred successive governments into action. The result has been the introduction of **Rural Advocates for the Countryside**, various **Rural Commissions** and has encouraged the **Countryside Agency** (now the **Commission for Rural Communities**) to support planning and development in rural areas under the '**Vital Villages**' scheme. HRH The Prince of Wales' **Affordable Housing Initiative (ARHI)** has also highlighted the problem of the lack of affordable homes and encouraged businesses to find solutions.

Low supply in market and affordable housing

Development has been constrained in villages since the Second World War and there has been a decline in **social housing** in rural areas. Social housing accounts for 13% of the housing stock in rural areas, compared with 22% in urban. For council housing the figures are even lower with only 4% of the stock in the most rural authorities now being in council ownership.

The traffic squeeze

> Expenditure pressures, carbon reduction targets, and a weak economic climate will all impact on transport planning.

There is widespread recognition that rural transport is not meeting the expectations of many rural residents. This is because mainstream transport policy does not readily recognise and translate into rural settings. Small scale rural needs require different transport solutions.

Case study: Solutions, strategies and schemes to help the rural poor into housing

- In the village of Buckland Newton in Dorset average house prices are £337 000. Concerned at the lack of affordable homes for locals the parish council established a community land trust to develop a housing scheme to provide affordable homes for locals on incomes of less than £20 000. Land was made available within the village through an exception site policy at below market development value. The affordability of the homes was assured by using low–cost, sustainable building methods, with timber–frame construction and straw bale insulation. Much of the building work was completed off–site using local suppliers and materials, helping to reduce time and costs. Ten new homes were eventually completed to either rent or to buy an equity share.

- A similar scheme at Great Massingham in Norfolk provided 12 two and three bedroom affordable homes to buy and to rent. The success of the first scheme in 2005 has now spurred on a further scheme.

- In South Devon the goal is to build 600 new affordable houses per year, thereby helping the rural poor gain a foothold on the housing ladder.

Planning Policy Statement 3: Housing (PPS3). This policy put in place in 2006 outlines the strategic housing policy of the government, which is to ensure that everyone has the opportunity of living in a decent home, which they can afford, in a community where they want to live. This policy is expected to have a dramatic effect on the rural housing landscape.

Case study: The rural poor in LEDCs

Village Enterprise Fund

The Village Enterprise Fund (VEF) works with subsistence farmers in remote, rural areas of Kenya, Tanzania, and Uganda. Most of the rural population struggle to feed their families. The VEF provide business training, funding, and mentoring to help them create and sustain income-generating small businesses and farming enterprises. Success ensures communities stay together. A lack of accessible water in many rural areas in LEDCs is having a serious effect on sustainability and development of such areas.

Rural transport in North Darfur, Sudan

For poor people trying to escape poverty, lack of mobility can push them further into physical and economic isolation. Work has been going on to help poor communities in North Darfur to increase their access to facilities and services, including markets that are essential to improving their lives, opportunities and economic growth. The development of intermediate means of transport (IMT) devices, such as animal drawn carts, has been promoted. Metal workers in North Darfur have been trained in the use of wheel-making sets and production of carts and trailers.

Building roads: training has been undertaken with local people. By the middle of 2005, more than 22 km of feeder roads connecting villages in North Darfur to markets and services had been built through community-building schemes.

Helping the rural poor in LEDCs

Various schemes have attempted to support the rural water-poor. **Water-Aid** is probably the most high profile in terms of portable water in Africa and Asia. In terms of irrigation the Food and agriculture Organisation (FAO – part of the UN), and in particular their **Water Development and Management Unit**, have done much to support LEDC farmers.

Technology adapting to change

Around the country more than two million people in rural areas have inadequate broadband.

It is estimated that 166 000 people are left behind in 'broadband deserts' or 'not-spots' around the UK and more than two million people in rural areas have inadequate broadband. Furthermore, those in rural areas have much less choice of broadband provider, are likely to get slower speeds and pay a different price. With superfast broadband just over the horizon (relying on optical cabling) the digital world is of the opinion that things are about to get worse for rural UK.

Why is broadband so important for rural areas? Since 2005 broadband internet has become the '**fourth utility**' for most of the UK population.

The incidence of rural home working is as much as three times greater than for urban areas. Most rural businesses are also small and medium sized enterprises (SMEs) which are a key source of innovation and rural wealth creation. Many would be forced out of business, or have to relocate, if adequate and competitive broadband provision was not available.

Much of the demand in rural areas is driven by the need for **online shopping**, **banking** and **communication**. Approximately three-quarters of rural internet users say they use the internet for transactions – higher than the UK average of 69%.

Additionally the Campaign for Rural England (CRC) has identified a number of key areas which they believe must be tackled as a priority.

- **Business sustainability** – rural SMEs need broadband and mobile telephone coverage to grow their businesses or to just maintain competitiveness.
- **Education and lifelong learning** – where employees are restricted in the degree of 'on-the-job' learning, businesses cannot nurture local or existing talent through e-learning. Students, in both full and part-time education, are disadvantaged if they cannot get adequate broadband at home, whilst an ageing population needs access to broadband to maintain their skills and continue to engage fully in society.
- **Equitable access to services** – increasingly, access to government services and to the full range of social benefits is available through various interactive services, e.g. cheaper bills, healthcare diagnostics and public service efficiencies created through increased online service delivery.
- **Social and community cohesion** – using broadband and mobile technology to communicate and use of the internet enables communication between isolated rural settlements.

PROGRESS CHECK

1. Why is affordable housing so central to the vitality of the countryside?
2. Why might villagers be upset when "exemption status" is afforded to housing?
3. Why is broadband so central to the re-development of the countryside?

Answers: 1. Housing retains the local population in the area ensuring its vitality, that schools are full and that services are fully utilised 2. Exemption status means that planning law is just about superseded and building goes ahead with minimal interference from the planning authorities 3. Because it allows the rural "disconnected" to relate to the rest of society, to be educated, to shop and most importantly in our high-tech world it allows industry to operate and be successful in the countryside.

8.5 Rebranding urban areas

LEARNING SUMMARY	**After studying this section, you should be able to understand:**
	● the apparent need for rebranding cities
	● that a range of options can be utilised to rebrand urban areas
	● that not all rebranding schemes are successful

Urban rebranding

AQA	**A2 U3**
OCR	**AS U2**
Edexcel	**AS U2**
WJEC	**AS UG2**
CCEA	**AS U2**

Many consider urban **rebranding** as a **reinvention** of the city. Urban areas can be thought of as 'products'. They provide labour, land, premises and industrial infrastructures to businesses. They also offer housing, shopping, leisure and other amenities, and a social environment for residents. Increasingly, cities and places within them compete against each other to attract new investment, tourists and visitors.

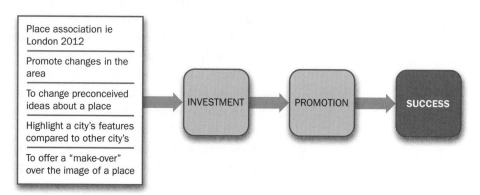

The aims of urban rebranding

Time to rebrand cities

Urban rebranding gives cities an opportunity to make up for past failures, industrial decline and social problems. Urban decline within cities has led to media portrayals of them being associated with severe economic and social deprivation, homelessness, high levels of crime, vandalism, public disorder, dirtiness and a death of civic amenities. A city can be repositioned in a better market sector, the new brand image can be used to convey the message that a wider and different range of place products are now on offer to business investors, visitors and tourists. Many of the highest profile place rebranding exercises have been completed by cities such as Glasgow, Manchester, Leeds or Newcastle. Previously they were old industrial centres that have needed to rebrand themselves as centres of leisure and amenity rather than of heavy industry.

> Brand and image is something that exists in the collective mind of the consumer, visitor or investor.

> Rebranding invariably starts with slogan raising: 'Birmingham: The Global City at the Heart of England!'; 'Glasgow's miles better'; 'Kingston-upon-Hull: The pioneering city'; 'Uniquely Manchester' and so on...

The rebranding process in city areas includes one or all of the following processes.

● **Environmental rebranding** – making improvements to buildings and built iconic structures.
● **Social rebranding** – schemes to reduce poverty and the cause of poverty.
● **Economic rebranding** – to increase jobs and opportunities.

So the response to the problems faced by urban areas has been to establish projects, such as the Leeds and Norwich examples, which hope to change the local and national perception about a place and to bring job opportunities and further investment to the area. The focus of many projects has been to enhance CBDs and the city as a whole.

Case study: Rebranding Leeds, UK

The strategic aim of rebranding Leeds was to 'go up a league'. It was in fact a city not making the most of its strengths. Marketing investigations suggested that Leeds needed to be more flexible in the way it operated if it was to attract the private finance it needed to move forward.

Since 2008, marketing has concentrated on raising the profile of Leeds on national and international levels, promoting a strong business image and a wider culture of the city. Rebranding has brought infrastructure improvements and range of cultural activities and institutions. New skills have become embedded in the city, and many finance and insurance companies now exist in numbers. There are also reported improvements in social cohesion and a reduction in crime.

Case study: Norwich – 'a fine City'

Rebranding has occurred in Norwich city centre to keep it ahead of a number of burgeoning out-of-town estates, but also in its bid to become Britain's first City of Culture in 2013 (in July 2010 it was announced that Norwich came second after Derry-Londonderry, but had beaten 27 other UK contenders). The methods used in Norwich are commonly utilised by other 'rebranding cities'.

Establishment of a town centre management and marketing team	<ul><li>City Centre Partnership (CCP) vacant shop campaign brightens Norwich's city centre streetscape</li><li>business pride campaign</li><li>city centre toilet provision</li><li>anti-litter campaigns and recycling help</li><li>Norwich In Bloom campaign.</li></ul>
Encouraging cultural quarters within the city area	Four quarters, shopping, cathedral, historic and entertainment. All with various slogans, e.g. 'Welcome to Norwich Lanes... the vibrant heart of Norwich life'
Building new 'flagship' shopping	<ul><li>Christmas shopping encouraged with the "Norwichristmas" campaign.</li><li>Norwich National Retail Skills Academy established.</li></ul>
Cleaning buildings and the restoration of cultural heritage	The Norwich 12 is the UK's finest collection of individually outstanding heritage buildings spanning the Norman, medieval, Georgian, Victorian and modern eras.
Encouraging the redevelopment of old buildings or industrial sites	A redevelopment of 42 acres of former industrial and railway land at the edge of the city centre, *Riverside* includes a food superstore, seven retail units, a multiplex cinema, a nightclub, restaurants, bars and health club, residential units, affordable housing, a swimming pool and a fitness centre.
Improving public transport	Norfolk County Council operates six purpose built Park and Ride sites more to improve the air quality in the city centre.
Pedestrianising of the city	St. Stephens master plan which aims to pedestrianise the final major shopping street by 2013.
Safer city	<ul><li>Business Crime Initiative, Safer Neighbourhood Action Panel, Radio Security System Scheme</li><li>Secure Incident Reporting & Community Engagement System (SIRCS)</li><li>Taxi Marshall Scheme.</li></ul>
Industrial redevelopment	Norwich is the largest economy in the eastern region.

The problem with rebranding

Rebranding rarely has a city-wide effect. At best small areas of cities can be re-invented which means there are always losers, gaps and areas of continued isolation in rebranded cities. The Glasgow example demonstrates this.

Case study: Miles better? Life in Glasgow

'Miles better' is the rebranding slogan for Glasgow. Glasgow is a vibrant city, but rebranding has only really scratched the surface of its many problems.

Health: UK average life expectancy for men is 77 years. In Glasgow it is 63–66 years.

Wealth: The UK's three poorest constituencies are all in Glasgow.

Commerce: Employment in Glasgow broke the 400 000 barrier in 2002; tourism employs 58 000 people; at its height, shipbuilding employed 38 000.

Crime: In 2004 murders rose by 5.4% in the last year in the Strathclyde region. Serious assaults rose by 2.4%.

Culture: 1988 Glasgow Garden Festival; 1990 Glasgow was the European City of Culture; 1999 Glasgow was the UK City of Architecture and Design; Commonwealth Games, 2014.

8.6 Rebranding rural areas

LEARNING SUMMARY	**After studying this section you should be able to understand:**
	• the apparent need for rebranding rural areas
	• that a range of options can be utilised to rebrand rural areas
	• that not all rebranding schemes are successful

Rural rebranding

OCR	**AS U2**
Edexcel	**AS U2**
WJEC	**AS UG2**
CCEA	**AS U2**

Rural rebranding happens when a rural location has to reinvent itself to halt a decline and provide for a rich future. Many villages suffer from social, economic and environmental problems; the so-called spiral of rural decline. Features of this decline include: a lack of skilled local jobs, low wages, declining farm work, long commuting journeys and few local services (as the population departs the shops/doctors are not adequately supported). Houses are bought up by rich suburban dwellers so younger villagers can't afford the house prices.

> Just as in urban areas the rural response has been to establish projects to help to change ideas and perceptions and to bring an end to the rural area's problems.

The response and strategy to rural decline

Deliberate rebranding strategies fall into the following broad categories.

- The **promotion of rural tourism** to increase visits to attractions and increase tourist visits to the country.
- **Using technology** to increase the attractiveness of the rural area. Businesses are attracted and not being tied to urban areas freelance/self-employed workers can move to the country as the internet is available.
- Rural areas **change and diversify** to cope with changing demand from the local, new and suburban population.
- **Value** can be added to what is on offer in the countryside e.g. farmers markets and organic production has moved the countryside forward.

Measuring the success of rural rebranding

Economic benefits

- New, different and unique jobs for those in rural areas.
- Increases in average wages.
- Plenty of disposable income to spend in and on local services.

Social benefits

- Skill levels changed and improved.
- New shops and services appear in the rural area.
- Transport schemes support the rural area.
- Rural areas feel attached to the village and locality.

Environmental benefits

- Old buildings and farms are regenerated and renovated.
- The inhabitants of rural areas want to improve and protect the environment.

How successful is rural rebranding?

The resentment felt by many in rural areas rebranding is well documented.

- On tourists: they bring pollution, parking problems and congestion and can be insensitive to the local environment.
- On services: the new 'rural-dwellers' demand premium products and the effect is to increase prices for many products for locals.

Overall the opinion of many in rural areas is that rebranding is merely window-dressing.

Case study: Rebranding Cornwall, UK

Cornwall is a popular holiday destination, with 4 million visitors per year. Tourists may only see the attractions of Cornwall, but full-time residents of Cornwall also see the decline in the rural economy.

In farming

- Supermarkets drive down farm prices.
- EU subsidies by 2013 will be withdrawn.
- Imported food is cheaper.

In fishing

- The UK's adherence to quotas and their redistribution to other EU countries has adversely affected the industry.
- Fishing stocks have dropped.

In mining

- Reserves are exhausted.
- Prices have collapsed.

How to rebrand Cornwall?

Principally as a tourist destination. Cornwall's economy is a seasonal one – there is a need to diversify and to create a year-round economy.

The Eden Project (opened 2001) is a diversification project that consists of two specially designed conservatories containing many plant species. It has been a huge success, transforming the local environmental landscape. However, it has also brought pollution and congestion. Even so, the benefits that the scheme has given the St Austell area far outweigh any problems. Local farmers have seen an increase in demand for their products, more jobs have been created, tourism has increased and interest in many other attractions has been generated.

PROGRESS CHECK

1. Why does the city need to be rebranded?
2. Why do some urban rebranding projects fail?
3. What is the spiral of rural decline?

Answers: 1. Collapse of businesses, lack of employment and social decline 2. They deal only superficially with issues, money runs out, no whole city rebranding projects have been undertaken 3. This is when social, economic and environmental factors all conspire together to drag rural areas down

Exam practice questions

1 **(a)** Describe the land use of any named LEDC city. [7]

(b) Explain the limitations of the application of any model of urban morphology to the city you have described in (a). [7]

(c) With reference to poorer residential areas of LEDC cities:

(i) Outline the environmental, social and economic problems that can be experienced in poorer residential areas. [9]

(ii) Suggest ways in which:

either

- the local population can help themselves to overcome such problems or
- city decision-makers can solve their problems for them. [7]

2 **(a)** Using Figure 1 describe and explain the differences between urban and rural environments. [10]

Figure 1

(b) For what reasons are many rural settlements in MEDCs exhibiting the characteristics of urban settlements? [10]

9 Population and migration issues

The following topics are covered in this chapter:

- Population growth and distribution
- Population structure
- Population migration
- Business of migration
- Population control

9.1 Population growth and distribution

LEARNING SUMMARY

After studying this section, you should be able to understand:

- why and how populations have grown and changed
- the concept of optimum population
- a population density and distribution

World population change

AQA	AS U1
OCR	A2 U3
Edexcel	AS U1
WJEC	AS UG2
CCEA	AS U2

World population in 2010 = approximately 7 billion.

For thousands of years the global population changed very little, probably reaching 250 million about 200 years ago. Population remained steady until about 200 years ago, and growth was linked initially to the industrial revolution in Europe.

World population reached one billion in the early nineteenth century, two billion in the 1930s and three billion in the 1960s. In effect the world population doubled between 1800 and 1930 and it doubled again up to the 1970s. Theoretically it will double again by 2040. The world's population is increasing at an **exponential rate**.

Population has not increased evenly over the world – some countries, especially in Africa and Latin America, since 1950 have grown exponentially. Asia during all of this period has maintained a very high half of the world's population! The growth rates of some North African countries are in the order of 3% (enough to double the population in 25 years). This compares with many European countries that have experienced drops in population in recent years. It is estimated that 95% of growth will occur in Africa, Asia and Latin America over the next 25 years.

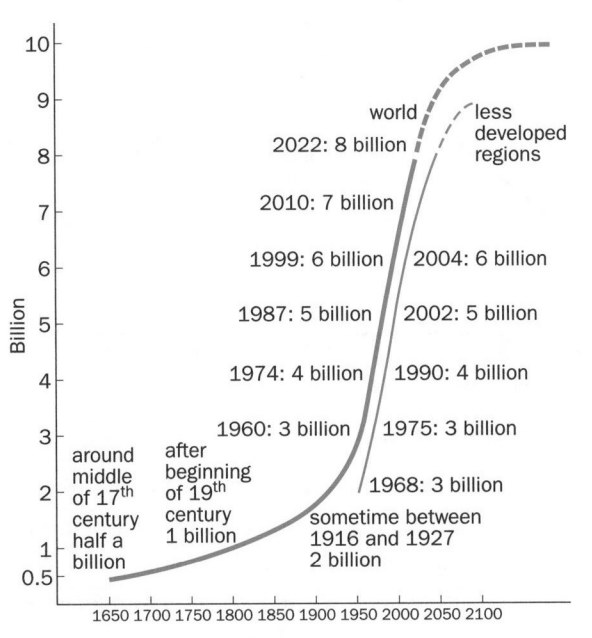

Growth of world population

Population growth or decline, 2008–2050 (estimates)	
Largest percent decline	
Country	**%**
Bulgaria	−35
Swaziland	−33
Georgia	−28
Ukraine	−28
Japan	−25
Moldova	−23
Russia	−22
Serbia	−21
Belarus	−20
Romania	−20
Bosnia-Herzegovina	−20
Largest percent increase	
Country	**%**
Uganda	+263
Niger	+261
Burundi	+220
Liberia	+216
Guinea-Bissau	+205
Congo, Dem. Rep.	+185
Timor (East Timor)	+179
Mali	+169
Somalia	+166
Angola	+155

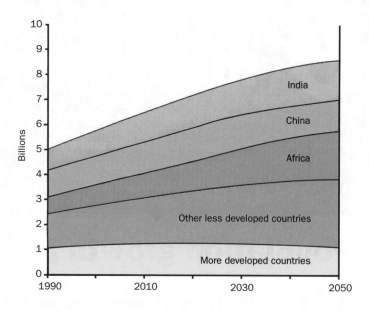

Africa and other developing regions make up an increasing share of the world population

Alternative view and scenario of population growth in the future

The Millennium Ecosystem Assessment (MA) developed four global scenarios exploring plausible future changes in ecosystems, ecosystem services and well being. The scenarios are:

- Global orchestration
- Order from strength
- Adapting mosaic
- Techno garden.

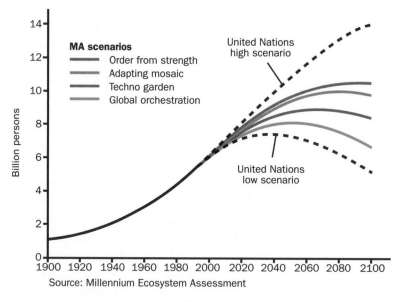

Source: Millennium Ecosystem Assessment

World population scenarios

The probability of any one of the scenarios being the real future is low: the real future is likely to be some mix of the scenarios. The future could be far worse or far better for the world's population than any of the individual scenarios, depending on the choices made by decision makers as well as on unforeseeable events.

Defining characteristics of the four scenarios

Scenario name	Dominant approach for sustainability	Economic approach	Social policy focus	Dominant social organisations
Global orchestration	Sustainable development; economic growth; public goods.	Fair trade (with a reduction in tariff boundaries), with enhancement of global public goods.	Improved world; global public health; global education.	Transnational companies, global NGO and multi-lateral organisations.
Order from strength	Reserves; parks; national polices; conservation.	Regional trade blocs; mercantilism.	Security and protection.	Multinational companies.
Adapting mosaic	Local-regional co-management; common property institutions.	Integration of local rules regulating trade; local non-market rights.	Local communities linked to global communities; local equality is important.	Cooperatives and global organisations.
Techno garden	Green technology; eco-efficiency; tradable ecological property rights.	Global reduction of tariff boundaries; free movement of goods, capital and people; global markets in ecological property.	Technical expertise valued; follow opportunity and competition; openness.	Transnational professional associations; NGOs.

Case study: A view on population growth

Environmentalists think that with declining fertility rates in many countries population growth will slow. At present, half the world now has fertility rates below the replacement rate of 2.3 children. Women in Iran were giving birth to eight children in the 1980s, but now on average give birth to fewer than two. In Bangladesh, where many mothers are poor and badly educated, women have an average of three children. Birth rates have fallen to 2.8 in India and two in Brazil, despite the influence of Catholicism.

Low birth rates in Taiwan, Hong Kong and Singapore have led to the governments paying couples to have children. Among patriarchal societies, such as Yemen and the Palestinians and Orthodox Jews in Israel, birth rates are rising. Mass migration and an ageing population are future challenges. In fact an older population could be 'wiser and greener', living life at a slower pace with less desire for high-speed consumption. Environmentalists feel we should be more worried by a population crash rather than a population explosion.

Negative **natural population** growth means there are more deaths than births or an even number of deaths and births.

A zero growth future!

There are twenty countries in the world with negative or zero natural population growth, including Germany, Czechoslovakia, Poland, Italy and Greece.

This figure does not include the impacts of immigration or emigration. The country with the highest decrease in the natural birth rate is the Ukraine at 0.8% each year. The Ukraine is expected to lose 28% of their population between now and 2050 (from 46.8 million to 33.4 million). Japan is the only non-European country in the list and it has a 0% natural birth increase and is expected to lose 21% of its population by 2050 (from 127.8 million to 100.6 million).

Measures of population growth

How fast is the world's population growing? The world's current growth rate is about 1.3%, representing a doubling time of 54 years. The UN predicts it will increase to 10.6 billion by about 2050.

You need to remember the trends depicted in these maps and tables.

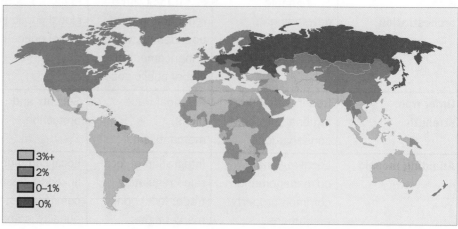

3%+
2%
0–1%
-0%

Actual population growth rate as % (2010)

Differences between developed and developing countries can be stark (see below)!

Try to remember a few of these figures!

Key demographic Indicators	Ethiopia	Germany
Total population	77.1 million	82.3 million
Population <15	33 million	11.9 million
Population 65 and older	2.2 million	15.3 million
Annual births	3.1 million	672 000
Annual deaths	1.2 million	821 000
Annual infant mortality	236 000	2 600
Life expectancy at birth	49 years	79 years

Population growth rates and doubling time

The rate of national growth is expressed as a percentage for each country, commonly 0.1% to 3% annually. Two percentages are associated with population – natural growth and overall growth. Natural growth represents the births and deaths in a country's population and does not take into account migration. The overall growth rate takes migration into account e.g., Canada's natural growth rate is 0.3%, while its overall growth rate is 0.9% due to Canada's open immigration policies.

We can expect the world's population of 6.5 billion to become 13 billion by 2067 if current growth continues.

The growth rate can be used to determine a country's population doubling time, e.g. Canada's population will double from its current 33 million to 66 million. The world's current growth rate is about 1.14%, representing a doubling time of 61 years.

Many Asian and African countries have high growth rates. Afghanistan has a current growth rate of 4.8%, representing a doubling time of 14.5 years! If Afghanistan's growth rate remained the same, then the population of 30 million would become 60 million in 2020. Increased population growth generally represents problems for a country – it means increased need for food, infrastructure and services.

The trends and concepts this map attempts to convey need to be understood. It is a common ploy in A-level exams to ask you to analyse and comment on such maps.

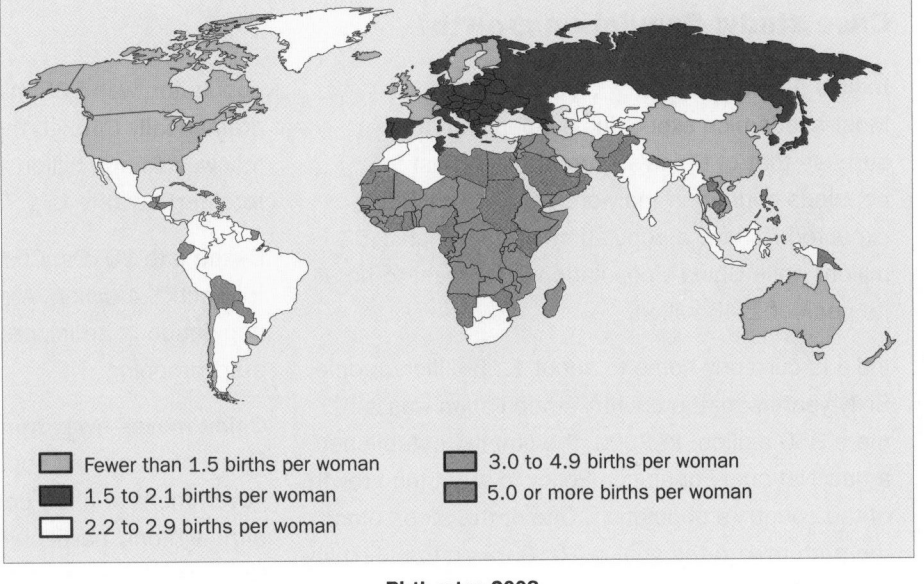

Fewer than 1.5 births per woman
1.5 to 2.1 births per woman
2.2 to 2.9 births per woman
3.0 to 4.9 births per woman
5.0 or more births per woman

Birth rates 2008

Case study: Sub-Saharan Africa: the population emergency

Sub-Saharan Africa has been experiencing population growth since the beginning of the twentieth century. The region's population increased from 100 million in 1900 to 770 million in 2005 and 1.5–2 billion inhabitants are projected for 2050.

Half of Sub-Saharan Africa's population will be urban dwellers by 2030. Intra-regional migration is severely disrupted by the conflicts and crises affecting several neighbouring countries. Population growth in sub-Saharan Africa will be a major handicap to economic and social development.

Case study: Global census

A census is the procedure of systematically acquiring and recording information about the members of a given population. It is a regularly occurring and official count of a particular population.

The 2001 UK census
The UK Census, which happens every ten years, collects population and other statistics essential to those who have to plan and allocate resources (i.e. health, housing and education). Enumeration Districts (EDs) are the areas within which the data is collected and Output Areas (OAs) are the base units for Census data releases. The UK 2001 census cost £259 million to administer and was dubbed the **Count Me In** census. The basic population summary statistics are that the UK population is ageing and that growth has been slower than predicted. The population on census day in 2001 was 58 789 194.

The Canada 2001 census
This was carried out in May 2001 and the total population count of Canada was 30 007 094. This was a 4% increase over the 1996 census of 28 846 761.

The 2006 Nigerian census
The Nigerian census was in 2006. The population was put at 149 229 090. Nigeria has recently undergone the start of a population explosion due to higher fertility rate. Its population growth rate has been put at 1.9% to 3.2%. Nigeria's census is notoriously unreliable. It is thought there was a massive undercount of people in 1991 at 89 million instead of the predicted 112 million. Even before the 2006 census was taken its 'outcomes' were being questioned! Many developing countries experience such issues.

Case study: Population growth

India's population

Most academics expect India's population to surpass that of China's, currently the most populous country in the world, by 2030. India is expected to have a population of more than 1.53 billion, while China's population is forecast to be at its peak of 1.46 billion.

India is currently home to about 1.15 billion people. Sixty years ago, the country's population was a mere 350 million. In 2000, the country established a new National Population Policy to stem the growth of the country's population. One of the steps along the path toward the goal in 2010 was a total fertility rate of 2.6 by 2002. This was not achieved and fertility remains high at 2.8. India's high population growth will result in increasingly impoverished and sub-standard conditions for growing segments of the Indian population. As of 2007, India ranked 126th on the United Nations' Human Development Index. Population projections for India anticipate that the country's population will reach 1.5 to 1.8 billion by 2050. India's growth rate of 1.6% represents a doubling time of under 44 years.

China's population

With just over 1.3 billion people, China is currently the world's largest and most populous country. The world's population is approximately 6.7 billion; one in every five people on the planet is a resident of China. China's population growth has been slowed by the one-child policy. As recently as 1950, China's population was 563 million. The population grew dramatically through the following decades that followed to one billion in the early 1980s. China's total fertility rate is 1.7.

By the late 2010s, China's population is expected to reach 1.4 billion. Around 2030, China's population is anticipated to peak and then slowly start dropping.

China moves away from one child policy

Fears of an ageing population has led the Chinese government to allow couples to have two children – although only particular kinds of people can apply. Couples who were both only children, which includes most of the city's newly-weds, are allowed two children. Couples are allowed to have two children if both partners have PhDs, are disabled, come from a rural area, or in some cases if their first child is a girl. There are exceptions for when a widow or widower, or a divorcee, marries someone childless.

However the one-child-policy remains in place in most parts of the country. The main focus of the one-child policy has been in the countryside, where farmers traditionally liked to have large families, especially sons. Middle class Chinese in the cities have fewer children by choice. By 2050, China will have more than 438 million people over 60, with more than 100 million of them 80 and above.

Causes of change

China will have only 1.6 working-age adults to support every person aged 60 and above.

If the number of births (birth rate = BR) exceeds the number of deaths (death rate = DR) in any one year then populations tend to increase. This is called **natural increase**. In addition, if the number of immigrants that join a country exceeds emigration, populations will rise.

> **KEY POINT**
>
> Three factors influence population change:
> Fertility + / − Mortality + / − Migration = + / − Growth

Reasons for change

AQA   **AS U1**
OCR   **A2 U3**
Edexcel **AS U1**
WJEC   **AS UG2**
CCEA   **AS U2**

The influence of falling death rates

Through the twentieth century, Africa has seen death rates halved, particularly amongst children. This has been brought about with improvements in sanitation and improved medical techniques, medicines and general provision. **Crude death rate** (the number of deaths/1000 inhabitants/year) is the most common measure of mortality. However, as it takes no account of population structure it is of little use. A better measure is infant mortality, the number of deaths of infants under the age of one year old/1000 live births.

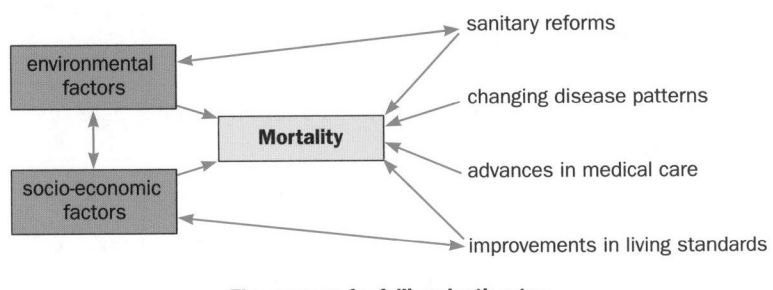

The reasons for falling death rates

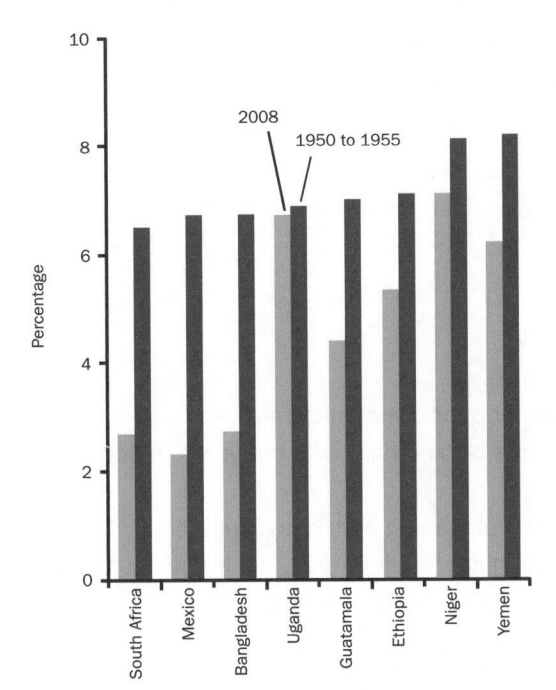

The northern hemisphere generally has the lower death rates.

The influence of birth rates

Birth rates in the more economically developed countries (MEDCs) have fallen, while in the least economically developed countries (LEDCs) high birth rates have persisted, but are falling in some. This will, as outlined above, result in these countries' populations doubling over the next half century. At the same time, MEDCs' populations will shrink.

Since the mid 1980s there has been a significant drop in fertility among developing countries. In Bangladesh, the total fertility rate dropped from 6.7 lifetime births per woman in the early 1950s to 2.7 in 2008, aided by a strong government commitment to population policies and successful community-based family planning programs.

Fertility also fell dramatically in Guatemala, from 7.0 to 4.4 children per woman. Mexico saw an impressive decline, as the country developed economically and embraced the idea of smaller families. Ethiopia, Niger, and Uganda show much more modest declines, helping to explain why Africa's population growth continues to outstrip that in other regions.

Birth rates decline in MEDCs because of:
- availability of family planning
- increased education and literacy
- better health and fewer child deaths
- more employment opportunities
- later marriage
- migration to the cities
- better deals for women
- more income and rising living standards.

Total fertility rates

The reasons for the continuing high birth rate in LEDCs (the number of live births/1000 people/year) is a complex problem to untangle. Reasons include:
- the importance placed upon child-bearing in some countries
- large families are seen as an insurance for the future
- women are disempowered
- more children equates to more workers.

In the past it was said that MEDCs with 25% of the world's population used 80% of resources and LEDCs with 75% of population used 20 per cent of resources. In 2010 the mantra still relates to resources, though indirectly through emission figures. The richest 500 million (7%) of the world's population release 50% of world emissions, while the poorest 3.5 billion (50%) of the world's population release just 7% of world emissions.

Generally population increase equates to high birth rates, e.g. in Africa the birth rate is in excess of 40/1000 while in Europe it's less than 15/1000!

Implications of increased population

If population increases the following increases will also occur:
- fuel consumption
- climate change resulting in flooding, salinisation and rising sea levels
- acid rain, waste and pollution
- agricultural activity, to feed the growing world population, causing greater use of pesticides and fertilisers
- use, and therefore depletion, of world minerals and hardwoods.

The ability of the Earth to sustain human life is limited. Resources can only be exploited to a certain level before they are exhausted, a limit known as the **carrying capacity**. Even though MEDC populations are stable, they are the biggest consumers of global resources, as LEDCs strive to industrialise their consumption will increase and so will pollution and environmental damage. The world can probably sustain a huge population for a short time, but medium to low population growth rate would enable a more sustainable future.

The concept of optimum population

AQA	AS U1
OCR	A2 U3
CCEA	AS U2

The concept of **optimum population** involves an ideal population living and working in a given area – numbers of people in balance with resources, maximising gross national product (GNP).
- If population is > than resources, it is said to be **overpopulated**.
- If population is < than resources, it is said to be **underpopulated**.

The best A-Level students will provide accurate definitions of optimum, over and under–population.

Though the idea of an optimum population is an interesting one, it is the extremes, over population and under population that draw most attention today.

Under-population

The causes of under-population include:
- physical disadvantages, e.g. climate
- inaccessibility to remote locations or poor communications
- historical reasons, e.g. Australia has worked hard over the years to boost its population, i.e. the £10 ticket
- types of economy, i.e. intensive manufacturing and/or agricultural
- small indigenous populations, e.g. in Brazil, where 92% live in the south-east and there are lots of natural resources as yet untouched.

The consequences of under-population include:
- resources developed by foreign countries
- regional disparity
- high urbanisation
- high standard of living
- high immigration.

Under-population does not imply a country is poor or has a low population density.

Over-population

With regard to over-population there are two opposing views.
- The **Neo-Malthusian** approach, that increasing population leads to environmental degradation, which limits population growth.
- The **Boserupian** approach (after Ester Boserup, a Danish economist), that necessity is the mother of invention. Increasing population drives agricultural productivity, which allows further increases in population.

The best students will be well aware of these views on under and overpopulation, and have good case studies to support them.

The consequences of over-population include:

- starvation and malnutrition
- poor health
- lack of jobs
- slow economic growth.

Case study: Population theories

The Reverend Thomas Malthus, 1798

Writing in 1798 the basis of Malthus' theory was that food is necessary for human existence and that population tends to grow faster (geometrically 1-2-4-8-16, etc.) than the power in the earth to produce subsistence (arithmetically 1-2-3-4-5-6, etc.). He felt for a stable population the effects of these two unequal powers must be kept equal. He felt that since maintaining equality between the two was probably impossible that it would lead to war, famine and disease. Once this had been reached he felt that further growth in population would be prevented by negative and positive checks. **Negative checks** include abstinence/postponement of marriage which lowered the fertility rate (worth noting is that he proposed this only for the poor and working classes!). **Positive checks** were ways to reduce population size by events such as famine, disease or war thereby increasing the mortality rate and reducing life expectancy. He called this his **population crash model**.

Was he right? It is debatable whether Malthus could have foreseen the massive technological advances, including hugely successful irrigation techniques and reducing population growth, as countries move through the demographic transition model (DTM). Supporters of Malthusian theory point to Africa with its explosion of population and its repeated famines, wars, food crisis, environmental degradation, soil erosion, crop failure and disastrous floods. The world population in 1798 was at nine million people. We have now passed the six and a half billion mark. So was he right?

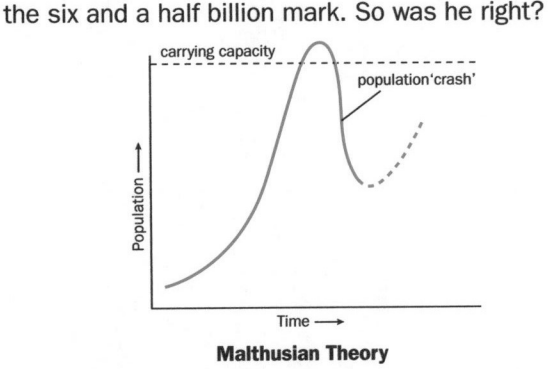

Malthusian Theory

Case study: Ester Boserup, 1965

Boserup believed that people have the resources of knowledge and technology to increase food supplies. Her view was directly opposite to those of Malthus – she suggested that population growth has enabled agricultural development to occur. Demographic pressure (population density) promotes innovation and higher productivity in the use of land (irrigation, weeding, crop intensification, better seeds) and labour (tools, better techniques). Boserup argued that the changes in technology allow for improved crop strains and increased yields, e.g. GM crops and the **Green Revolution**. She also recognised that overpopulation can lead to unsuitable farming practices which may degrade the land, e.g. desertification in the Sahel. Her ideas were forward thinking. The one point she missed in her assumptions was that she assumed a 'closed' society, which is not the case in reality, with easy "world" migration.

Paul Ehrlich, 1967

In December 1967, biologist Paul Ehrlich wrote in the New Scientist, about the **Population Bomb**, that the world would experience famines sometime between 1970 and 1985 due to population growth outstripping resources. He stated that in the 1970s and 1980s hundreds of millions of people will starve to death in spite of any crash programs embarked upon now. Ehrlich concedes that he did overstate his case, underestimating the effects of the green revolution, but that part of the reason that there have not been such serious famines has been due to a reduction in birth rates.

Since Ehrlich stated his case over 4.5 billion people have been added to world population. Furthermore, billions are undernourished and predictions about disease and climate change were essentially correct.

The Club of Rome, 1972

A group of industrialists, scientists, economists and statesmen from ten countries published *The Limits to Growth* in 1972. It stated that if the population of the world continues to grow, the limits to growth on this planet will be reached sometime in the next 100 years. The outcome of this would be declines in industry and uncontrolled population decline. The Club of Rome did not consider how adaptable and innovative man can be, for instance HYV seeds to prevent starvation in parts of Asia and so on.

Julian Simon, 1981

Julian Simon's book *The Ultimate Resource* is a criticism of what was then the conventional wisdom on population growth. He challenged the notion of an impending Malthusian catastrophe — that an increase in population has negative economic consequences, that population is a drain on natural resources and that we are at risk of running out of resources through over-consumption. Simon argues that population is the solution to resource scarcities and environmental problems, since people and markets innovate.

Population density and distribution

AQA	**AS U1**
OCR	**A2 U3**
Edexcel	**AS U1**
WJEC	**AS UG2**
CCEA	**AS U2**

- **Density** – a measure of the average number of people/unit area. This measure should not be used to compare countries in terms of overpopulation, as different countries have different carrying capacities.
- **Distribution** – relates to location based mainly on economic and physical factors. It is difficult to measure as it is a spatial indicator.

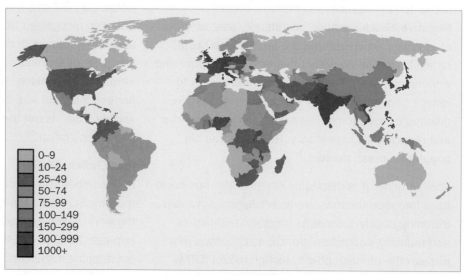

	0–9
	10–24
	25–49
	50–74
	75–99
	100–149
	150–299
	300–999
	1000+

World population/km²

Is Britain crowded?

The map (right) shows data from the 2001 Census (people per hectare). The map shows that the south and London have the greatest population pressure upon them and that pressure continues to increase. Those in the north–west and north-east of England and around Glasgow have actually seen population pressure falling between the 1991 and 2001 census.

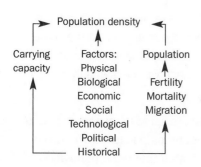

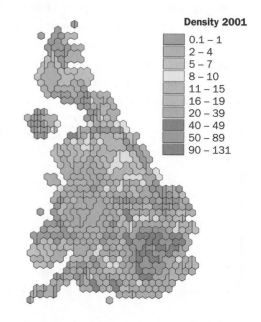

Density 2001

	0.1 – 1
	2 – 4
	5 – 7
	8 – 10
	11 – 15
	16 – 19
	20 – 39
	40 – 49
	50 – 89
	90 – 131

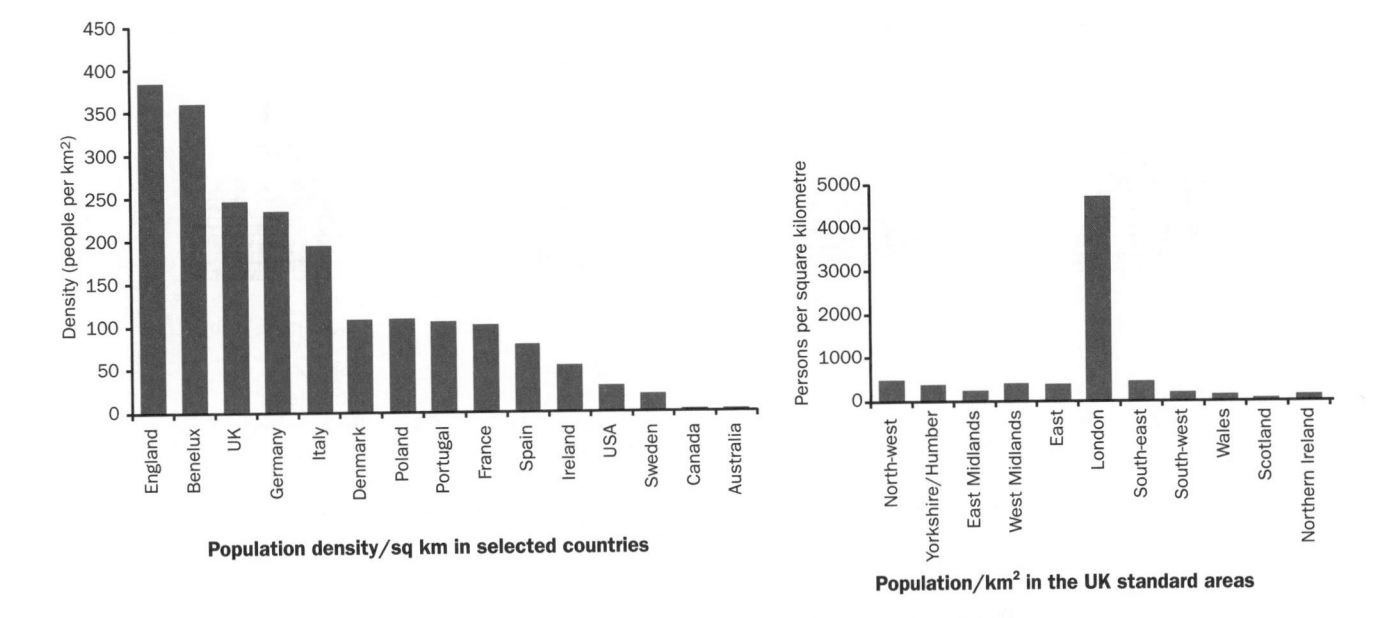

Population density/sq km in selected countries

Population/km² in the UK standard areas

If you can remember some of the data in these maps and charts great, but what you really have to get over to the examiner is that you understand and can verbalise the trends and patterns they show.

PROGRESS CHECK

1. Why did the millennium goals scheme see different population outcomes?
2. Why does population growth represent problems for a country?
3. What is carrying capacity?
4. What did Paul Ehrlich underestimate?

Answers: 1. This scheme shows a series of different scenarios as to how the future might develop in terms of population. It is likely to be a mix of scenarios that contributes to overall population 2. Issues with food, infrastructure and services. 3. Resources can only be exploited to a certain level before they become exhausted. 4. The effects of the Green Revolution.

9.2 Population structure

LEARNING SUMMARY

After studying this section, you should be able to understand:

- the use of population pyramids to analyse age and gender structures
- how these can be used to study the proportion of the youthful, mature and elderly populations
- how the demographic transition model uses population structures to assess development

Population pyramids

AQA   **AS U1**
Edexcel   **AS U1**
WJEC   **AS UG2**
CCEA   **AS U2**

Learn the population pyramid shapes and be able to apply them to any given country.

Population pyramids are a graphical method of representing the age and sex structure of a population at one point in time. The shape of a country's pyramid is the result of past fertility, mortality and migration within the population. It is sensitive to 'baby booms', wars, epidemics, population planning policies. On the pyramid each group is represented by percentages, therefore comparisons in the age–sex ratios can be made between countries. Four main types of population pyramid can be distinguished: stationary; progressive; regressive; composite.

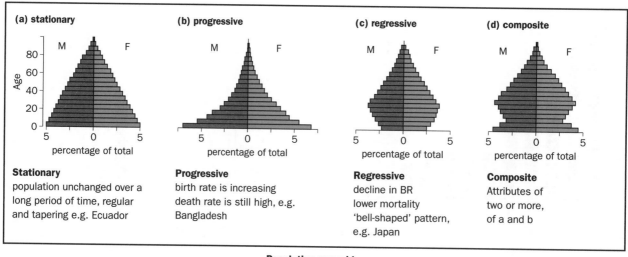

Population pyramids

Dependency

One of the most common measures derived from population pyramids is the **Dependency Ratio**. For convenience, the age structure is divided into three broad age bands.

- **Youthful** dependants aged 1 to 14.
- **Elderly** dependants aged 65 and over.
- **Working** population aged 15 to 64.

> Current problem, therefore important at AS!

> Dependency Ratio = Young people (0–14) + Old people (over 65) / Working age population (15–64)

To calculate the proportions of dependants, divide the proportions of young and old by the proportion who are economically active, to yield a ratio. If the dependent group constitutes a relatively high proportion, the ratio will be high. The greater the proportion in the working age group the fewer the dependants. Dependency ratios are higher for developing countries.

Pyramids for countries may combine attributes of different types. These are termed **composite**. Points to bear in mind include:

- Is a young dependant more or less of a burden on a working person's resources than an elderly dependant?
- Many over 15s continue to study, so this age group is now incorrectly classified in the working population.
- In many LEDCs children start jobs before they are 15.
- Students, housewives and the unemployed are not represented as dependants.

Trends in population structure in LEDCs and MEDCs

KEY POINT

LEDCs	MEDCs
General	**General**
• Birth rate is still high, but reducing.	• Increased numbers of elderly.
• The numbers of those who are 15 or fewer is still very high.	• The upper end of the population structure is increasingly fit and healthy.
• There are more old people.	• Reduced birth rates.
	• Decline in the working population.

KEY POINT

LEDCs	MEDCs
General effects on the population structure	**General effects on the population structure**
• Population will continue to increase; anti-natal population policies will be imposed to deal with increased population.	• Tax burden on workers increases.
	• Greater Government spending.
	• Fit OAPs retire later.
• Pressure on the countries' economies.	• Career paths close because of the above.
• Children need to work.	• The state is unable to fully provide for OAPs.
• Migration increases.	• The elderly are moneyed and mobile.
• Unrest and instability within the country.	• Growth in 'grey' investment/economy, holidays and retirement homes.

The 'greying' world population

> The world's population is 'greying' fast and this affects all of us.

As countries become 'developed' they experience a static population with low birth rates (BR) and death rates (DR). It is the lowering of the DR that is most important as, by 2050 1:7 of the world's population will be over 60 years.

- **Supporting the elderly** – in the MEDCs increasingly the old will have to look after themselves in terms of pension provision.
- **Impact on young** – the 'greying effect' impacts on the young, in many countries as it leads to unstable structures and economies. Clearly the situation differs in LEDCs and MEDCs.

Case study: A comparison between the UK's and Europe's population structure

Age

90
80
70
60
50
40
30
20
10
0

500   250   250   500
Male population   Female population
(thousands)   (thousands)

Age structure of England and Wales mid 2008

UK population structure

This population pyramid shows the age-sex structure of the population in England and Wales in 2008. The dent in the pyramid at around age 63 represents those born during the first half of the Second World War, when fewer births took place than usual. In contrast, the spike in the pyramid at ages 54–57 reflects the large number of births occurring in the late 1940s, often referred to as the post-war baby boom. The large bulge in population for those in their late 30s and early 40s is a result of the high number of births that occurred during the 1960s. Similarly, the smaller bulge around ages 10 to 20 represents the children of the large number of women born in the 1960s. The pyramid does follow the general pattern for a developed country – a stable or declining base to the pyramid and a large number of persons aged over 65.

The population structure of Europe

Numbers are low at the base of the pyramid (over the page) of the age structure of Europe's population. **Birth rates** being low at the end of the 1990s and at the start of the twenty-first century. The structure of the population demonstrates that birth rates have fallen since the 1960s with the size of the cohorts becoming progressively smaller, by age group from age 35 to 39 down to the under-fives. The **bulge** at ages 35–39 corresponds to the **baby boom** of the mid to late 1960s and is followed by smaller cohorts at older ages. This reflects the low fertility during the Second World War and in the 1930s, as well as the impact of

mortality at older ages. There is a **gender imbalance** at older ages due to the lower mortality rates and higher longevity of women.

Similarities and differences

The UK pyramid does have some similarities to the European pyramid (e.g. the bulge at ages 35–39), but it also demonstrates some clear differences. In the UK there is a noticeable blip for people in their late 50s

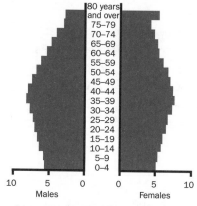

Age structure of Europe's population mid 2008

in 2006 – this does not show in the European pyramid. This corresponds to a high **birth rate** in the UK after the end of the Second World War. At ages between 0 and 29 in Europe, it appears that the **fertility rates** declined between the early 1970s and 2006, but this did not happen in the UK in such an obvious way. **Low fertility rates** in the UK during the late 1970s, resulted in 2006 in small groups of young adults in their mid to late 20s. The UK pyramid shows larger cohorts of people in their teenage years in 2006 than had been the case five to ten years earlier. High fertility rates since the 1970s and the large numbers of women born in the 1960s became mothers during the 1980s and 1990s and this was the cause of larger cohorts of teenagers in 2006. Even so, fertility rates were low in the UK throughout this period relative to the 1960s.

Case study: Three differing developing population structures

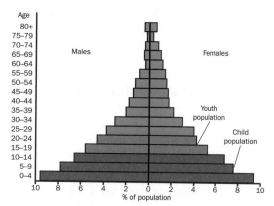

Mali – a young population

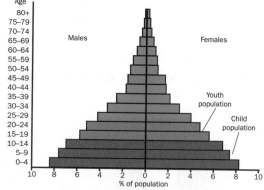

Zambia – youthful populations

Women in Mali:

- Marriage common by age 15.
- Early childbearing.
- Desired family size – up to 6 children.
- Low school attendance.
- Spousal abuse accepted.

Women in Zambia:

- Less access to education.
- Marriage more frequent.
- More children during teen years.

- Desired family size – up to 4 children.
- Spousal abuse accepted.
- High HIV prevalence.

Urban women in Peru:

- Well educated.
- Young marriage is infrequent.
- Few have children as teens.
- Desired family size – 2 children.

Rural women in Peru:

- Teen childbearing more common in rural areas.
- Poverty and childbearing.
- Unsafe abortion and maternal death.

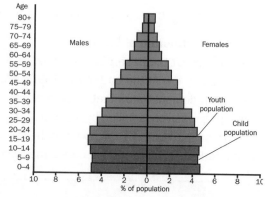

Peru – population in transition

The demographic transition model

AQA	AS U1
WJEC	AS UG2
CCEA	A2 U1

The differences between the two ratios of **Crude Birth Rate** (CBR) and **Crude Death Rate** (CDR) are known as the rate of **Natural Increase**. Utilising these two simple indices it is possible to analyse the idea of demographic transition and change in the rate of natural increase over time.

Stage 1: high stationary	Stage 2: early expanding	Stage 3: late expanding	Stage 4: low stationary
Pre-industrial society	*Early industrialisation*	*Later industrialisation*	*Developed country*
Typical of the UK in the eighteenth century and LEDCs today, e.g. Ethiopia and Bangladesh. • DR high – little medical care. • BR high – no birth control and children an economic advantage. • DR fluctuates due to plagues and famines.	Typical of the UK in the nineteenth century and Nigeria, Peru, Sri Lanka, Kenya today. • DR declines – medical developments improved nutrition and sanitation. • BR remains high – children remain an economic advantage since urbanisation and mechanisation at an early stage and births are seen as desirable. • Increasing difference between BR and DR.	Typical of the UK in the late nineteenth and early twentieth centuries and China, Cuba, Australia today. • DR low and slowly decreasing – continued medical and nutritional developments. • BR starts to decrease rapidly – improved education and availability of contraception and decreased economic value of children due to increasing urbanisation and use of technology. • Decreasing difference between BR and DR.	Typical of the USA, Canada, Japan and UK today. • DR remains low and slowly falling – continued medical progress and enhanced welfare provision. • BR declines to just above DR – economic independence of women, improved contraception and changing views on desirability of births in highly urbanised society.
high proportion are young	very high proportion are young	increasing numbers surviving to old age	high proportion of population are ageing

> There may be a further stage of demographic development as outlined in *Nature* (August 2009) by Myrskyla, Kohler and Billari.

The original **demographic transition model (DTM)** has just four stages. However, some population theorists feel that a fifth stage is needed to represent countries that have **sub-replacement** fertility (that is where death rate exceeds the birth rate), e.g. Italy and Germany. Many European and East Asian countries now have higher death rates than birth rates.

Case study: Applying the DTM to the UK

- 1700–1760: birth and death rate are both high and oscillating. This is broadly consistent with stage 1 of the DTM.
- 1760–1880: death rates are generally falling, but birth rates are remaining high and even increasing. This is broadly consistent with stage 2 of the DTM.
- 1940–1980: birth and death rates are low. This is broadly consistent with stage 4 of the DTM.

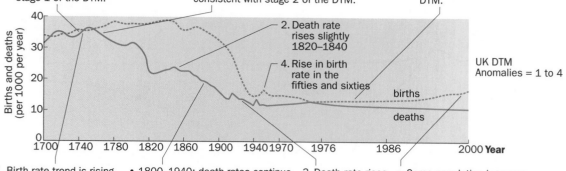

UK DTM
Anomalies = 1 to 4

2. Death rate rises slightly 1820–1840

4. Rise in birth rate in the fifties and sixties

births

deaths

1. Birth rate trend is rising 1700–1840, whereas it would be expected to be approximately constant.

- 1800–1940: death rates continue to fall and birth rates decline. This broadly consistent with stage 3 of the DTM.

3. Death rate rises around 1916 and 1940.

- Some population increase into stage 5. This is consistent with the DTM.

Why did mortality decline?
- Improved sanitation and hygiene.
- Improved food supply.
- Reduction in disease impact.
- Medical advance.
- Rise in living standards.

Why did fertility decline?
- Legislation to do with female and child labour.
- More women in the work force.
- Contraception.
- Declining infant mortality.

To LEDCs

Most are bunched in the second stage, unable to achieve economic and social progress to enable them to move on. Population increases and ecological problems further hamper progress. However, there have been notable decreases in mortality and fertility in the last 100 years.

Why has mortality decreased?
- Malaria and other tropical diseases have been eradicated or controlled.
- Improved health care.
- Stronger economies.
- Better nutrition.

Why has fertility decreased?
- The age of marriage has increased.
- Increase in the use of contraception.
- Urbanisation/Western values have been taken on.
- The status of women has improved.

PROGRESS CHECK

1. How do you calculate dependency ratios?
2. What are the challenges of a youthful population?
3. Why has a fifth stage been developed for the DTM?

Answers: 1. Young people (1 to 14) and old people (over 65) divided by the working age population (15 to 64). 2. Less access to education, marriage more frequent, spousal abuse, few opportunities. 3. Where countries have sub-replacement fertility.

9.3 Population migration

After studying this section, you should be able to understand:

- the causes and characteristics of migration and its profound effects
- models used to describe migration patterns

The causes of migration

AQA	**AS U1**
OCR	**A2 U3**
Edexcel	**AS U1**
WJEC	**AS UG2**
CCEA	**A2 U1**

Technology and economic progress increase mobility and permit increased migration. Migration is usually defined as a 'change of residence of substantial duration'. Migration can be classified in several ways.

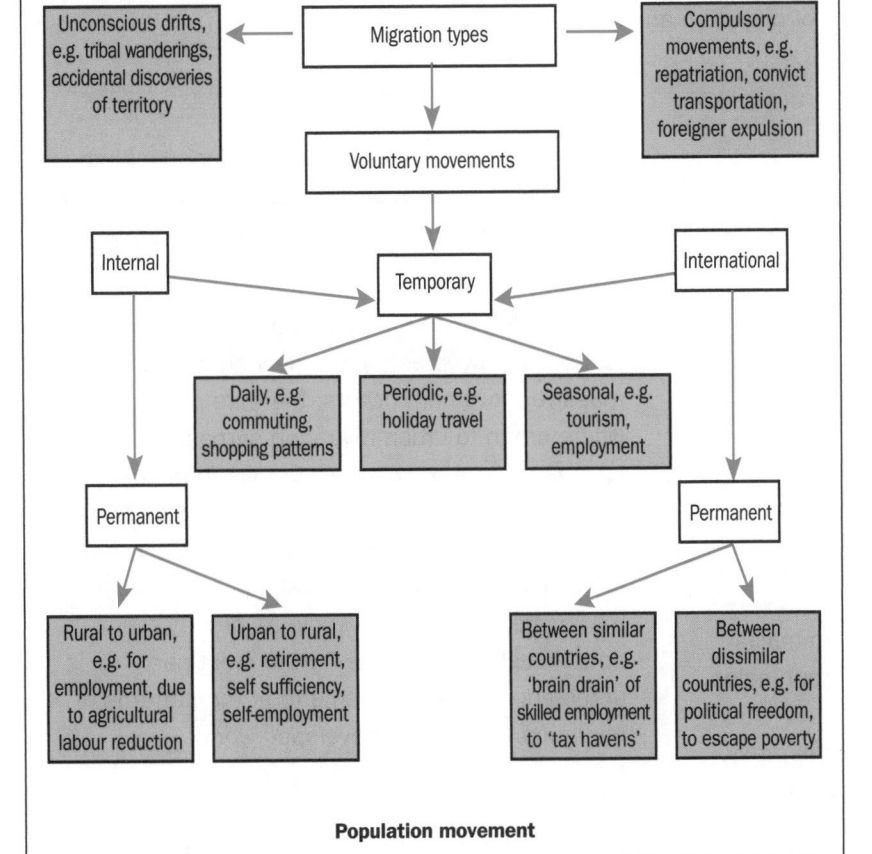

Population movement

Timescale

- Seasonal in nature, e.g. Mexican fruit pickers in California.
- Temporary, e.g. asylum seekers.
- Periodic, e.g. forced out by conflict.
- Permanent (over 1+ years), e.g. emigration to New Zealand.

Distance

- Internal, e.g. within cities.
- External, e.g. a move abroad.
- Inter-regional, e.g. a job move.
- International, e.g. emigration or immigration.

Causation

- Forced, e.g. Darfur, Sudan.
- Spontaneous.
- Free, e.g. could be politically, socially or economically motivated.
- Planned.

Migration does tend to be highly selective, but most migrants have a combination of traits.

- **Age** – the majority of migrants are between 18 and 35, often moving for a first job. Increasingly people migrate on retirement.
- **Gender** – in general in MEDCs males and females migrate in roughly equal amounts. In LEDCs it tends to be the young males.
- **Marital status** – in MEDCs most migrants have in the past been single. Nowadays family migrations are more common.
- **Occupational groups** – professional migrants tend to predominate. Occupational migrations tend to be selective in terms of race, nationality and education.

Characteristics of migration

The conditions that cause migration can involve both 'push' (usually at the place of origin) and 'pull' (usually at the destination) factors.

Push factors include:
- increased mechanisation in agriculture
- low wages
- highly political, racial or oppressive governments
- natural disasters

Pull factors include:
- marriage
- family
- employment offers
- retirement.

Modelling

AQA	AS U1
OCR	A2 U3
Edexcel	AS U1
WJEC	AS UG2
CCEA	A2 U1

A series of models study **migration typologies**. They look at the distance travelled by a migrant, internal and international movement, the permanence of migration, the causes of migration and its selectivity.

KEY POINT

Ravenstein's Laws of Migration

These were developed on the basis of migration for the UK between 1871 and 1881. Ravenstein outlined eleven laws as follows.
- The majority of migrants go only a short distance.
- Migration proceeds step-by-step.
- Migrants going long distances generally choose a great centre of commerce or industry.
- Each current of migration produces a compensating counter-current.
- The natives of towns are less migratory than those of rural areas.
- Females are more migratory than males within the country of their birth, but males more frequently venture beyond.
- Most migrants are adults; families rarely migrate out of their country of birth.
- Large towns grow more by migration than by natural increase.
- Migration increases in volume as industries/commerce develop and transport improves.
- The major direction of migration is from agricultural areas to centres of industry and commerce.
- The major causes of migration are economic.

Peterson's typology of migration

This model identifies five classes of migration: primitive, forced, impelled, free, and mass move-ments, each with an activating force and initiator. Each class of migration is further sub-divided into conservative migrants and innovating migrants.

Zelinsky's mobility transition model

This is a five-phase model. In phase one there is just cyclic movement; in the second phase massive movement occurs; the third phase is where urban to urban migration surpasses rural to urban migration; phase four migration has levelled off; phase five the only real movements are temporary inter-urban movements. Zelinsky's model closely mirrors the DTM. Zelinsky argued that migration is on the whole an orderly event and, though his theory is for the most part untested, it certainly seems to fit the patterns of migration seen in the developed world.

Lee's laws of migration

This outlines why groups choose to migrate and summarises 'push and pull' ideas.
- There are factors linked to the destination of the migrants.
- There are factors associated with the origins of the migrants.
- Some intervening obstacles exist between origin and destination.
- Personal factors come into play.

Factors in the choice of destination include:

- cost of moving
- presence of friends and relatives
- employment
- amenities
- features of the physical environment
- assistance and subsidies
- information available
- lack of alternative destinations.

Socio-economic factors influencing migration include:

- technological change
- changes in economic practice and organisation
- propaganda
- regulations on migration
- healthcare and education
- population pressure
- climate and vegetation
- natural barriers
- size of countries.

Push and pull factors. Learn them!

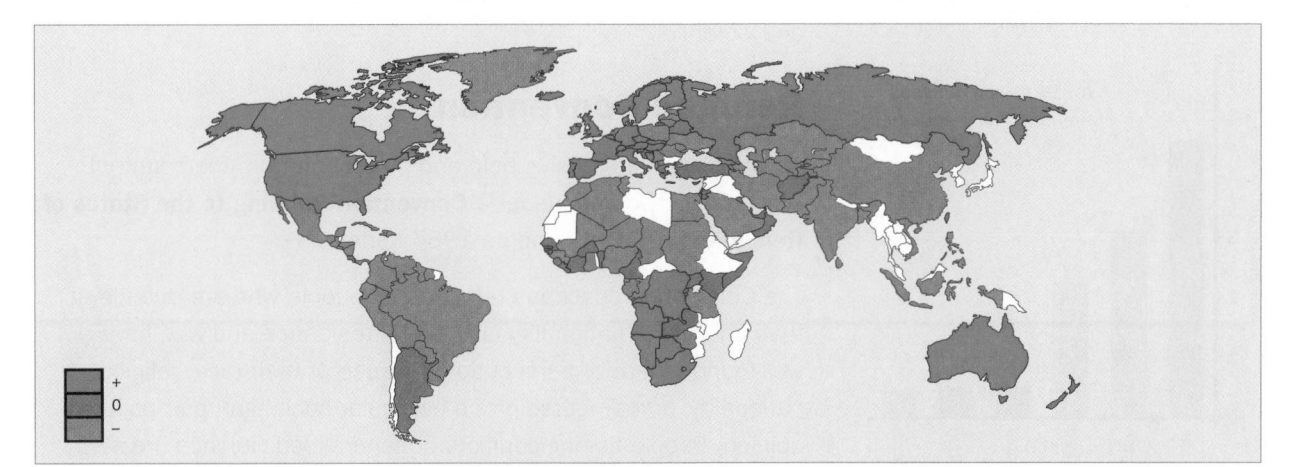

Positive and negative international migration rates around the world for 2009

The causes of migration

AQA	**AS U1**
OCR	**A2 U3**
Edexcel	**AS U1**
WJEC	**AS UG2**
CCEA	**A2 U1**

There are three main reasons why people leave their homes:
Political – repression or war forces them out;
Economic – they are forced to seek a livelihood elsewhere;
Disasters – natural or man-made disasters cause them to leave.

Migration is at an all-time high. In 2005 there were 191 million people on the move (3% of the world!). Those who cross national borders usually move to nearby countries, e.g. from Mexico to the United States. The flow of migrants is principally from less developed to more developed countries. In 2005, 62 million migrants from developing countries moved to more developed countries, but importantly almost as many (61 million) moved from one developing country to another, such as from Indonesia to Malaysia. People also move from one industrialised country to another, e.g. from Canada to the United States, and move from more developed to less developed countries e.g. Japan to Thailand.

Generally the world (and the UN) wants international migration to be voluntary, and has tried to limit **forced migration**, whether motivated by persecution or economic deprivation at home.

KEY POINT

Migrants: People who move from their home to another place. The migration may be internal or international. Migrants can return home if they wish to. There is no danger to their lives.

Refugees: Cannot return home immediately, or are in fear of their lives if they do so.

Asylum seekers: People seeking a place of safety after being persecuted in their own country.

Internally displaced persons: These migrants remain in their own country, but have been forced from their homes, usually by war.

Returnees: Refugees who return to their countries.

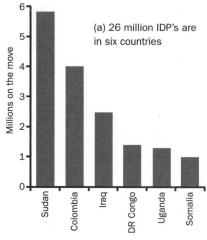

(a) 26 million IDP's are in six countries

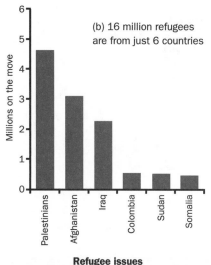

(b) 16 million refugees are from just 6 countries

Refugee issues

Persecution and conflict

There are currently 42 million people around the world that have fled armed conflicts and are searching for safety. They are children, women, and men living in temporary shelters, camps, or shanty towns, struggling to survive in new and often hostile environments. Those who have sought refuge in another country are **refugees**, a status which entitles them to certain rights under international law.

Those who are seeking refuge within their own countries are officially called **internally displaced persons (IDPs)**. They have fewer rights than refugees, yet make up almost two-thirds of the people around the world today who are seeking safety from armed conflict and violence.

Refugee Convention

Refugees should receive help and protection from the country to which they have fled, under a **Convention Relating to the Status of Refugees** from 1947 and its 1967 update.

The Convention describes refugees as people who are outside of their country of nationality or habitual residence and who have a well-founded fear of persecution because of their race, religion, nationality, or membership of a particular social group or political opinion. People fleeing conflicts or generalised violence are also generally considered as refugees and are therefore entitled to refugee rights.

KEY POINT

United Nations High Commission for Refugees (UNHCR), Geneva

- Set up in 1951 to address the problem of 1.2 million Ethiopian refugees left homeless after the Second World War.
- Works closely with governments and NGOs to implement humanitarian aid.
UNHCR aims to:
- Protect and assist refugees, i.e. ensure basic human rights.

- Seek lasting solutions , i.e. return refuges, arrange asylum or resettlement.
- Pay particular attention to the needs of children and the rights of women and girls.
- UNHCR's involvement may be long-term, as in Vietnam, or brief as in Mozambique.

The UNHCR is responsible for ensuring that these rights are respected and for finding solutions to refugee crises.

Refugee rights

- Humanitarian assistance, such as food, water, shelter, medical care and protection from violence.
- Refugees have the right to either return home when it is safe, to stay in the 'new' country or to resettle in another country – Liberians and Sierra Leoneans returned home following many years of civil war.

Internally displaced persons

The majority of people 'moved' by conflict are **internally displaced persons** (IDPs). For whatever reason they are unable make it to another country to find safety, seeking refuge within their own countries, in camps and shanty towns for instance. The Sudan, Colombia, and Iraq carry the most IDPs. Other countries with high number of IDPs include Somalia and the Congo. IDPs have no international agency to protect and assist them. The UNHCR has a limited role in terms of IDPs – they often end up being the target for rebels and have limited humanitarian help.

Case study: Migration issues

Operation Murambatsvina, Zimbabwe

Zimbabwe has experienced massive rural to urban drift since independence in 1980. Black people had not been allowed to 'move' to the city when the country was ruled by the white minority. With independence, demand for jobs and housing very quickly outstripped supply.

Knowing that Zimbabwe's rural to urban solutions depend upon a return to the land, the government has started to plan for a rural future for its people. To encourage the rural move the government started a clean-up campaign. In 2005 Operation Murambatsvina bulldozed houses and market stalls – estimates of the displaced varied from 300 000 to over a million and hundreds of thousands of people lost sources of income. With nowhere to live and an inflation rate of 1200%, people have struggled to make ends meet. The solution was to make a fresh start in the rural areas. There were tensions between the foreigners and the communities that the displaced are moving to. The new arrivals built homes and businesses on land that is not their own. There were many incidents of culture clashes and cases of theft, prostitution and public violence. White farmers were driven from their 'ancestral' farms by force to allow the Black settlers to build a new life.

The endless conflict, Afghanistan

During the Soviet occupation of Afghanistan in the 1980s and the civil war that followed, more than six million Afghans were displaced. Two million refugees have returned home since the Soviets left in 1989, while another 2.2 million remain in Pakistan and Iran – including 340 000 who have fled since the Taliban seized power in 1996. The latest conflict between coalition forces and the Taliban has seen a similar number move, remain and return to/from Iran, Pakistan and Tajikistan. As many as 8 million have been on the move at any time.

With the slightly more stable Karzai (the PM) government and conditions in Afghanistan the biggest problems the new regime faces are the returnees and deportees!

The Arab–Israeli conflict

Since 1948, some 3.2 million Palestinians have been at the centre of the Arab–Israeli conflict. Treaties, such as the Madrid Peace Process, the Oslo Process, the Jordanian–Israeli Peace Treaty and the Alon Plan, have all failed to find a solution. The wars, such as the Six-Day War, the Yom Kippur War and the Lebanese civil war, also all failed to find solutions to the conflict. For the Palestinians the 'right of return' is a central tenet, but the Israelis counter this by asserting they bear little moral responsibility for the plight of the Palestinian's and that de facto they accepted Jewish refugees from the Arab world.

Developments within the area over the last fifty years are unlikely to impact on the way any return is handled. Many refugees may well leave one camp to be sent into another! The Israelis are keen to continue a form of demographic security.

Like all refugee situations the Arab–Israeli conflict is a complex issue, Furthermore, it is unusual in that normally the poorest countries of the world carry the burden of refugee situations. The refugee challenge stemming from the second half of the twentieth century seems likely to continue long into the twenty-first century. The hope is that **conflict-driven** migrants/refugees will decrease. It will require a change in the UN charter, however, for the UN to intervene within countries before people are forced to move out.

Having facts at hand is useful when discussing refugees as there is so much mis-information about the subject.

> **KEY POINT**
>
> **Facts about refugees in the UK**
> - Asylum seekers account for only 3% of net immigration to the UK. The UK hosts 2% of the world's 10 million refugees.
> - Eleven refugees from the UK have won the Nobel Prize for science.
> - There are more than 1500 refugee teachers in England. It costs £10 000 to prepare a refugee doctor to practise in the UK. Refugees have the right to apply for protection in the UK.
> - A third of refugees do voluntary work in the UK.
> - The UK's refugee population amounts to 0.6% of the total UK population.
> - In Leicester over 30 000 jobs have been created by Ugandan Asian refugees since the 1970s.

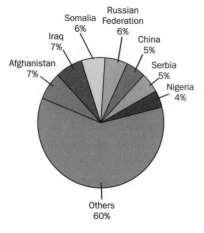

Main nationalities of asylum seekers, 2009

Asylum seekers

For the first time in many years the number of asylum seekers in industrialised nations was stable in 2010. The overall number of asylum seekers remained the same with 37 000 applications. Afghans top the list of asylum applicants with their submissions representing a 45% increase over 2009. Iraqis dropped to second; the Somalis moved to third position.

The USA is the main destination country for asylum seekers where there has been a marked rise in Chinese asylum seekers. Second was France, due to increasing claims from Serbians, originating predominantly from Kosovo. Canada, at third among receiving countries, saw a dramatic decrease in applicants. In the UK applicants have dropped, while Germany has seen increased claims for asylum. The five top destination countries received 48% of the total claims recorded in 2009.

Case study: Australia shuts the door on asylum

Australia has suspended refugee and asylum applications from Afghanistan. They are of the opinion that life for Afghanis has improved so much that they no longer need to seek protection elsewhere. Most Afghanis and Sri Lankans have arrived illegally by boat.

Many oppose the current Australian **fast-track** method of dealing with asylum seekers and Australia's human rights campaigners feel the Australian Government decisions are misinformed.

The reasons why people move

AQA	**AS U1**
Edexcel	**AS U1**
WJEC	**AS UG2**
CCEA	**A2 U1**

Reasons to move – economic

Increasing **globalisation** (the relative ease with which people can travel in the modern world) together with the wide and growing gap between rich and poor, is the catalyst, in many instances, for migrations based on the need for economic security. Within the world there are at least three established interfaces, where most economic migration occurs. There are a number of new interfaces developing.

Case study: Interface 1: USA and Mexico

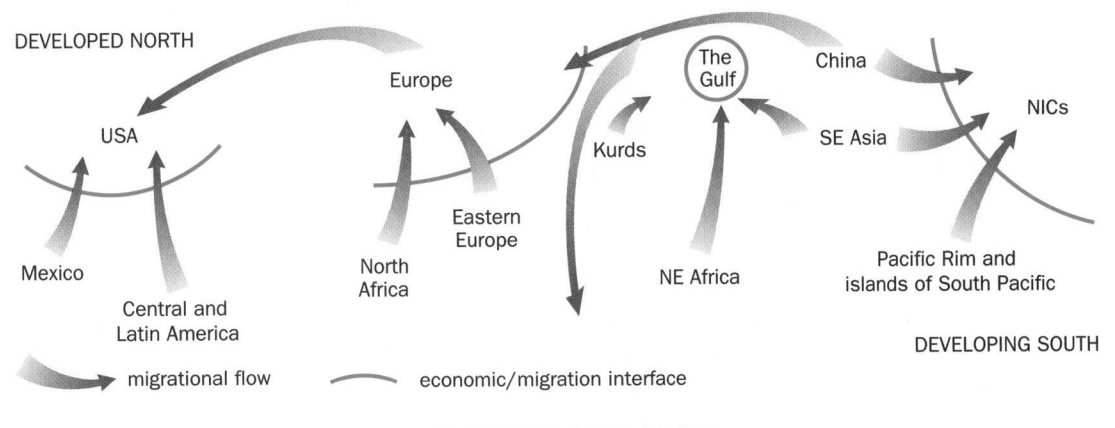

Economic and migration interface

> All these case studies are well documented. Read papers and browse the Internet.

Nearly 1:7 urban dwelling Mexicans are thought to be in the USA. Typically working in low paid jobs, they send half of their average wage of $400 home to Mexico each month. Remittances to Mexico exceed $30 billion a year. The 25 million Mexican-origin workers in America create a larger gross product than Mexico itself!

Mass migration from Mexico to the US is a relatively recent phenomenon. It has grown through century old social networks linking specific immigrant communities in the US to their hometowns in Mexico (e.g. working in the citrus fields of California, the mines and railroad systems of the north-east). Most of these networks have their roots in rural Mexico.

Remittances do help to boost Mexican living standards, but they change the character of Mexican life. In some towns with a long history of migration, leaving home to work in the USA has become a rite of passage for young men, often in place of completing school. Many of these towns are 'short' on men and dominated by single-parent households. The money flowing in reduces local incentives to work and fuels inflation.

The trip into the USA is dangerous and can be expensive (if an illegal entry is contemplated) so workers have to remain for some time in the USA to pay for their transit.

The USA has also realised that they can keep Mexicans in Mexico by setting up factories in the border areas. Here they produce cheap goods for the American market.

A growing number of East African immigrants have begun to travel the routes into the USA that traditionally the Mexicans have taken. Once at the border the asylum seekers look for the authorities and ask for asylum (Somali refugees say they are fleeing repression by armed militias). To get to the border may well have cost the Africans in excess of $10 000. There are about 87 000 Somalis in the country who have settled in cities like Minnesota and Seattle. A new and emerging refugee from Mexico is the 'narco-war' refugee; usually middle-class and wanting to escape the drug cartels and corrupt authorities.

Case study: Interface 2: Europe

The British government estimated immigration from the countries that joined the EU in 2004 at somewhere between 5000 and 13 000 people per year. By 2006 there were around half a million Eastern Europeans officially registered to work in the UK.

In the UK, Polish immigrants are seen as either the death knell for British values or as a boost to the industry as they attempt to expand with help from the hard working Poles.

Eastern Europeans, armed with higher wages and previously unheard of disposable incomes are learning to exercise their purchasing power. They have contributed to the British economy. Polish, and other Eastern European immigrants, are having a positive economic impact by making the UK workforce younger, more flexible and economical. This has also helped an otherwise waning buy-to-let market.

But there are issues. Peterborough, in north Cambridgeshire, claims it is being stretched to breaking point by the influx of Eastern Europeans, attracted to the area by the promise of high wages and decent living conditions in exchange for manual labour. But the city of Gdansk wants its workers back. Its shipyards are struggling without their skills.

Leaders in Gdansk have visited Peterborough to plead for the return of their workers.

The Polish economy has performed well since it joined the EU, helped by billions of pounds in EU grants for roads and other big infrastructure projects. Poland now offers better opportunities than the UK where unemployment is rising, wages are falling and there is growing resentment against foreign workers.

Case study: Interface 3: The Gulf

A large population of Indians live and work in the 'Oil Rich' Middle East. Most moved to work as labourers and in clerical jobs. They retain their Indian passports since most of the countries in the Gulf do not provide citizenship or a permanent residence for the Indians. The main reason they head to the Gulf is the incomes it can provide. In 2005, about 40% of the population in the United Arab Emirates were of Indian descent.

The majority send back considerable amounts to their dependents in India. It is estimated to be in the order of $10 billion/year.

The years 2002–06 were great for the property market in Dubai. Census records at the time indicated that 17% of the population of Dubai were UAE nationals, while the remaining population was distributed into the immigrants from neighbouring countries, expatriates (1.62 million) and temporary residents with work related or other visas.

With increasing numbers of businesses and multinational companies (MNCs) starting operations in Dubai, the inward migration of professionals and their families was increasing exponentially. The tax-free salary in Dubai was one of the major attractions for professionals from all over the world. Additionally, the wealth of job opportunities and leisure activities saw massive overseas investment and foreign tourism cashing in on Dubai.

With the recent global recession, everything has changed in Dubai. Increasing numbers of well-paid executives working in property (60% fall in property prices is predicted) and finance are heading home as they are made redundant and lending by the banks dries up. Many expatriates are abandoning their homes as their value has dropped. In the race to skip the country, and to avoid prison for defaulting on loans, they abandon their luxury cars at the airport (3000 every month).

Case study: Interface 4: Migration into the NICs

In Tonga, MIRAB provides half the national income!

Countries such as Japan, Korea, Taiwan, Singapore and Malaysia have highly restrictive immigration policies, though they are becoming more dependent on foreign migrants. In such countries, 35% of the foreign labour force are from south-east Asia. Migration from south-east Asia to Taiwan is referred to as **MIRAB** (migration, remittances, aid, bureaucracy).

In 2006, 1.5% of the population of Taiwan were registered as migrant workers, mainly from the Philippines, Vietnam, Thailand and Indonesia. Most are unskilled, are exploited, paid low wages, have bad living conditions and are referred to as 'resident aliens' by the Taiwanese. They tend to compete for '3D' jobs – 'dangerous, demeaning and dirty'.

During the 1980s, Taiwan experienced a 'brain drain', with many people emigrating to the USA. Despite this, Taiwan has become a major international power and its growth has encouraged emigrants to return home to help develop its manufacturing industry. The return of educated Taiwanese has also led to an up-and-coming technology industry.

Natural disasters, drought, sea level rise, melting glaciers, hurricanes and their effect on migration

Anthropogenic (man-induced) climate change will continue to show itself, through an intensification of environmental processes such as drought, sea level rise and the melting of glaciers. Climate change will modify rain patterns geographically and temporally, shifting the start of rainy seasons, as well increasing precipitation in some temperate areas and decreasing it in other regions like the tropics. This decline in rainfall may cause aridity and more frequent drought (migration as a way of coping in drought conditions). Significantly one third of the world's population is vulnerable to drought.

In north Mexico 60% of arid, or semi-arid, land suffers some form of erosion. The Mexican government estimates that 900 000 people/year have left these areas since the 1990s. Similarly in north-east Brazil there have been huge spikes in out-migration following poor crop yields after years of very severe drought.

It is claimed that anthropogenic climate change is also affecting sea levels. The greatest impact is likely to be in the very densely populated coasts of south-east Asia, e.g. from Tuvalu almost 30% of the population has already migrated to New Zealand. Approximately 50% of the Caribbean islands populations who live within 1.5 km of the coast are at particular risk from sea level rise.

> When you get scarcity of resources, population movement and poverty, you almost inevitably end up with conflict and destabilisation. It will make the world a more conflict-ridden place.

The melting of glaciers is causing problems in the Andes Mountains, e.g. with reductions in water availability for agricultural and the population, the numbers of fires during the dry season has increased and changes in rain patterns could mean greater flooding during the rainy season.

Climate change is also attributed to increased occurrences of certain natural hazards. The tropics have most of the developing countries of the world, but a combination of political, economic and social factors lower their populations' resiliency, ability and capacity to respond effectively to disasters.

Reasons to move – natural world

Compared to migrations brought about by political and economic reasons, displacements of population caused by natural disasters are small in number, but increasing. Typically floods, earthquakes and volcanoes, drought, famine and climate change have moved people.

Case study: Under the volcano, Montserrat

> Being able to link the natural environment into your understanding of migration is vital.

The residents of Montserrat, affected since July 1995 by the continually erupting Soufriere Hills Volcano, are settling into rebuilding a society in the safe northern corner of their tiny eastern Caribbean island. Two-thirds of the island remains out of bounds, still threatened by the volcano. Abandoned communities have become ash-ridden ghost-towns or have been buried under millions of tonnes of volcanic rubble. The capital, Plymouth, has been destroyed.

At the worst moments during the eruption, many Montserratians believed that the UK was deliberately seeking to 'depopulate' the island, either completely or sufficiently enough to reduce Her Majesty's Government's financial burden. The British always denied this, maintaining that given its priority to protect lives it had to make contingency plans for a worst-case scenario. Many Montserratians did take advantage of lucrative relocation monies and move away from the island.

Reasons to move – failure of agrarian economies

Lack of resources can also drive people to migrate. Deforestation and soil damage have forced many Africans to alter their agrarian life style and go in search of waged labour to supplement the family income. This is also a problem in India.

Case study: The Punjab diaspora

In an attempt to stave off starvation in India during the 1960s the state launched the **Green Revolution** which was spearheaded by Punjab. The Green Revolution specialisation of wheat-paddy crops brought tremendous growth in food production. The Punjab became the 'food-basket' for India.

Recently the Punjab has faced an acute crisis in agriculture that has manifested itself in stagnating productivity, rising costs of production, shrinking income, employment indebtness and ecological damage.

The production of wheat and rice has resulted in over-utilisation of natural resources which have had adverse environmental consequences with regard to the long-term viability and sustainability of the agrarian economy. Reduction of domestic and overseas demand of these crops, lower quality of production and declining international prices of these crops have made them uncompetitive in the global market.

What is needed now is change and investment based on diversification of agriculture and organic farming to save the Punjab's agriculture.

Reasons to move – development causes migration

Infrastructural projects may also cause massive displacement/migration/re-location. Globally millions of people a year continue to be 'moved' by such new projects.

Case study: Dam resettlement programmes

Human rights violations associated with the displacement of people for the construction of dams are growing. An estimated 30 to 60 million people worldwide have been moved from their homes to make way for major dam and reservoir projects. These refugees tend to be poor and politically powerless.

To build the Three Gorges Dam in China 1.14 million people were moved (at a cost of $7.9 billion) from their family homes. It is expected that 300 000 people living near the dam will be relocated to protect the environment by installing buffer belts to improve the water quality of the Yangtze River streams.

9.4 The business of migration

LEARNING SUMMARY

After studying this section, you should be able to understand:
- people have replaced drugs as the most lucrative trade item for criminals

Moving people for profit

AQA AS U1
Edexcel AS U1
WJEC AS UG2
CCEA A2 U1

People want to be 'safe' and many want a better life. As a result there is a growing trade in people **trafficking**. Hundreds of thousands want to 'escape' and are prepared to pay sums of money to achieve this. They may have to take a clandestine route to safety because of the exclusionary measures that have been put in place in the host countries: draconian visa regimes and penalties imposed on airlines and lorry drivers harbouring immigrants. For many, illegal entry into other countries is the only option. The trafficker becomes the migrants' lifeline!

Often the people who deal with 'the paperwork' are ruthless criminal gangs happy to arrange passage for the migrant by the dozen in the back of a refrigerated truck or even in the wheel bays of jet airliners!

With criminal gangs involved, failed immigrations are common, often leaving the migrant penniless and in debt. Few 'home' countries worry about this illegal trade however as 'illegals' contribute to **GNP** by sending money home. Those that fail to secure employment in the 'host' country frequently end up involved in crime, usually to pay back money for tickets and passage.

> Could it be that criminals are reacting more quickly to changing patterns of labour demand, ahead of governments?

On a global scale, the victims of human trafficking are used in a variety of situations, including forced labour (bonded labour or debt bondage), child labour (for purposes which include labour, military, adoptions, and commercialised sexual exploitation of children), sexual slavery, commercialised sexual exploitation and other forms of involuntary servitude. Virtually every country in the world is affected by these crimes. The challenge for all countries, rich and poor, is to target the criminals who exploit desperate people and to protect and assist victims of trafficking and smuggled migrants, many of whom endure unimaginable hardships in their bid for a better life.

KEY POINT

Only the UK, Europe, Canada and Australia have signed up to a protocol to protect the victims of trafficking. The most common destinations for victims of human trafficking are Thailand, Japan, Israel, Belgium, the Netherlands, Germany, Italy, Turkey and USA. The major sources of trafficked persons include Thailand, China, Nigeria, Albania, Bulgaria, Belarus, Moldova and Ukraine. In some areas, such as Russia, Eastern Europe, Hong Kong, Japan and Colombia, trafficking is controlled by large criminal organisations.

Case study: Mail-order trafficking into the UK

It is estimated that four million people are moved annually around the EU, generating a potential $5–7 billion for traffickers.

Individuals have been quoted from $4000 for transportation from Turkey to Dover and quotes of $25 000 for transportation from China to the UK. The trade in humans is a multi-billion dollar global industry, with many of the most popular locations within the EU.

Fujian province in China has had a criminal tradition since the 1970s with so called 'snake-head gangs' smuggling people into the UK, mostly to work in restaurant kitchens for the Hong Kong Chinese community. Alongside the illegal workers and gang-masters who make the headlines, a hidden parallel trade was emerging in girls recruited into the Chinese brothel network creeping across the UK. In 2009 £93 million was seized by UK authorities from an account set up to return money back to China.

Case study: The challenge of immigration in Europe

Many Africans are prepared to cross the Sahara in search of a job. Their goal is Ceuta, a Spanish enclave tucked into Morocco's Mediterranean coastline – a journey of thousands of miles. They hire the services of a trafficker, who takes their money, puts them in a lorry and dumps them at the border fence. Spain hung onto Ceuta and its twin enclave, Melilla, on entry into the EU. By 2015 it is estimated between 15 and 20 million migrants will have made a bid for Western Europe via these two Spanish territories.

There are social, political and economic factors which push people away from one country, towards another. Often the only way people can make this transition is by seeking the help of illegal traffickers. Conversely there are equally many factors which will pull migrants towards certain, often western, countries.

Dealing with clandestine migration is high on the EU's agenda. It is estimated that 1.5 million migrants enter the wealthy European states every year and that for every other migrant in the developed world another migrant is there illegally. The situation on the EU's frontiers has become so bad it is intended to implement Europe wide laws that will allow illegal immigrants 18 days to make arrangements to remove themselves from the country. After this point they can be put into detention centres for up to 18 months and then forcibly removed. For the EU the problem appears to be insurmountable. Frontiers are becoming harder to penetrate, but it is unlikely the trade in humans will stop. As long as there is money to be made a trafficker will be there to take it from unwitting 'migrants'.

9.5 Limiting migration

LEARNING SUMMARY

After studying this section, you should be able to understand:

- what can be done to prevent destabilising movements of refugees and immigrants
- that most people want to stay in their own countries if there is no persecution and a reasonable standard of living

Restrictive measures

AQA	AS U1
Edexcel	AS U1
WJEC	AS UG2
CCEA	A2 U1

Tighten asylum rules and fortify Europe?

Tightening asylum is hard to implement against a backdrop of open borders in Europe, but it is a policy that many favour.

In Europe in 2007 there were 249 600 asylum applications, in 2008 a further 283 700 and 286 700 in 2009. The main target for these applications was France followed by the UK and Germany. Many asylum claims are rejected. Several governments have come to power in Europe with tacit support from an electorate worried about rising immigrant numbers and potential home unemployment!

Continued adaptations of the migration and asylum acts will also begin to severely limit the UK's attraction.

If it is accepted that large-scale cross-border legal/illegal migration is a fact of life and here to stay, it is likely that the only effective method of managing the situation is to reduce the motivation to migrate in the first place. The UK, for instance, is increasingly putting 'obstacles' in the way of unfounded claims for benefit payments from immigrants and asylum seekers. The UK is increasingly less attractive to potential migrants.

Why return to Kosovo, Rwanda, Somalia, Sudan, Iraq and Iran?

Edexcel **AS U1**
WJEC **AS UG2**
CCEA **A2 U1**

Imagine the essay question! Do this with all your work/topics. Perhaps Why do displaced populations find it hard to return to their homes?

Resolving conflicts

Displaced migrants, probably want to live in a peaceful environment. War is one of the biggest causes of displacement and one of the hardest to resolve.

Patterns of prolonged displacement through forced migration have taught the world some lessons about the standards necessary for people to return to their homes – and to return for good.

- **Security** – refugees say that the most important condition for their return is safety. In any post-war situation there are many different security threats. Weapons, ammunition and explosives are rife and a disorganised society creates space for all kinds of criminal acts.

- **Documentation** – closely linked to the security question is the issue of personal documentation and the documentation of property. Many refugees are deprived of their personal documents before being forced out of the country.

- **Shelter** – housing has to be made available as many houses are destroyed during conflicts.

- **Infrastructure** – fighting and sabotage destroys infrastructure which is slow and costly to rebuild.

- **Income and social** – even when security, documentation, shelter and infrastructure are in place, people cannot return to a place where there is no likelihood of economic survival.

- **Prospect of lasting peace and commitments** – if there is little prospect of a lasting peace, few people will opt for repatriation.

- **Reconciliation** – possible internal conflicts among the returning population can also create problems.

- **Motivation and information** – correct, reliable, concrete and comprehensive information must be made available to refugees, IDPs and all those involved in relief and support efforts. The existence of accessible, independent media can make a decisive difference compared to a situation where IDPs refugees and returnees are left victims of propaganda, misinformation and prejudice.

- **Time** – the longer people are away from home, the more they integrate with and get accustomed to their new environment.

- **Distance** – the further away refugees have been settled from their homeland, the less chance there is for voluntary repatriation.

- **Overcoming trauma** – going back to the place where atrocities occurred is extremely difficult for some people. They need help.

- **Reducing migrationary pressure** – sustainable development that includes and encourages education, healthcare and sanitation, along with the provision of jobs, would greatly reduce migrationary pressure. Restoration of democracy is also important.

- **Increasing tolerance** – if 'host' countries can be encouraged to accept that immigration strengthens countries, illicit trafficking would be lessened and refugees more easily rehoused.

Law and order is increasingly difficult to accomplish in lawless societies and developmental aid does not assist the process.

The biggest migrationary problems are the result of complex historical, racial and political conflicts and issues. Migrational crises will continue until stable government is universal and economic migration is halted.

9.6 Population control

LEARNING SUMMARY	After studying this section, you should be able to understand:
	• why population policies are necessary and what they hope to achieve

Population policies

AQA **AS U1**
WJEC **AS UG2**
CCEA **AS U2**

When population projections are viewed they often spur governments into action. Policies influence growth, mortality, fertility, distribution and migration.

Anti-natalist policies

Anti-natalist policies encourage population control and include:
• providing contraception advice
• legalised abortion and late marriages
• economic and social measures to discourage large families, e.g. in China.

Pro-natalist policies

Pro-natalist policies encourage population growth. Generally governments of pro-natalist states believe that there is strength in numbers and the economy will prosper. Various countries have adopted such policies over the years.
• France with its 'La famille est prioritaire' scheme (Liberté. Egalité. Fertilité!)
• Malaysia and Germany (the 'Give the Chancellor a child scheme') are two of many countries that offer maternity benefits and tax concessions for those families that enlarge.

> Many are uneasy with the German scheme as it has similarities with the Nazi 'Give the Führer a child scheme'.

A final thought… in the time it took you to read this section 700 children have been born in India alone!

> ### Case study: Japan's 'Angel Plan' population policy
>
> In 1989, Japan's fertility rate hit a record low of 1.57, below the replacement rate of 2.08. In an attempt to raise their country's birth-rate the Japanese introduced the Angel Plan in 1994. The Japanese were worried that with decreasing fertility the government would need to allocate a lot more money on pensions, nursing homes and other programs to help the elderly.
>
> The Angel Plan offered support to working parents, such as counselling services and child day care centres. Sadly, the Angel Plan has not been successful. Most Japanese feel that child support was not reassessed within the scheme to help ease the expenses of raising a child. Japan is rapidly becoming the oldest human population in the history of mankind. Many now believe that the only possible solution is to encourage immigration into the country.

KEY POINT

- Pro-natalism is expansionist policy while anti-natalist is a control policy.
- Overpopulation is when the ability of an area to support a population is exceeded.
- Exponential growth is a population explosion.

PROGRESS CHECK

1. What are the typical traits of a migrant?
2. What are the three main reason people move?
3. Why was Polish migration made easier to the UK?
4. Why have Poles started to return to Poland?
5. What has gone wrong in the Punjab? Why are people leaving?
6. Japan introduced the 'Angel Plan'. Why?

Answers: 1. Aged 18 to 35/males and females move in MEDCs and males predominantly in LEDCs; most are single; professional migrants still predominate. 2. Political, economic and disasters. 3. Schengen Agreement and EU membership 4. Building work has declined, fewer jobs, changes in how agricultural practice is carried out, pound is weak. 5. Rising agricultural production costs, environmental damage, depleted ground water, reduction in agricultural demand etc. 6. Want more babies because the Japanese are worried that pensions would not be supported by small populations and that the labour force had diminished.

Exam practice questions

1 Study the photo which shows a shanty town on the periphery of Manila, the Philippines.

With reference to cities in LEDCs:

(a) Suggest what problems a new migrant might face on arrival in a large city. **[5]**

(b) Assess the solutions that could be employed to reduce the rate of urbanisation and thus reduce the problems created by it. **[5]**

10 Development issues

The following topics are covered in this chapter:

- Describing and measuring development
- Spatial differences in development
- Causes of disparity
- Aid and trade as forces for change
- Inequalities in MEDCs

10.1 Describing and measuring development

> **LEARNING SUMMARY**
>
> After studying this section, you should be able to understand:
>
> - that there is a wide development gap between the Northern and Southern hemispheres
> - that development is increasingly associated with human welfare and not just economic gains and this is reflected in the criteria chosen to describe a nation's wealth/state or stage of development

The development gap

AQA	A2 U3
OCR	A2 U3
Edexcel	A2 U3
WJEC	A2 UG3
CCEA	AS U2

Many of the globe's populations struggle to develop the resources that lead to a dignified and productive life. This struggle is often called '**development**'. This struggle is also viewed as a process of change operating over time. You will be aware of the fact that the countries of the world have not developed equally. The focus of much of the development work carried out in the latter part of the last century, and well into this new century, will be to ensure that regions develop equitably, with adequate food supplies, medical services and educational opportunities, under regimes that value social justice and political and economic freedom and quality of life. Sustainable development, too, will be high on the agenda.

> Development is difficult to define. Is it about human welfare, environmental sustainability or economic growth?

Development is a complex process driven by many variables. On the one hand the problem is **economic**, while on the other it is **political**, **environmental** and **social**. It is a process operating in the context of growing disparity between nations and within poor nations, the so-called **development** gap.

Evidence for the development gap

- Food and malnutrition issues (malnutrition and infant mortality).
- Resources imbalance (food insecurity).
- Health issues (safe water availability and access to disease control).
- Educational opportunities and issues (illiteracy rates).
- Demography issues (high birth rate (BR) and high infant mortality).
- Child exploitation (child labour).
- Gender issues (women in low paid work).

- Economic migration (brain drain).
- Increases in conflict (e.g. Darfur).

Closing the gap

The Millenium Development Goals (MDGs) are eight development goals that all 192 United Nations (UN) member states and 23 international organisations signed up to improve social and economic conditions in the world's poorest countries. There are eight goals with 21 targets to be achieved by 2015.

To accelerate progress towards the MDGs, finance ministers met at Gleneagles, Scotland, for the **Gleneagles Summit** in 2005 and reached an agreement to provide enough funds to the World Bank, the International Monetary Fund (IMF) and the African Development Bank (ADB) to cancel a $40–55 billion debt owed by members of the **Heavily Indebted Poor Countries (HIPC)**. This was later to be called the **Multilateral Debt Relief Initiative (MDRI)**. The MDRI supplements HIPC by providing each country that reaches the HIPC completion point 100% forgiveness of its multilateral debt.

> Solutions to the world's development problems are being addressed.
>
> **Goal 1:** Eradicate extreme poverty and hunger.
> **Goal 2:** Achieve universal primary education.
> **Goal 3:** Promote gender equality and empower women.
> **Goal 4:** Reduce the child mortality rate.
> **Goal 5:** Improve maternal health.
> **Goal 6:** Combat HIV/AIDS, malaria and other diseases.
> **Goal 7:** Ensure environmental sustainability.
> **Goal 8:** Develop a global partnership for development.

Describing development

AQA	**A2 U3**
OCR	**A2 U3**
Edexcel	**A2 U3**
WJEC	**A2 UG3**
CCEA	**AS U2**

The problems of describing development data dates from the Cold War. At this time the USA and its allies were termed the **First World** and Russia and China as the **Second World**. The so-called **Third World** were the southern countries.

Nations with high living standards are said to be developed – they have gone through the trials of development. Those that currently negotiate the barriers to development are said to be developing. In the near past the poor nations have also been 'labelled' the Third World, the under-developed world and so on! The latest derivation is MICs and LICs (more and lower income countries).

In 1980 Willy Brandt, Chancellor of West Germany, was tasked by the UN to both clarify and to suggest solutions to global inequality. In his report *North-South: a Programme for Survival* an urgent plea was made for change. The report was updated in 2001 in a report called the *Brandt Equation*; his original report established the North *vs* South divide of the world.

> To Western observers industrialisation holds the key to development, hence the tag **less economically developed country** (LEDC). The industrialised West was tagged the **more economically developed countries** (MEDCs).

Trends since the Brandt Report

> The Brandt Report suggested that 80% of the world's wealth exists north of the Brandt line and 20% beneath it.

> The positives of Brandt's map are that it simplifies and mirrors gross domestic product (GDP) data. Some generalisations in that some countries above the Brandt line are LEDCs.

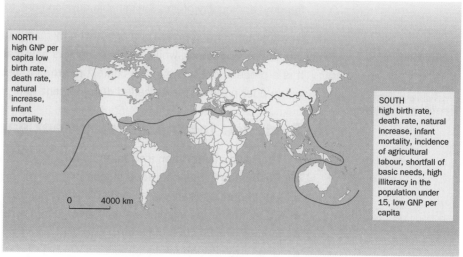

NORTH high GNP per capita low birth rate, death rate, natural increase, infant mortality

SOUTH high birth rate, death rate, natural increase, infant mortality, incidence of agricultural labour, shortfall of basic needs, high illiteracy in the population under 15, low GNP per capita

0   4000 km

The Brandt line dividing the North from the South

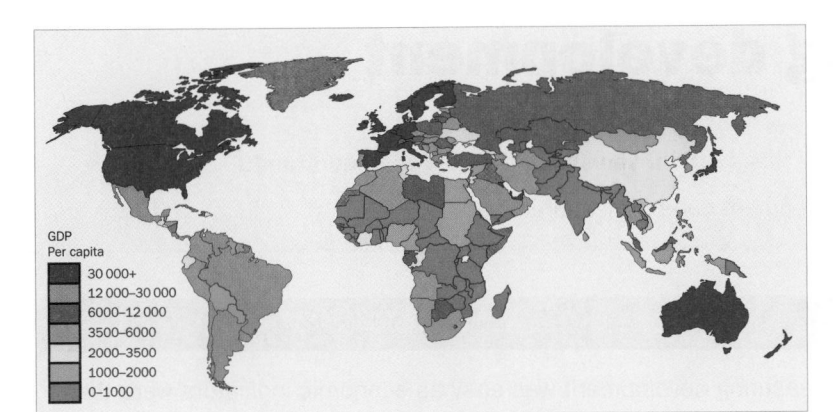

GDP per capita by countries, 2009

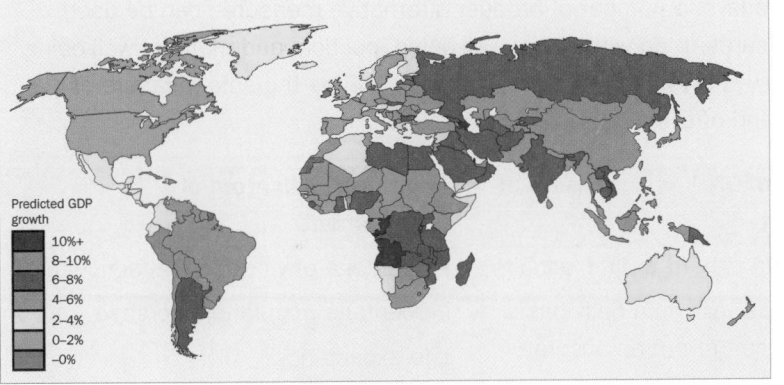

GDP predicted growth

Much has changed since the 1980s. The top map shows GDP in 2009. The bottom map shows GDP predicted growth over the next five years. It is not the countries of the North that are showing the growth. It is the countries of the South! Why have things changed?

Positive

- Many countries have increased production and standards of living have improved.
- The Cold War has ended.
- Military expenditure and arms sales in/to the South have declined.
- Democracy is supported around the world. Colonialism has all but vanished.
- Apartheid has been abandoned.

Negative

- Absolute poverty has increased: 1.4 billion people live on $1/day.
- Children continue to die through hunger.
- Population growth is not under control.
- Environmental threats are becoming more alarming.

PROGRESS CHECK

1. What is the development gap?
2. What is the evidence for the development gap?
3. What is the reasoning behind the millennium goals?
4. In the past developing countries have been termed Third World and LEDC's. What is the latest derivation?

Answers: 1. Disparity between nations and within nations. 2. Malnutrition issues, resource imbalances, health issues, child exploitation and conflict. 3. To accelerate movement towards important economic and social conditions by 2015. 4. LICs – lower income countries (at the other end of the scale HICs are higher income countries.)

10.2 Measuring development

LEARNING SUMMARY	After studying this section, you should be able to understand:
	• how development can be measured

Measuring development

AQA   **A2 U3**
OCR   **A2 U3**
Edexcel **A2 U3**
WJEC   **A2 UG3**
CCEA   **AS U2**

In the past measuring development was easy as economic indicators were the sole indices used. **Gross National Product (GNP)**, the total value of goods and services produced by the country divided by population, was the principal indicator. Nowadays, a number of broader alternative measures can be used including measures to do with social well-being, political and material well-being. The table below charts the limitations of using GNP as the sole measure of development and offers a range of alternatives.

> Within nations, individuals have access to wealth, power and privilege.

> Costa Rica is an LEDC based on its GDP. By any other measure it's not!

> Measurements made with this scheme ignore unpaid, subsistence and informal work.

> Other problems of using these measures: accuracy of measurement, figures hide regional variations and using the US$ hides exchange rate issues.

Limitations of GNP as a measure of development	Additional indicators of development
Accurate data is hard to find/acquire.	Percentage of workforce in agriculture.
Currency rates vary on a daily basis and it is difficult to put an accurate value on goods.	Percentage population increase.
	Life expectancy.
	People : doctor ratios.
Subsistence industries and farming production do not enter the market economy and are therefore not measured.	Percentage with access to safe water.
	Percentage unemployed.
	Calories per capita per day.
Per capita GNP hides the internal distribution of wealth in a country.	HDI (Human Development Index).
	Literacy rates.
GNP is always given in $US.	Percentage of school enrolment as a % of relevant age group.
GNP reveals nothing about quality of life and well-being (ditto GDP).	

Clearly no single measure, such as GNP, can adequately indicate development. Combination or composite indices are far more illustrative of development.

> Don't worry about learning figures exactly, although great if you can! But know the trends.

Measures of quality of life across countries			
	Burkina Faso, West Africa (LEDC)	Taiwan, East Asia (NIC)	UK (MEDC)
Demographic indicators			
Life expectancy	46	76	77
Population density (km^3)	57	834	254
Economic indicators			
GDP ($)	1304	31 834	34 619
Poverty (below % line)	46	<0.5	<0.5

	Burkina Faso, West Africa (LEDC)	Taiwan, East Asia (NIC)	UK (MEDC)
Social indicators			
Adult literacy (%)	24	98	99
Calorie intake	2360	3060	2940
Environmental			
Carbon dioxide emissions (tonnes/person)	0.1	8.2	9.4
Political indicators			
Homicide/1000	4	3	2
Composite indicator			
HDI (higher = better)	.3	.72	.94

Do remember that development is not just about economic development in the world's countries!

KEY POINT

Ways of measuring development and quality of life include:

- **Quantitative measures** – demographic, economic, social, political and combined measures.
- **Qualitative measures** – by classification (e.g. north and south) as stairway groups (e.g. MEDCs, etc), as groups (e.g. Organization of the Petroleum Exporting Countries (OPEC)).

Composite measures of development

The Physical Quality of Life Index (PQLI)

The Physical Quality of Life Index (PQLI) was devised in the 1970s by the Overseas Development Agency, as a means of determining 'need', aid requirements, etc. The value is the average of three statistics: basic literacy rate, infant mortality, and life expectancy at age one, all equally weighted on a 0 to 100 scale. The higher the number the more developed the country.

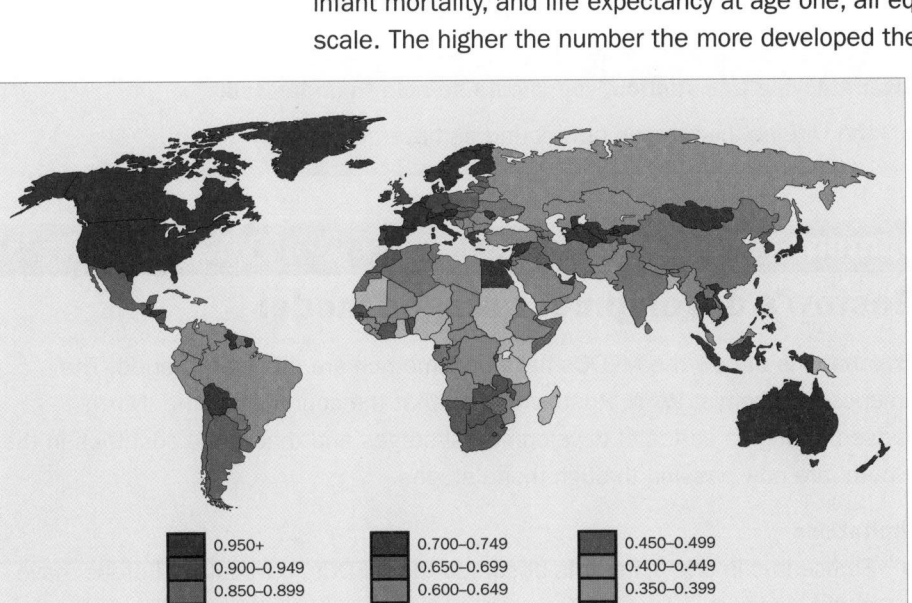

Physical Quality of Life Index

0.950+	
0.900–0.949	
0.850–0.899	
0.800–0.849	
0.750–0.799	
0.700–0.749	
0.650–0.699	
0.600–0.649	
0.550–0.599	
0.500–0.549	
0.450–0.499	
0.400–0.449	
0.350–0.399	
–0.350	
Unknown	

Human Development Index (HDI)

Devised by the United Nations (UN) in 1990, the HDI uses three development indices to assess development:

- real per **capita income**,
- a measure of **adult literacy**
- life expectancy at birth. Countries are placed on a scale from 0 to 1 – the nearer they are to 1 the more developed the country is.

Alternative methods: the World Bank uses the development diamonds approach to assess levels of development in countries.

Problems with social measures of development include:

- they do not reflect inequalities in income distribution
- there is no agreement over which social indicator to use
- rights of women and freedom of speech cannot be included
- development is about inter-dependence between countries – how can this be measured?

There are other measures of development; how these contribute to the branding of the North/South divide will have to be seen. The rebranding now calls the divide the **development continuum**. It is felt that this will close the gap between the rich and the poor. The HDI does this effectively. This is well exemplified in **Rostow's model**.

So-called **development diamonds** portray relationships among four socio–economic indicators for a given country relative to the averages for that country's income group (low-income, lower-middle-income, upper-middle-income or high-income). **Life expectancy** at birth, **gross primary** (or secondary) **enrolment**, **access to safe water**, and **GNP per capita** are presented, one on each axis, then connected with bold lines to form a polygon. The shape of this 'diamond' can easily be compared to the reference diamond (see examples left), which represents the average indicators for the country's income group, each indexed to 100%. Any point outside the reference diamond shows a value better than the group average, while any point inside signals below-average achievement.

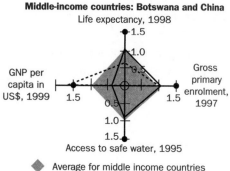

Low-income countries: Ethiopa and India

◆ Average for low income countries
---- Ethiopa — India

Middle-income countries: Botswana and China

◆ Average for middle income countries
---- Botswana — China

Development diamonds for selected countries

10.3 Spatial differences in development

LEARNING SUMMARY

After studying this section, you should be able to understand:

- the various theories on how countries have developed

Theories of development

AQA	A2 U3
OCR	A2 U3
Edexcel	A2 U3
WJEC	A2 UG3

Rostow's development stages model

Explanations of how the MEDCs have become rich are many and varied. The American economist W. W. Rostow argued that the countries of the 'North' passed through a series of developmental stages and that those countries in the 'South' are now passing through these stages.

Limitations

- Eurocentric in origin, i.e. it is assumed the MEDCs, particularly Europe, have all the answers.
- How accurately can the economies of the twentieth century North be compared effectively with, e.g., Guyana today?
- Cash offers and injections have failed to bring about take-off in many LEDCs.
- Late developers do not have the resources that were available to the developed countries when they industrialised.

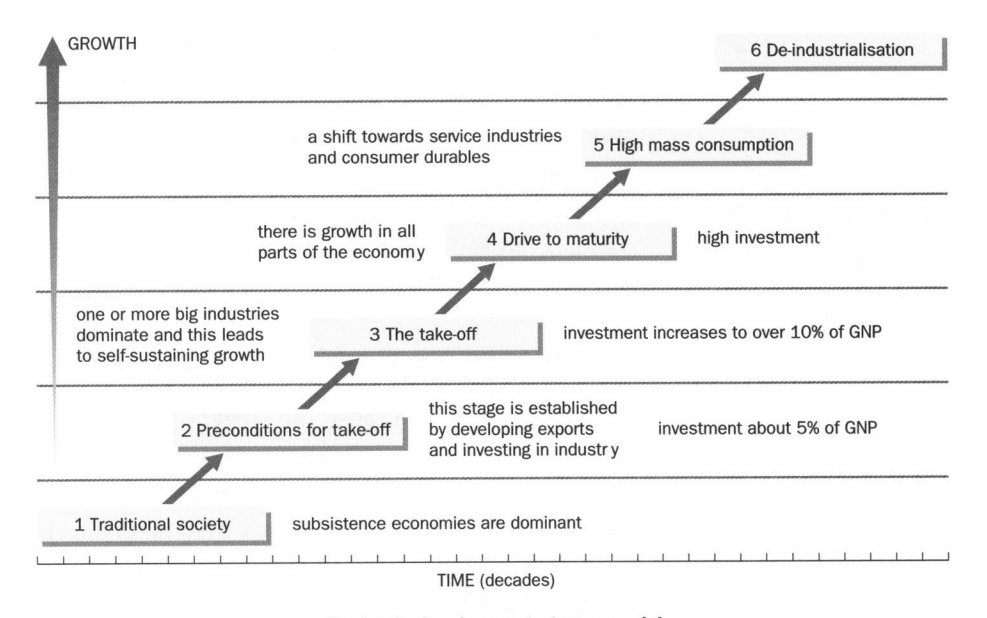

Rostow's development stages model

Kondratieff waves

In the Kondratieff system five stages of economic development are recognised:

- K1 = early industrial development
- K2 = the age of steam power
- K3 = electrical engineering
- K4 = introduction of mass production
- K5 = IT and biotechnology and globalisation.

This system identifies development and dereliction stages in all industrial developments. Government involvement is identified throughout the process.

Frank's dependency theory

The **development of underdevelopment** is a well-worn statement in development circles, one recognised by André Frank, who believed the MEDCs (of the North) have insidiously bled the LEDCs (of the South) by exploiting the people and the resources. His argument originates from the sixteenth century (the so-called colonial period) and explains why the developing world became poor and has stayed poor.

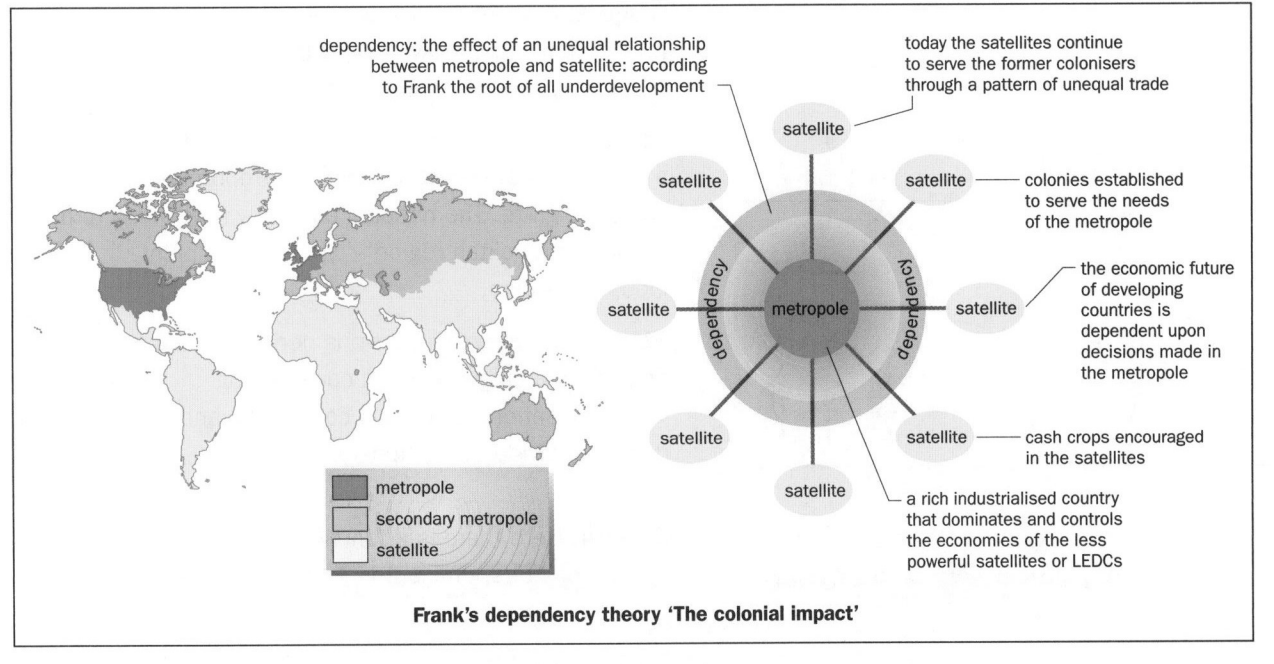

Frank's dependency theory 'The colonial impact'

Blaut and neo-colonialism

James Blaut used the term neo-colonialism to describe the relationship between developed and developing countries at the present time in his book *The Theory of Development*. To his mind although LEDCs are **politically independent** of the MEDCs, they are still **economically dependent**. Banks, investment, aid, exports and trade are all controlled by the MEDCs. In some countries the MEDCs still play a big part, e.g in Sierra Leone, the UK, in Rwanda – the French and Dutch and so on.

> Benefits are brought by colonialism, i.e. roads, rail, health services and schools.

Economic dualism – Myrdal and Friedmann

The colonial era encouraged the growth of 'modern' expanding and capitalist economies that formed cores of commercial activity within the much larger traditional economy. As the core areas enlarged, the economically weak traditional areas shrank in importance. Both Myrdal and then Friedmann attempted to explain this phenomenon. Myrdal attempted to explain and decipher the causes of regional inequality.

> Models are representations of reality. Examiners love them!

rich area (traditional area expands and develops) → poor area (rapid change)

growth area

area of great attraction and influx

core site area

comparative advantage offered by the original site (fuel, materials and labour)

PERIPHERY

new industry attracted from the periphery

feedback from growth area = backwash effects

spread effects make original site grow, it becomes attractive to the periphery

Myrdal and Friedmann's economic dualism

Stage 1: A series of independent local centres with no hierarchy. Each town lies at the centre of a small region – typical pre-industrial structure.

Stage 2: A single strong core emerges, together with a periphery from which potential entrepreneurs and capital migrate. The national economy is reduced to a single metropolitan region – typical of incipient industrialisation.

Stage 3: There are now strong peripheral sub-cores as well as a single national core. During industrialisation, secondary cores form and in so doing subdivide the periphery.

Stage 4: Here there is a functional interdependent system of cities, efficient in location to give maximum growth potential.

Friedmann's development model

Case Study: Venezuela

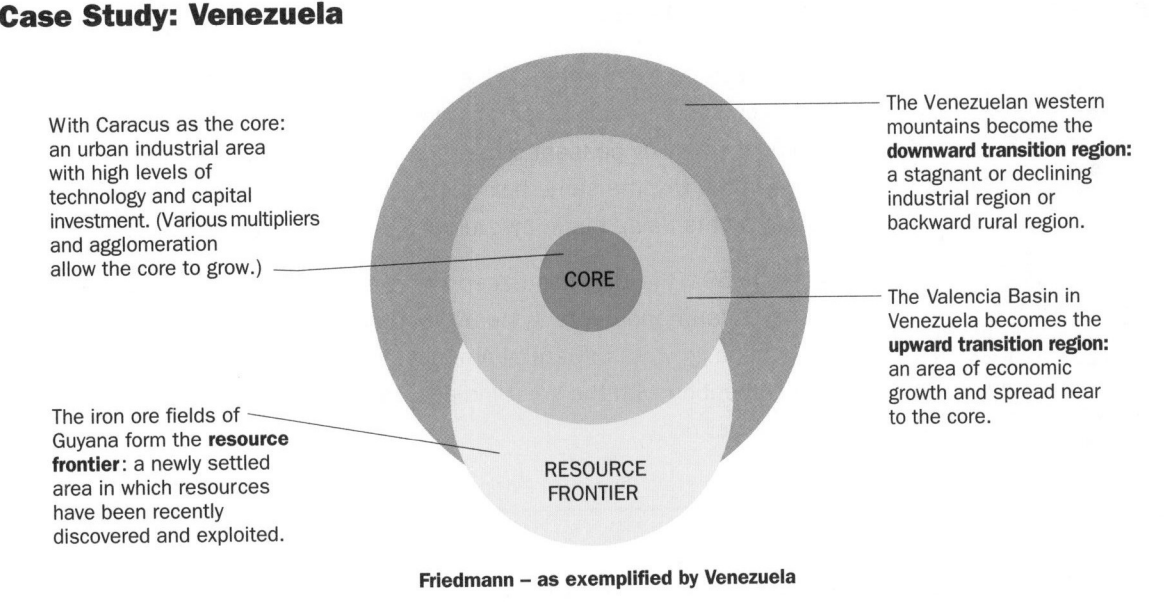

With Caracus as the core: an urban industrial area with high levels of technology and capital investment. (Various multipliers and agglomeration allow the core to grow.)

The Venezuelan western mountains become the **downward transition region:** a stagnant or declining industrial region or backward rural region.

The Valencia Basin in Venezuela becomes the **upward transition region:** an area of economic growth and spread near to the core.

The iron ore fields of Guyana form the **resource frontier**: a newly settled area in which resources have been recently discovered and exploited.

Friedmann – as exemplified by Venezuela

Stage 1: Cities are few, scattered and small. Most roads run North to South.

Stage 2: Oil is discovered on and near the coastal strip.

Stage 3: Caracas grows as the single national capital.

Stage 4: Migration flows from the periphery (a backwash effect).

Stage 5: The core becomes attractive to investors and dominates the country while the periphery stagnates.

Stage 6: The core's popularity spills over into the periphery as oil development continues; sub and secondary cores are established.

Stage 7: Policy change to develop the periphery.

The Marshall reparation plans for MEDCs in Europe in the 1940s became the blueprint for LEDCs in the 1980s. However, pumping large sums of money into LEDCs has not helped their plight. This is called a 'top-down' strategy.

Governments have largely accepted the concept of core against periphery. It is recognised that regional imbalances exist and a range of socio-economic reasons have to be addressed.

The cycle of poverty theory

The cycle of poverty theory suggests that LEDCs will never reach the higher levels of development as they are too overly dependent upon MEDCs to do so. However, experience suggests that many of the large scale schemes, aimed at relieving problems in LEDCs, actually make problems worse, e.g. dam building in Brazil, has caused malaria to increase and any benefits that have accrued have been small and focused on the wrong groups of people.

Dependency theory

Dependency theory suggests that the colonialistic tendencies of the past are perpetuated in the economic and political power that can today be exerted by MEDCs through agreements such as the General Agreement on Tariffs and Trade (GATT) or through the edicts of the World Trade Organization (WTO).

Case study: The secret life of a banana

Only 7% of bananas brought into Europe come from the Caribbean.

Unfair trade can be exemplified by the situation on the West Indian island of Saint Vincent which has one main export commodity – bananas. The bananas are grown almost solely by peasant farmers, who receive about £2 for a 40 lb box of fruit. Once in the UK, the same box of bananas could sell for more than £50, resulting in profits being made by everyone except the farmers.

The USA filed complaints, through GATT, against the EU schemes giving banana producers exclusive access to Europe's markets.

During the 1950s, the UK encouraged the West Indies to grow bananas and established a tariff and quota system to protect the small banana farmers against banana growers in Central America. But the WTO has ruled that it is unfair to deny competition with the West Indians, meaning they lose out whilst the North continues to get rich.

10.4 Causes of disparity

LEARNING SUMMARY	After studying this section, you should be able to understand: • that the political and economic relationships between rich and poor countries reinforce the dependency of LEDCs on the MEDCs • that capitalism increases poverty and inequality in LEDCs

Historic origins

AQA	A2 U3
OCR	A2 U3
Edexcel	A2 U3
WJEC	A2 UG3
CCEA	AS U2

Capitalism

Capitalism has led to inequalities. The capitalist world relies upon a constant requirement to produce capital – money. The nature of most capitalism means that countries that have a **comparative advantage** (the developed countries) to accumulate money through 'taking' capital from the least advantaged (the developing countries of the world).

Comic Relief, Sport Relief and a range of other money raising activities achieve about £25 million each year. This is about half the amount paid back in interest payment every day!

KEY POINT

Put simply ...

In the 1960s the USA spent more than it earned → printed more $s → value of $ fell → oil is priced in $s therefore export value of oil dropped → 1973 oil prices were hiked up considerably → large sums invested in world banks worldwide → interest rates plummeted as money was lent out too quickly → to re-coup losses the economies of the developing world are targetted → lavish amounts are lent, at rates below inflation, and with no thought as to how it was to be used or repaid → some was used to repay other debts, about one-fifth went on arms, much was spent on development projects that proved of little value. By the mid–1970s, encouraged to grow the same cash crops, the LEDCs found that they weren't getting the prices they expected for their exports → interest rates rose → oil increased in price again → the trap was sprung! LEDCs were earning less, paying more on loans and had to borrow more to pay off interest → effectively the LEDCs become bankrupt!

Abbreviating your notes in this way and then highlighting them is a must.

The map below shows some of the world's most highly indebted poor countries (HIPCs).

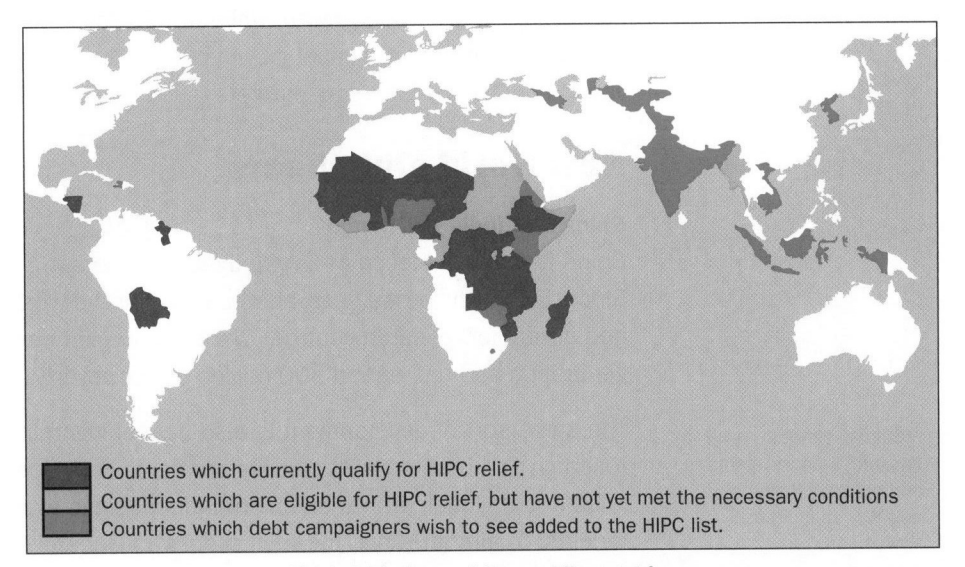

Countries which currently qualify for HIPC relief.
Countries which are eligible for HIPC relief, but have not yet met the necessary conditions
Countries which debt campaigners wish to see added to the HIPC list.

The indebtedness of the world's countries

Why globalise?
- To save labour costs in LEDCs.
- To gain access to new markets.
- To meet the demands of investors.
- Tariffs blocks favour home products.
- New technology helps.
- Transport has improved – containers; deregulation.

Globalisation

Globalisation is possibly the number one cause of global inequality. Globalisation has the effect of increasing social and economic gaps as it needs economies to adapt in a very fast way and this almost never happens equally. Some nations will always grow faster than others. Rich countries exploit developing countries to the point where developing countries become dependent on the rich ones. So globalisation only serves to perpetuate the divisions between rich and poor.

The diagram shows flows of trade around the world – what is important to be aware of is the lack of trade flowing back from the developing world.

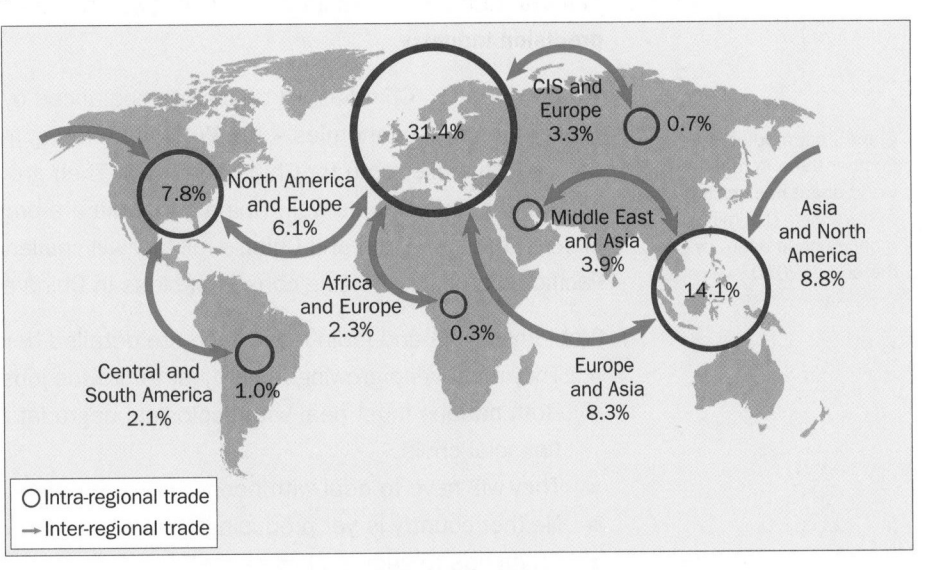

Inter-regional flow

Flows of trade around the world

AQA	**A2 U3**
OCR	**A2 U3**
Edexcel	**A2 U3**
WJEC	**A2 UG3**
CCEA	**AS U2**

The dominance of MEDCs is changing – the **two-speed world** now rules the global economy.
- Developed economies → slow growth.
- Developing economies → fast growth.

The two-speed world describes advanced economies losing growth momentum, while developing economies are flourishing.

In a report titled *Preparing for a Two-Speed World: Accelerating Out of the Great Recession* (2010), it was stated that developed economies would experience a continuing slowdown in growth, while developing economies would maintain rapid expansion. It projected annual growth rates of less than 2% for the USA, Europe and Japan until 2015. Emerging markets, such as China and India, will return 6–8%.

New world globalisers

China and India – 'Chindia'

China is a global mecca of knowledgable workers in an information economy. Since 1990, China has been able to boast of the world's biggest industrial zones and dozens of research centres. China has tripled per capita income in a generation and has eased 300 million out of poverty.

> Motorola, Hewlett-Packard, Cisco Systems and other tech giants now rely on India to devise software platforms for next-generation devices.

The same sort of development is also happening in India where office towers and research and development centres are being developed. Indians play invaluable roles in the global innovation chain. India designs everything from car engines, to forklifts, to aircraft wings for such clients as General Motors and Boeing Co. In 2010 such **outsourcing** work is expected to quadruple to $56 billion a year. Bangalore is rapidly becoming India's Silicon Valley.

For two decades, China has been growing at 9.5% a year, and India by 6% and they are expected to continue to grow at about 7–8% range for decades. In the next 30 years India will have overtaken Germany as the world's third biggest economy and by 2050 China should have overtaken the USA as No. 1. By then, China, India and the USA will account for half of global output.

China and India complement one another. China will stay dominant in **mass manufacturing** whilst India is a rising power in **software**, **design**, **services** and **precision industry**.

> China's car market is number three in the world and it is expected to have over 700 million mobile phone users by the end of 2010.

What's good for 'Chindia' will be counter–balanced by layoffs and lower pay for workers in other economies – the West is suffering from 'future shock'. The West will have to make room for China and India. Their growing economic might well carry into geopolitics as well. China and India are pressing their interests in the Middle East and Africa and China's military will challenge US dominance in the Pacific. In addition, a rising consumer class in China and India will drive innovation.

Problems China and India might face are detailed below.

- They must keep growing rapidly just to provide jobs for the expanding workforce.
- Both nations must deal with ecological degradation, social strife, war and financial crisis.
- They will have to deal with boom and bust cycles.
- Neither country is yet producing companies with world-beating products.
- China has to address low wages.
- China's working-age population will peak at 1 billion in 2015 and then shrink steadily. China then will have to provide for a greying population.
- China is surprisingly weak in innovation – they have yet to yield many commercial breakthroughs.
- There is a lack of intellectual property protection in China.
- China also is hugely wasteful – its factories are not energy efficient.
- India has nearly 500 million people under age 19 and higher fertility rates.

> By mid-century, India is expected to have 1.6 billion people and 220 million more workers than China. That could be a source for instability.

Can India reproduce the Chinese miracle?

India has had to develop with scarcity. It gets little foreign investment and has no room to waste fuel and materials like China. India also has Western legal institutions, a modern stock market and private banks and corporations. Indian corporations have achieved higher returns on equity and invested capital in the past five years in industries from cars to food products.

Export manufacturing is one of India's best hopes of generating new jobs. India has sophisticated manufacturing knowhow, but what holds India back is red tape, labour laws, and the inability to build an infrastructure fast enough.

In the coming decades, China and India will disrupt world workforces, industries, companies and markets. How these Asian giants integrate with the rest of the world will largely shape the twenty-first century global economy.

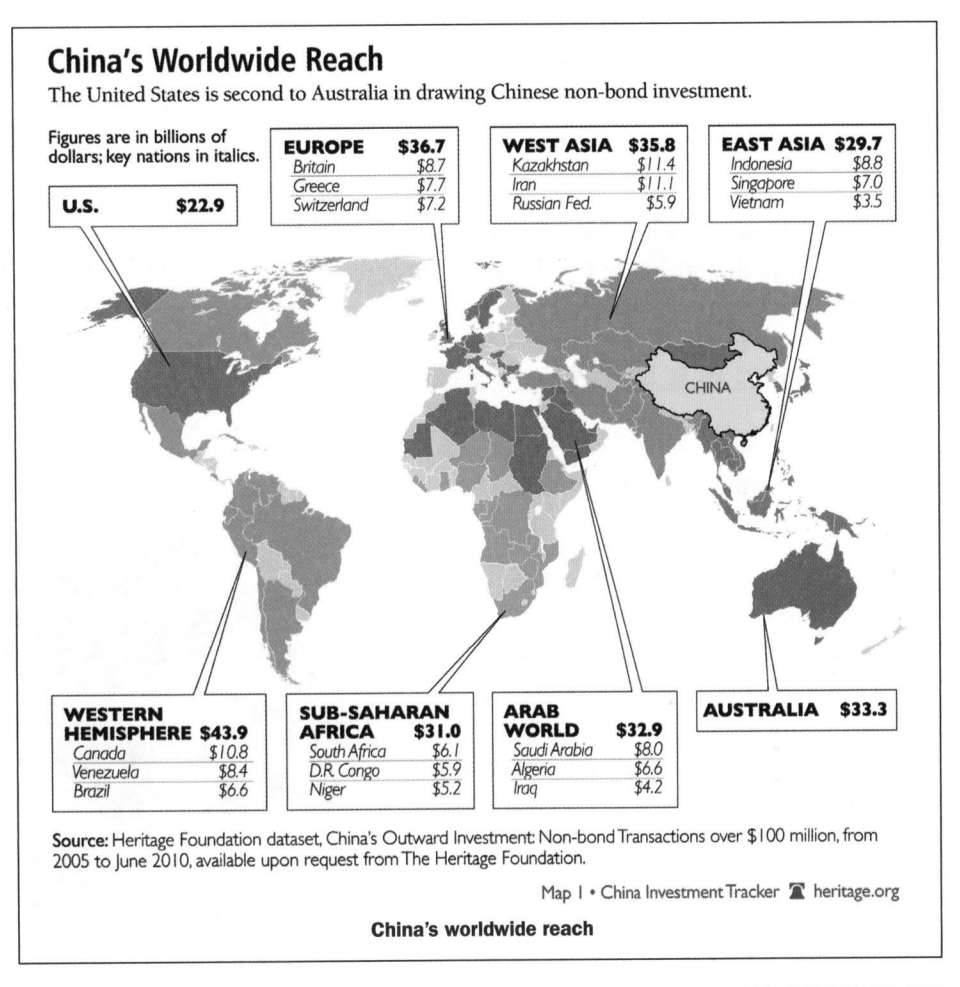

China's Worldwide Reach

The United States is second to Australia in drawing Chinese non-bond investment.

Figures are in billions of dollars; key nations in italics.

U.S. $22.9

EUROPE	$36.7
Britain	*$8.7*
Greece	*$7.7*
Switzerland	*$7.2*

WEST ASIA	$35.8
Kazakhstan	*$11.4*
Iran	*$11.1*
Russian Fed.	*$5.9*

EAST ASIA	$29.7
Indonesia	*$8.8*
Singapore	*$7.0*
Vietnam	*$3.5*

WESTERN HEMISPHERE	$43.9
Canada	*$10.8*
Venezuela	*$8.4*
Brazil	*$6.6*

SUB-SAHARAN AFRICA	$31.0
South Africa	*$6.1*
D.R. Congo	*$5.9*
Niger	*$5.2*

ARAB WORLD	$32.9
Saudi Arabia	*$8.0*
Algeria	*$6.6*
Iraq	*$4.2*

AUSTRALIA $33.3

Source: Heritage Foundation dataset, China's Outward Investment: Non-bond Transactions over $100 million, from 2005 to June 2010, available upon request from The Heritage Foundation.

Map 1 • China Investment Tracker ☎ heritage.org

China's worldwide reach

Case study: Chinese overseas investment

Why are the Chinese investing so much money abroad?

- It is thought the Chinese are engaging in building influence for the government, by buying up significant stakes in companies that have influence in western governments.
- They have targeted companies that have traditionally invested heavily in China.
- It is hoped that through investment the government can influence the policies of multinational companies and protect China's interests in international spheres.

- The Chinese remain a passive investor to avoid political difficulties in overseas markets.
- It is suggested that China invests in natural resources to increase its strategic reserves.
- The Chinese have invested/acquired shares in high-tech companies to help China rapidly close the gap with leading industrialised nations.

It is also important to remember that China has accepted the biggest amount of inward investment of any country in the world! It is receptive to agricultural, transportation and energy projects and any raw materials projects help!

Case study: The Chinese exploitation of Africa's iron ore

The Belinga iron ore deposit in the north-east of Gabon was discovered in 1895 and is regarded as one of the last uptapped iron ore deposits. The reserves are estimated at one billion tonnes. The project is the country's largest investment. Exploitation of the deposit is dependent on building the Belinga iron ore facility, a deep-water port in Santa Clara and 560 km of railroad track to export the iron ore. It will also include two hydroelectric dams to provide power. The Belinga project is being financed by the China Exim Bank. But there are issues related to the exploitation of the Belinga iron ore resource.

Social

- The Gabonese government has given contracts related to the project to friends and family.

- The project is tainted by government corruption.
- Will the 30 000 proposed jobs go to Africans or to Chinese workers?

Environmental issues

- The dam will cause flow volume problems.
- Resources will be lost.
- The 560 kms of new railway and the new ports will endanger ecosystems.
- The dam must be located at Tsenué-Lélédi to limit environmental damage.

Human rights issues

- Bans imposed on Gabon's environmental activists.
- Local communities are moved against their will.

Case study: Is it all 'win – win' for Africa?

China has worked with Africa to ensure its economic boom does not halt, that its raw material doesn't run out and to ensure a market for its goods.

Positives

- China has restructured or cancelled loans for many African countries.
- China has implemented long-lasting cooperation in improving local infrastructure by building roads, bridges, schools, hospitals and other public undertakings.
- Most Africans believe the Chinese bring what Africa needs – investment and money for governments and companies.

Negatives

- China exploits African workers.
- China neglects the environmental damage it causes.
- China makes no efforts to transfer skills and knowledge to Africans.
- It is Africa biggest arms supplier.

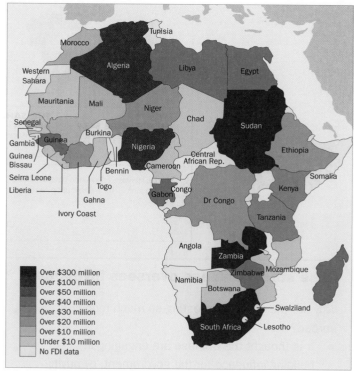

China's foreign direct investment in Africa (2005)

New world globalisers

Brazil and Russia

Just like China and India, Brazil and Russia have fast growing economies and fast emerging markets. Brazil by 2025 is expected to almost exclusively be supplying the world with soya bean and iron ore and Russia will be supplying massive amounts of oils and natural gas. These two countries have effectively become a commodities suppliers to both India and China.

They were coined the BRIC's in 2001 by economists at Goldman Sachs Bank in the USA.

Brazil and Russia, along with India and China (the BRICs), are expected by 2050 to be four of the six most dominant economies in the world.

> **PROGRESS CHECK**
>
> **1** For many years GNP (Gross National Product) was used as the way of measuring development. What other measures are used today?
>
> **2** What do Rostow's and Kondratieff's models compare?
>
> **3** What is the "two speed world"?
>
> Answers: 1. Life expectancy, population density, adult literacy, CO_2 emissions, HDI 2. Stages of economic development 3. Advanced economies are losing momentum whilst developing economies are growing.

The principle cause of disparity – debt

AQA	**A2 U3**
OCR	**A2 U3**
Edexcel	**A2 U3**
WJEC	**A2 UG3**
CCEA	**AS U2**

The history of debt in developing countries

Approximately a quarter of the world's population lives in poverty, but why is this?

The history of the developing world's debt is the history of siphoning off the resources from the most deprived people. Money was lent to developing countries in the 1970s and 1980s for development projects. Export revenues from raw materials (cocoa, coffee, tea, palm oil) and crops help fund the repayment of interest on loans. However, with prices falling for commodities, oil prices rising and higher interest rates making it harder to repay loans, many developing countries struggle to repay the interest on the loans and have to take out more loans. New loans are often tied to **structural adjustment programmes (SAPs)** which place strict spending conditions on public services such as healthcare, education, sanitation and housing.

Odious debt

Each year, LEDCs pay the MEDCs nine times more in debt repayments than they receive in grants.

Odious debt is unfair debt. Money lent was invariably used to oppress the people or embezzlement. Many poor countries have to repay heavy debt burdens as they strive to start their independence process. South Africa is paying for debts incurred during the apartheid era (approximately $62 billion) and people die in this young state because of debt repayment. In countries where odious debt exists they are oppressed by the regimes propped up by the loans, impoverished by the cost of servicing the loan and they are oppressed by the penalties imposed if their country reneges on the loans. Currently, Indonesia estimates it has paid $151 billion relating to odious debt – twice the level of its recorded debt.

Each person in LEDCs owes £250 to the MEDCs! Africa spends four times as much on debt repayments as on healthcare.

Mismanaged lending

Abbreviating your notes in this way and then highlighting them is a must.

In summary, in the 1960s the USA spent more than it could afford → more dollars printed → oil prices pegged to the dollar decrease → oil producing companies/countries increase the cost of oil → more money earned → interest rates fall → banks increase lending to avert a crisis → Mexico and other countries default on their borrowing → the IMF step in → loans and SAPs put in place → the poor suffer, as education, health are cut to pay off debt.

Are poor countries subsidising the rich?

Payment is often made by asset stripping resources as their goods are not needed!

Loans come with conditions, such as preferential exporting which in effect means more money goes out of the country than comes in. Furthermore, the production of **over-excess** in the developed world means that the developing world find it near impossible to compete.

Help with globalisation

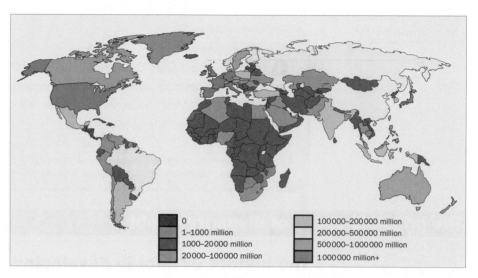

0	100 000–200 000 million
1–1000 million	200 000–500 000 million
1000–20 000 million	500 000–1 000 000 million
20 000–100 000 million	1 000 000 million+

The amount of debt owed to others in dollars

Responses to the debt crisis

The political response to the debt crisis has been varied and none of the initiatives have been adequate to deal with the whole debt crisis.

- **The Brady Plan** – according to this plan from 1989, Brady suggested that banks should reduce the remaining debt of large debtor countries by writing off, or re-scheduling, their debt through conversion schemes. Usually converting it into sellable bonds!

- **Trinidad/Naples Terms** – this is where a cancelling of half the debt of the poorest countries is contemplated. When the plan was finally agreed in 1994, a figure of about 67% was suggested and agreed at the G7 summit. This whole scheme hasn't worked as the creditors have been reluctant to offer debt relief and few countries would sign into the SAPs to reduce debt.

- **The Gleneagles Summit: 'Make Poverty History'** – this campaign revolved around the G8 meeting at Gleneagles in 2005. The three demands of the campaign were **trade justice**, **drop debt** and **more and better aid**. None of these aims were new (there were many attempts over the preceding decades to promote them), but the scale of the 2005 campaign dwarfed previous efforts.

- **HIPC Initiative/Mauritius Mandate** – the highly indebted poor country (HIPC) initiative, from 1996, was a shift by the International Monetary Fund (IMF) and the World Bank to cancel debts owed to them and to be financed by the sale of IMF gold, through a trust fund. HIPC initiatives have been limited in effect. Christian Aid estimates that only 6.4% of debt of the 41 poorest countries will be tackled. The Mauritius Mandate encouraged the key creditor countries of Japan, USA, Germany, France and the UK to put fresh impetus behind the HIPC initiatives.

To date only about £7 billion of the £70 billion promised has actually been delivered. It seems that the 2000 debt relief target hasn't been met by any of the creditor countries.

Case study: In the shadow of the elephant i.e. rich South Africa

Mozambique has set an example to all other poor countries. Once ranked as the world's poorest country, following years of civil war and post-colonial poverty, the progress made in Mozambique means it is now benefitting hugely from debt reduction and increased foreign investment. Mozambique has benefitted from the HIPC Initiatives. The government has made considerable efforts to stimulate the economy by supporting its industries globally, committing to its labour intensive agricultural

system and offering incentives to inward investors such as Maputo Iron and Steel. The most important development has been Mozambique's commitment to linking its economic progression to human development. The Poverty Action Planning is having a huge effect throughout the entire country, particularly in health, water and sanitation, and in education – in the 1970s, 90% of Mozambique's population was illiterate; this had been reduced to 45% of men and 73% of women by 2007.

Barriers to tackling debt relief

Poor countries have to prepare Action Plans or **SAPs** (**SAP = Special Adjustment Policy**), which have to be approved by the IMF and World Bank. Most HIPCs lack the skills to draw up such proposals. IMF and the World Bank demand that all areas of an HIPC economy and development are tackled as they are reluctant to write cheques for money that might be spent on weapons rather than adjustment programmes. Creditor governments and their opponents, in time-honoured fashion, rebuff and dispute the releasing of money to the IMF. Even if a country is offered relief it doesn't always help! Tanzania has had 20% of its debt written off. However, the £59 million it still has to pay back is more than the annual spend on education. (A similar situation exists in Mali, Burkina Faso, Mozambique, Zambia and Malawi.)

> SAPs mean spending less on health, education and social services and devaluing currency. Food subsidies are removed. Jobs go. Subsistence agriculture dominates.

SAPs impose such onerous restraints on HIPCs that some have refused to co-operate. To meet IMF requirements countries may have to impose increased interest rates. Riots ensued in the Dominican Republic after food prices were doubled and medicine prices quadrupled (112 died and 500 were wounded in 4 days of rioting). In Argentina and Zambia, strikes and demonstrations have been common. In the Venezuelan capital of Caracas where wage levels collapsed and food subsides were cut, up to 1500 died in rioting in 1989.

> The G8 Okinawa meeting in 2000 actually backtracked on all previous agreements to relieve debt.

The effect of the imposition of SAPs, and the increasing unrest of individuals in the developed world with reference to the lack of progress with debt relief, has led to bloody confrontations in several developing world capitals, e.g. the riots at the 2001 Genoa G8 meeting.

PROGRESS CHECK

1. What is odious debt?
2. Outline three responses to the debt crisis.
3. What are SAPs?

Answers: 1. Debt resulting from dubious lending. 2. The Brady Plan; the Trinidad/Naples terms: *make poverty history* scheme. 3. Structural adjustment programmes (plans put together by LEDC's to repay loans).

Innovative ways to narrow the development gap

AQA	A2 U3
OCR	A2 U3
Edexcel	A2 U3
WJEC	A2 UG3
CCEA	AS U2

It is not all bad news. Debt is increasingly swopped or fair trade established – see the case studies below.

Case study: The largest ever debt-for-nature swap happens in Costa Rica

In 2007, the governments of the USA and Costa Rica, together with environmental groups, determined that $26 million of Costa Rican debt could be foregone in exchange for tropical forest conservation. The debt-for-nature swap comes under the Tropical Forest Conservation Act of 1998.

The agreement will target forest protection in some of Costa Rica's best known biodiversity hotspots. It is the first debt-for-nature swap in Costa Rica. Since 1998 the Tropical Forest Conservation Act has provided $163 million to 11 countries.

Case study: Made in Vietnam, cut in Cambodia

The USA consumes 20% of the entire world's coffee, making it the largest consumer in the world. The coffee is produced in what are described as 'sweatshops in the fields'. Most coffee farmers receive prices for their coffee that are less than the costs of production, forcing them into a cycle of poverty and debt.

Fair trade is the solution to this crisis, assuring consumers that the coffee they drink was purchased under fair conditions. To become fair trade certified, an importer must meet stringent international criteria, e.g. paying a minimum price per pound and providing technical assistance such as help moving to organic farming. Fair trade for coffee farmers means community development, health, education and environmental stewardship.

The **Costa Coffee Foundation** supports the coffee growing communities by building new schools, furnishing the buildings, investing in water supplies, developing land for families to grow crops, teacher accommodation, latrines, getting children to school, funding social worker salaries and ensuring all communities are represented.

Over the last few decades, forest in Cambodia has been reduced from 70% of land area to 30% due to warring factions and political parties trying to fund their military and political ideals. Illegal loggers depend on a market for their timber, a major section of which is the garden furniture trade, especially over the last few years. Retailers across Europe want garden furniture made in Vietnam meaning this is a big trade and one which is increasing. Garden furniture imports into Norway in 1998 were 95 times what they were in 1990 (in monetary terms).

The furniture is often marketed as being environmentally friendly – supposedly a new tree is planted for every one that is felled, but this does not seem to be the case. When consumers buy garden furniture made in Vietnam, they may be contributing to deforestation in Vietnam, Malaysia, Laos and Myanmar, but they may also be contributing to the enriching of military warlords and politicians in Cambodia.

Consequences of disparity

AQA	A2 U3
OCR	A2 U3
Edexcel	A2 U3
WJEC	A2 UG3
CCEA	AS U2

The 'Boomerang' effect on the developed world

The debt of the HIPCs also affects the rich world.

Brazilians cut 50 000 km²/yr of TRF to help service its US$112 billion debt.

- **The environment** – the easy solution to the debt situation is for HIPCs to milk their own resources, by over-fishing, by massive deforestation or by overusing soil to grow cash crops for the North. Plans set by SAPs to develop large dam projects and to drive industry with charcoal have also caused problems for the environment. All of this has consequences for world climate.

- **Dole queues** – in the MEDCs dole queues have lengthened as HIPCs under SAPs agreements have 'earned more, but spent less'. Before the debt crisis broke, Europe sold 20% of its exports to Africa. This has gradually reduced to less than 10%, despite the Lome Agreement, which was meant to support countries wishing to export to the UK and EU.
- **The Drugs Trade** – in an attempt to repay foreign debt HIPCs have increasingly raised currency by increasing the amounts of drugs, including cocaine and cannabis, they grow, e.g. 41% of Bolivia's workforce use the drugs trade for their livelihood.
- **War** – debt leads to and contributes to war. The escalations of conflict in various countries leads to MEDCs intervening, e.g. the USA in Somalia and the British in Sierra Leone.
- **The banks** – one of the winners in this entire debt problem has been the commercial banks. In the UK all the high street banks have lent to the HIPCs of the world. But through the bizarre system of debt exchange and tax relief in the secondary market all the banks have made substantial profits whilst debtor countries have gained nothing!

Effect on the less developed world

Effects of debt on children

There are thought to be 5000 'street kids' in Guatemala City. The so-called 'street cleansing' of 'undesirables' accounts for the deaths of up to 300 children in Honduras, similar numbers in Brazil – usually at the hands of the security forces!

Families in HIPCs have to cope with a great deal. Lack of health provision impacts on family planning availability, resulting in many unwanted children. Parental inability to feed, clothe and provide for their children leads to many being abandoned onto the streets. In Central America this situation has become a massive problem for the authorities. Robbery, prostitution and begging are rife, all an essential source of income for the 'street children'.

Effects of debt on housing issues

The move to the cities also increases pollution in city areas.

Uneven development results in migration, within and between countries, usually from rural to urban areas. The cities have attracted many millions of these migrants, exacerbating housing problems. Many thousands of shanty towns (favelas, bustees, ranchos) have sprung up to accommodate growing populations. Seen as symbols of underdevelopment they have been demolished and bulldozed by HIPCs to enhance their SAP arrangements. Most of the shanty dwellers have been moved into housing for which they cannot afford to pay rent, and as a result these areas are themselves rapidly falling into decline and so the cycle continues.

10.5 Aid and trade as forces for change

LEARNING SUMMARY

After studying this section, you should be able to understand:
- that aid may do more harm than good in the long-term
- that competition can result in economic growth in LEDCs

The burden of aid

AQA	**A2 U3**
OCR	**A2 U3**
Edexcel	**A2 U3**
WJEC	**A2 UG3**
CCEA	**AS U2**

What is aid?

- **Short-term aid** – supplied after disasters, includes tents, medicine and food.
- **Long-term aid** – usually financial assistance or equipment, advisors and technicians.

> Be able to show how the developed world helps the developing world.

- **Multi-lateral aid** – reaches the HIPC via organisations such as the IMF.
- **Bi-lateral aid** – goes directly from the donor country to the developing nation.
- **NGO aid** – derived from non-governmental agencies (NGOs) e.g. Oxfam.

Remember:

- Aid tries to correct the **imbalance** in world resources and wealth.
- Poor countries still back their own development. MEDCs finance less than 20% of development projects undertaken.

> In some cases less than 10% of the original aid reaches those that need it. In many cases smaller sustainable projects would be much more appropriate.

- Profits in MEDCs are greater than the aid that is offered to LEDCs.
- **Official aid** sends millions more to LEDCs than the NGOs.
- The aid that is offered sometimes does not reach the people who really need it. Some is used to pay for debt repayment, some is used to support the administration of aid. There is spending on weapons and prestige projects.
- **Aid distribution** is often tied to a particular government's **political or economic agenda**. It may also be tied to the purchase of donor country goods and services or to military access.

> Aid-based development can cause 'aid fatigue' in donor countries.

- Some people feel that aid actually **nurtures dependency**, and that trade should be encouraged rather than 'stop gap' aid being offered.

Case study: Damning the dams

The building of dams is a hot political issue in HIPCs. Generally, big dams are seen as misguided, inappropriate, technology-driven developments that have devastating effects on the environment. In light of this opinion the IMF and the World Bank looked into the 50 dam building projects they have been involved in. Based on independent research only 13 are identified as having problems. The Bayano in Panama and the Kariba in Zambia are said to have caused irreparable damage to the environment. The cost to the IMF of these 50 dams is US$7.4 billion, just 3% of IMF loans.

> The direct involvement of other MEDCs in dam building at the time saw a consortium withdraw from building the Bakun Dam on Sarawak. It too will finally be built at a cost of $6 billion with a completion date of 2015 – over 30 years after it was first proposed.

The direct involvement of the UK government in dam building in LEDCs ceased after adverse reports in the British media linking £234 million of British Aid for the Pergau dam in Kelatan Province in Malaysia, to the purchase of in excess of £1 billion of arms and machinery. Furthermore, five other schemes have been linked to such deals. The Pergau Dam was finally completed some 20 years later in 2003, for a country desperately in need of clean, reliable water.

PROGRESS CHECK

1. Why are SAPs seen as onerous by LEDCs?
2. What are some of the consequences of the disparity between rich and poor countries?

Answers: 1. Massive interest payments mean that less can be spent on health and education, for instance within the LEDC. 2. Over-fishing, massive deforestation, soil erosion, dole queues, drug issues, wars and conflict, abandoned children, shanty town growth.

Trade and the growth of NICs

AQA	A2 U3
OCR	A2 U3
Edexcel	A2 U3
WJEC	A2 UG3
CCEA	AS U2

Many see the best route forward for HIPCs is to trade, rather than rely upon aid. Since the 1970s some stronger economies have emerged from the developing world. These are linked either to oil discoveries or to the growth of **Newly Industrialised Countries (NICs)** who have gained a significant amount of world trade over the last 30 years. Those countries that have been successful generally have protectable markets and produce a product that has enabled them to expand abroad.

Common characteristics of NIC development

- Cheap labour.
- They had early foreign assistance and investment.
- They have co-operative governments who want to industrialise and develop.
- There is lots of investment abroad, to broaden and develop export markets.
- There is a strong work ethic, values and beliefs.

Case study: Strategy for the development of successful trade regimes and growth in HIPCs/LEDCs, South Korea

For South Korea, NICs have enabled export greater than imports. Their trading partners have increased and total corporate debt has dropped.

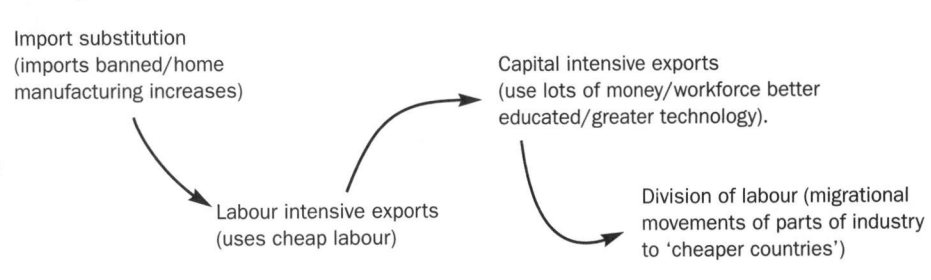

Import substitution (imports banned/home manufacturing increases)

Labour intensive exports (uses cheap labour)

Capital intensive exports (use lots of money/workforce better educated/greater technology).

Division of labour (migrational movements of parts of industry to 'cheaper countries')

South Korean development

Probably the strongest of the 'Asian Tigers'. By 1996 South Korea had in excess of US$8.4 billion invested abroad.

The USA initially heavily supported the economy of South Korea, but the advantages of deep-water anchorage, markets on the Pacific Rim and a cheap hardworking labour force and government backing also helped. Much of the economy of South Korea is now in the hands of giant corporations (or **chaebols**) such as Daewoo, Hyundai and Samsung. Initially the economy focused on textiles and heavy industry, but it has shifted of late to high-technology goods. It has also spread production abroad.

Case study: Taiwan – a contrasting 'tiger economy'

A **tiger economy** is where a country's economy undergoes rapid economic growth accompanied by an increase in the standard of living. Taiwan was formally a republic of China – the Island of Formosa. It is the second biggest Asian Tiger. Historically steel and ship building have been important, but Taiwan has gradually moved to more value added products, such as electrical products. The Taiwanese are innovators – 85% of goods are exported. Industry is either large scale state-owned enterprises or small family firms. Taiwan's agricultural output has increased thanks to mechanical and agricultural research. The government keeps Taiwan busy and competitive by offering competitively priced goods, by being ruthless in business and by being willing to take risks. High wages are expected by the Taiwanese so some offshore movement has occurred, mostly to other 'tigers'.

What future for the NICs?

Don't confuse NICs with TNCs in the exam room! TNC = Transnational Company.

The global shift and globalisation of economies will lead to LEDCs being looked upon differently over the next century. As **GATT (General Agreement on Tariffs and Trade)** spreads the message of economic liberalisation it seems unlikely that LEDCs will be left behind economically. NICs will continue to prosper and develop, despite competition from **TNCs (Transnational Corporations)**.

The threat of TNCs

AQA	A2 U3
OCR	A2 U3
Edexcel	A2 U3
WJEC	A2 UG3
CCEA	AS U2

All advantages and disadvantages need to be discussed in the exam room.

The TNCs and NICs have a huge influence over the world. This is a commonly examined area.

TNCs will be forced to take a responsible approach to world trade, so the NICs' future looks to be assured. Certainly the existence of TNCs look, on the whole, to benefit more than hinder HIPC/LEDC development.

The problems that TNCs will pose to the NICs in the future

TNCs operate and have ownership of assets in more than one country. They wield huge economic, political and financial power. Supporters of TNCs suggest they have fuelled development, whereas critics focus on their moral and social effects.

TNCs exist to:
- gain access to new markets
- avoid trading barriers
- diversify
- reduce costs
- exploit lax environmental rulings
- provide new export routes.

Advantages of TNCs

To host:
- multiplier effect
- income generator
- introduction of technology
- employment growth

and home:
- profit.

The disadvantages of TNCs

To host:
- local companies lose out
- loss of autonomy
- spread of technology is suppressed
- lax safety regulations
- lax environmental stewardship
- pricing is manipulated to reduce tax liability
- financial costs to the host
- goods produced are inappropriate to the country

and home:
- possible job losses.

Case study: Nike – the biggest sportswear manufacturer in the world

Nike's headquarters are based in Beaverton, Oregon in the US. It is the world's biggest sportswear manufacturer and largest shoe-maker in the world, selling in 160 countries.

Nike admitted to isolated cases of subcontractors employing children in the developing world. Children, as young as 10, were employed to make footballs, shoes and clothing in Pakistan and Cambodia.

Nike does have strict rules governing the employment of children, but when factories sub-contract work out children inevitably, and uncontrollably, get drawn into the manufacturing process. With 700 contract factories Nike now uses the Global Alliance to pick up and respond to child employment issues, and has worked to increase wages and improve conditions for all workers.

Case study: Toyota is the 12th biggest TNC in the world

Toyota is the largest Japanese car manufacturer with many factories worldwide. The countries in the European Union (EU) form the largest market for new car sales. By building cars in Europe Toyota avoid import quota payments.

The UK government gave Toyota financial assistance in building their factory in the UK. The UK is a large market for new cars and is in the centre of Europe. Workers in the UK are highly productive. Toyota chose Burnaston to build their factory as Burnaston was a large flat site of over 100 hectares. It had room to expand, is close to the motorway system, has a skilled workforce, benefits from cheap electricity and Derbyshire County Council was willing to support the scheme with £20 million investment.

Once the factory was built about 3000 jobs were created. The multiplier effect meant more money was being spent in Derby city centre. Public transport has improved. Toyota's suppliers have moved into the area creating 300 new jobs and 4700 new houses have been built in Derbyshire to cope with many new people moving the area. One negative aspect of the Toyota investment is that the site was a greenfield site, so more countryside has been destroyed. A brown-field site would have been better. There has been a huge increase in traffic and noise in the area.

'Alternative' strategies

AQA	**A2 U3**
OCR	**A2 U3**
Edexcel	**A2 U3**
WJEC	**A2 UG3**
CCEA	**AS U2**

'Bottom-up' policies in developing countries

'Bottom-up' policies are **smaller scale and more appropriate** for many local and national needs. Most development strategies have a developed-world origin. As developing countries have different aims and structures, alternative strategies to enable development and sustainable development have to be instituted. Usually this involves **appropriate and or intermediate technologies**.

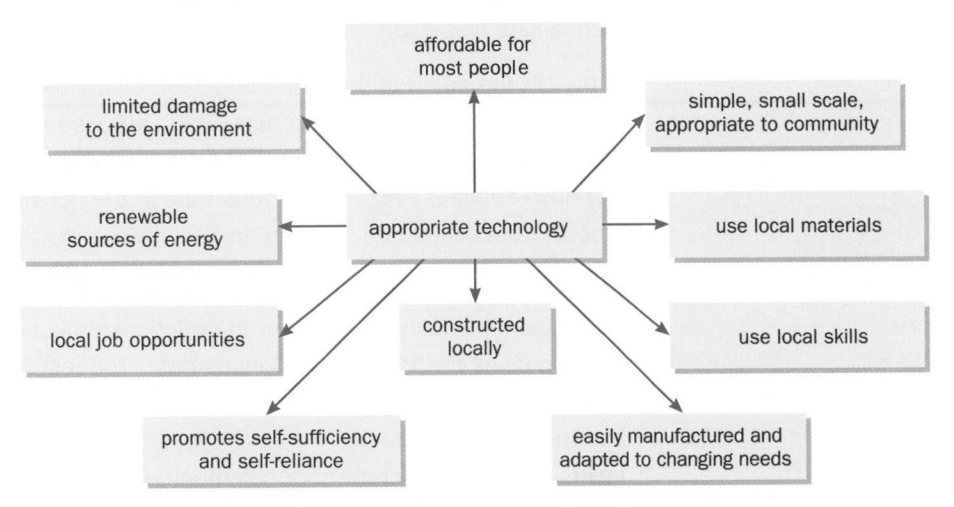

Advantages of using appropriate technology to meet local needs

Case study: Teddy Exports of Tirumangalam, Southern India

Teddy Exports employs 300 local people and produces a range of products such as massage rollers, cotton bags, shirts, lampstands and hair scrunchies. The business started in 1990 from a mud hut with five workers. Turnover is now about £1.5 million/year. Workers receive free medical care, subsidised food, housing loans and good wages. From the profits the Teddy Trust was formed and which has funded schools, evening classes, HIV/AIDS awareness programmes, vet camps and a 'teaching' farm. Many of their products come into the MEDC marketplace, e.g. Bodyshop.

Agenda 21 still has an important role to play in planning and many other decisions over 15 years beyond its inception!

Agenda 21

In 1992 the United Nations conference on Environment and Development was held in Rio de Janeiro in Brazil. Usually referred to as the **Rio Earth Summit**, one of the major outcomes agreed by 180 countries was a blueprint for sustainable development. This action plan, beginning in the 1990s, but projected forward into the twenty-first century, is called **Agenda 21**. It looks at environmental, social and development issues and how they inter-relate. Formulating, agreeing and implementing strategies for sustainable development is now a requirement for governments.

10.6 Inequalities in MEDCs

LEARNING SUMMARY

After studying this section, you should be able to understand:
- that every country has regional variations in wealth
- that regional development policies aim to spread wealth and development

Inequalities in the UK

AQA	**A2 U3**
OCR	**A2 U3**
Edexcel	**A2 U3**
WJEC	**A2 UG3**

Thus far we have focused almost exclusively on the development issues that face HIPCs and other major LEDCs of the world. Within countries of the MEDCs inequalities also exist.

In the UK the south is the richer area, containing the capital with its concentration of governmental and corporate power and a dense collection of rich towns. The towns on the whole have developed from the boom in producer services, business services and hi-tech industries. In the north cities are still trying to recover from de-industrialisation and decreases in manufacturing production.

Recent work, including *Life in Britain* commissioned by the Joseph Rowntree Foundation, looked exclusively at the relationship between poverty, affluence and location using the Millennium Census data. The results served to give an accurate picture of the UK today, one where generally the rich are getting richer and the poor much poorer. Furthermore, it uncovered what has been coined the **inverse care law** whereby poor communities have the least access to essential life chances and resources. A summary of findings of the above work is shown below.

- **Education** – areas with the greatest proportions of young people with no qualifications have the lowest number of teachers per head of population. The areas that are doing best have four times as many teachers. Young people are more likely to obtain good qualifications if their parents' generation (people aged 40–54) are also well-qualified.
- **Employment** – the UK is divided between 'work rich' and 'work poor' areas. In areas with low unemployment, the people who have jobs are more likely to be working very long hours. High-status jobs are concentrated in London and the south-east. With distance from London, the number of children in low income families increases. In London the greatest concentration of low income families are found in Central London.
- **Housing** – areas with high levels of overcrowded homes (including London, parts of the south-east, cities in the Midlands and the North, and Glasgow) tend not to have many under-occupied homes. Households without central heating is used as an indicator of the quality of housing stock and of levels of deprivation.
- **Transport** – there is a geographical divide based on access to cars. Access to cars is used as a measure of wealth – areas west of London have three or

more cars per family. The areas with the most families without cars are not only found in poor urban areas, but also in some rural areas. One side effect is the amount of carbon emissions in areas of the UK. The south emits the most – driven up no doubt by higher car ownership.

- **Health** – the proportion of people with life limiting long-term illnesses, and informal care providers, increases to the west and north of the UK, with the highest rates in the Welsh Valleys, parts of Scotland and areas around Tyneside and Merseyside. Areas with the highest levels of poor health have the lowest numbers of health care professionals living and working there. Only nurses, midwives and health visitors tend to be more concentrated in areas of higher need. The numbers of people with long-term and serious illness increases away from southern England.

Exam practice questions

Structured questions

1 **(a)** Explain what the Human Development Index (HDI) is. **[3]**

(b) (i) A Spearman rank correlation was calculated for the HDI and GNP%. Comment on the significance of this result. **[3]**

(ii) Account for the relationship between GNP% and the HDI. **[4]**

Country	HDI	GNP% $US
Austria	0.925	28110
Bangladesh	0.364	260
Colombia	0.836	2140
Ethiopia	0.227	100
Indonesia	0.637	1080
Jamaica	0.721	1600
Nepal	0.343	210
Norway	0.932	34510
Singapore	0.878	30550
USA	0.937	28020

Degrees of freedom = n −1 0.05	Critical values
8 0.	643
9 0.	600
10 0.	564
11 0.	523

Spearman rank correlation coefficient (rs) = 0.915

(c) Essay: Evaluate the options open to an LEDC government faced with substantial debt that wishes to improve the health of its people. **[10]**

11 Worldwide industrial change and issues

The following topics are covered in this chapter:

- **International patterns and trends**
- **Specific industrial examples**
- **Economic theories**

11.1 International patterns and trends

LEARNING SUMMARY

After studying this section, you should be able to understand:

- how manufacturing can be defined by the industrial process carried out
- the recent trends in manufacturing on a global scale

Classification

WJEC **A2 UG3**

Industry is work performed for an economic gain.

If you asked anyone sixty years ago what the word 'industry' meant they might have replied 'mining and manufacturing'. Ask the same question today and the response would be vastly different. 'Industry' is now one of many economic activities, which include transport, communications, services, tourism, mining and manufacturing. To make the study of economic activity a little easier, a traditional classification or grouping of activities is often used and referred to.

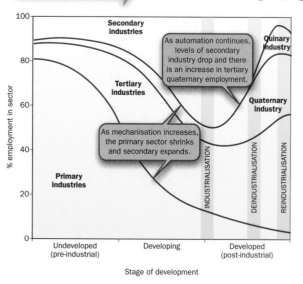

Stage of development

Fourastie's Model

- **Primary activity** – involves extraction, collection or the early processing of resources from activities such as quarrying, mining, forestry and farming. The activity is resource–based and is the basis of secondary industry.
- **Secondary activity** – involves manufacturing, processing and assembly. They change, or transform, the products of primary activity. The value of the raw material is increased by secondary activity and locations vary. Heavy industry adds bulk and light industry is smaller scale, such as making televisions.
- **Tertiary activity** – provides a service to customers and is market–orientated. This area of work includes transport, shopping, teaching, office, banks and doctors services.
- **Quaternary activity** – provides information and expertise. Many of these industries are located in universities research and development centres. Marketing and advertising and are other important areas.
- **Quinary activity** – includes education, government, health and research.

The use of computers and the internet means that these industries can be 'mobile' and can be located just about anywhere.

The diagram above is based on French economist **J.Fourastie's** model and relates the approximate sizes of economic sectors in 2010, based on their levels of economic development over many years. It also indicates the approximate time-frame of industrialisation, de-industrialisation and re-industrialisation.

Global themes and change in industry

AQA	A2 U3
OCR	A2 U3
Edexcel	A2 U3
WJEC	A2 UG3

A number of themes have influenced the type, location and distribution of industry since the mid-1990s. Secondary, or manufacturing, industry had formed the cornerstone for most MEDC economies, leading countries through phases of economic development and on to a good deal of prosperity. However, the requirements of these industries have changed, as have the requirements of consumers' expectations relating to environment and the need to be able to respond and react 'globally'. Fundamental changes in the location and types of industry will continue apace as de-industrialisation is completed (in many MEDCs) and rationalisation, restructuring, globalisation and re-industrialisation gather apace (the latter especially in India and China). The following are the routes taken by companies to avoid their products becoming obsolete and to ensure they continue to trade successfully.

- **Rationalisation and restructuring** – attempts to improve competitiveness and the productivity of a company. It is always better for a company to quickly reconfigure than to bring about wholesale change that can take many years.
- **Globalisation and internationalisation** – most big companies have broken into international markets. Many countries offer the home company a competitive and strategic advantage.
- **Flexible production (Just–in–time production (JIT))** – this is a form of lean production where components arrive just-in-time for the manufacturing process.

> This means huge numbers of parts don't have to be stored and there is a move away from mass production.

- **Environmental considerations** – for industries to avoid bad press, and also to meet government imposed targets on emissions, decision makers have had to be more aware of the potential to pollute.
- **Moves to the country** – because of constrained sites and planning policies in many countries and because of counter–urbanisation many industries have sought to relocate on the urban periphery.

In addition to the above points technological change in industry has speeded up since the mid-2000s.

- Increased rates of **innovation** (in micro-electronics, biotechnology and new materials). Innovation is also aided by completion and rivalry between producer countries and through research and development.
- The broader **application** of new technologies (in 2010 more than 85% of industrial activity utilises manufacturing processes that use micro-technology).
- Shorter **life-cycle** for products and the ability to be flexible to customer needs.
- There is pressure to **cash-in** as quickly as possible, so technology is shared between smaller and related companies.

> One issue here is that automation equates to job losses.

- **Automation** means that procedures can be completed quickly and more cheaply.

Case study: Bridging the technology gap – Brazilian laser production

The Brazilians have been quick to use shared photonics (laser) building knowledge and technology. In South America the cost of such technology, especially for use in medicine, was prohibitive. Brazil has been able to use research and development completed in other countries to come up with a product that is effective for medical use, but at a far more accessible price. The Brazilians have effectively captured for themselves a niche market and one that the Americans, the Germans and the Japanese find difficult to undercut!

De-industrialisation

Old industries, such as textiles, iron and steel manufacturing are the most likely to de–industrialise leading to job losses and plant closures.

De-industrialisation generally relates to the decline in jobs and production in the manufacturing sector of the employment structure (the **post-industrial society**). The characteristics of this phase are:

- shifts from traditional industry towards service industries
- the growing importance of multi–national/transnational companies (TNCs)
- the coincidental depletion of raw materials
- the rise of new technologies
- the lack of, or removal of, subsidies and investment moves this phase on, as do cheaper imports from competitors overseas.

Furthermore, overzealous trade union activity can speed the process of de-industrialisation which is coincidental with periods of depression and can be massive and sudden and may disrupt the local balance of payments. On the whole, de-industrialisation has been concentrated in the nineteenth century industrial cities of Europe and the USA, and though it has meant the shedding of literally millions of manufacturing workers, what has been left is a much more competitive manufacturing industry.

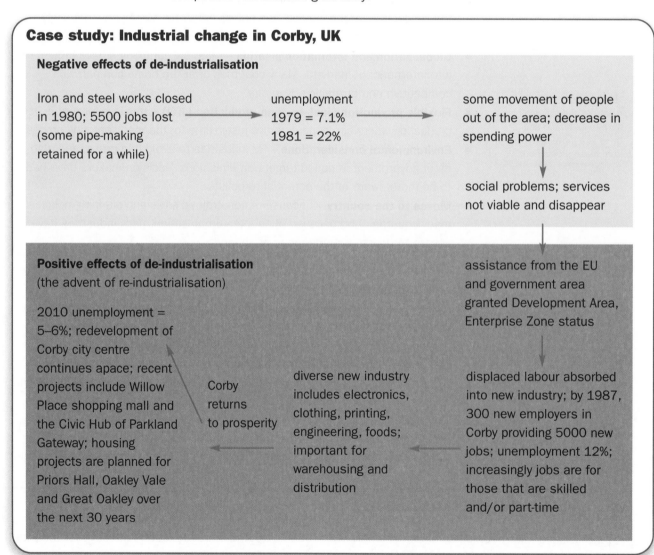

Case study: Industrial change in Corby, UK

Negative effects of de-industrialisation

Iron and steel works closed in 1980; 5500 jobs lost (some pipe-making retained for a while) → unemployment 1979 = 7.1% 1981 = 22% → some movement of people out of the area; decrease in spending power ↓ social problems; services not viable and disappear

Positive effects of de-industrialisation (the advent of re-industrialisation)

2010 unemployment = 5–6%; redevelopment of Corby city centre continues apace; recent projects include Willow Place shopping mall and the Civic Hub of Parkland Gateway; housing projects are planned for Priors Hall, Oakley Vale and Great Oakley over the next 30 years

Corby returns to prosperity

diverse new industry includes electronics, clothing, printing, engineering, foods; important for warehousing and distribution

assistance from the EU and government area granted Development Area, Enterprise Zone status ↓ displaced labour absorbed into new industry; by 1987, 300 new employers in Corby providing 5000 new jobs; unemployment 12%; increasingly jobs are for those that are skilled and/or part-time

De-industrialisation has on the whole been industry-specific, and invariably region specific in the UK, e.g. Corby and the surrounding area, Ravenscraig in Scotland, Durham and Tyneside. Most of these areas, thanks to government and European Union (EU) aid, are now experiencing the **post-industrial age**.

The shift to service industry

Service society	Industrial society
• Differentiated production	• Mass production
• Short series of production	• Long series of production
• Flexibility and complexity	• Standardisation
• Economies of scope	• Economies of scale
• Knowledge and creativity most important factors of production	• Capital most important factor of production
• Development, planning, management and marketing most important functions	• Goods production most important function
• Quality maximisation	• Cost minimisation
• Product competition (quality, service, adaptation to customer needs)	• Price competition
• Turbulent and segmented markets	• Stable and homogeneous market
• International markets	• National markets
• Individual consumption	• Mass consumption
• Automation	• Monotonous routine work
• High and diversified qualifications	• Standard qualifications
• Job enrichment	• Job specialisation
• Flexible employment	• Standardised labour market
• Flat hierarchies, network organisation	• Hierarchical organisation
• Subcontracting, externalisation	• Vertical integration
• Small and medium-sized enterprise, divisionalisation	• Large corporations
• International cooperation, local and regional self-government	• Nation states
• Division of labour among firms	• Division of labour among persons

Re-industrialisation

Three key areas of industrial growth are linked to re-industrialisation.

- The growth of the service sector (or **tertiarisation**).
- The rise of the small firm.
- The inward investment of large multinational companies (MNCs) through **globalisation**.

> Services industries include retailing and distribution; financial services, including banking and insurance; hotels and tourism; leisure, recreation and entertainment; professional and business services, including accountancy, marketing and law.

Around the world the service industry generates two-thirds of gross national product (GNP) and employs two-thirds of the work force. In the UK service industries employ upward of 6 million. The industry affects every part of the economy and de-industrialised economy, changing and adapting as different phases of the industrial process develop.

In the UK there is a high degree of inequality in terms of the amounts of service activity across the country. Similar inequalities exist in all developed economies. In the UK it's because of London's dominance, the pre-dominance of the south-east and decentralisation. Two different categories of activity are observed, as detailed below.

Consumer services, which deal with the general public – patterns of consumer provision have to match the locations and patterns of population. There are, for instance, fewer 'services' in rural West Norfolk than there are in and around Norwich. With regard to the location of such services, four important factors pervade.

> The advent of advanced ICT has been crucial to the recent relocation/location of service industries.

- Services have to be extremely **accessible**, in terms of transport, but also telephonic links, fax and web connections need to be of the highest quality.
- Invariably the **proximity** of the labour force can be a locating factor – they have to be close and there needs to be plenty of them! Service industries are major employers as many offer services 24 hours a day, 365 days a year!
- Huge capital **start-up costs** mean that **out-of-town** locations are preferred nowadays because these sites are cheaper.

On the whole, consumer services still tend to 'hug' centres of population.

- Being **footloose** ensures that consumer services can be fairly flexible in terms of their approach to location and potential market.

Producer services provide for other services

Producer services are concerned with **management**, **public relations** and **computer services**. In terms of location much of business today is face-to-face, by way of discussion, phone, fax or video-conferencing. The south-east of the UK in particular has seen a growth in this sector, being close to the capital's decision-makers and the research centres at the eastern end of the Thames valley. As expected, producer services locate near their customers' industry.

- It is feasible that a **low capital start-up** can occur with this industry.
- It is **labour intensive**, with the product being wholly dependent upon the standards applied by the staff and workers.
- Most jobs tend to be **white-collared** (i.e. non-manual).
- In a sense the user or purchaser of the service is an **integral part** of the production process.

Case study: Retail changes and developments in the UK

Changes tend to mirror the movements of the population, i.e. urban to suburban to re-urbanisation.

Traditional location	Present location	Future strategy
• High-order goods providers, in the CBD or on the 'high street', probably family run. • Low-order goods providers, in shopping parades and 'local neighbourhood stores'.	• Corporate firms dominate and superstores are common, usually on brownfield sites, though some greenfield developments occur. • Retail parks, clusters of DIY, electrical and furniture stores.	• Shopping villages and outlet centres (selling labelled and over-run goods). • Consumer targeting (airport shops, etc.). • Further development of the internet and teleshopping. • Affected by government policy, the return of the 'centre' – the mall.

Benefits include:

- 'Change' perceived as advantageous for customers by the companies involved.
- The fall in the total numbers of shops, but there is an increase in floor space.

Change is explained by:

- Small households, making small shops uneconomic.
- Increased affluence towards the suburbs.
- Car ownership and accessibility.
- Costly city-centre locations.
- Space for future development.

Retailing contributes about 50% of the 'service' GDP; the rest comes from banking and finance, the producer services.

Case study: Banking, UK

In the producer service industry, banking has gone through significant changes lately due to recessionary pressure brought on by poor banking strategy, but also due to the advent of new technology/telephone/internet-banking/increased use of ATMs and the advent of the credit card. Competition from new providers, for example Virgin Direct and Sainsbury's have a banking 'wing'.

The move from building society to bank status, e.g. Abbey National was bought out by Santander, and various other mergers have affected the industry. Many clerical tasks are now completed in less costly peripheral city sites and abroad.

The rise of the small firm

Re-industrialisation has spawned a new and vibrant growth of new small firms around the developed world. Their growth can be explained in a number of ways.

- When large firms are broken up they divide up their production, research and administration sections, e.g. into a variety of smaller and more efficient **subsidiary** companies.
- Quite often the skilled work force, instead of moving out of an area, will pool their expertise and establish competing, and often more successful, subsidiary firms.
- Rapid **technology change** and **innovation** always favours smaller and more reactive firms.
- **Enterprise culture** is the notion that an organisation employs people who are imaginative and creative, who challenge existing ways of doing things and come up with new ideas and solutions to the benefit of society as a whole. Most successful businesses in this country in recent years typify the 'enterprise culture'.

> This involves moves away from a 'dependency culture' where people continually expect others (often the government) to sort out problems for them.

Case study: Examples of enterprising firms

In the 1990s Peter Cruddas set up CMC businesses with £10 000 of capital having spotted a gap in the financial services market and by using the latest technology at that time – the internet. CMC were one of the first companies in the world to carry out financial transactions online. CMC took the risk, got in first and quickly took the lead. Business people refer to this as 'First Mover Advantage'. CMC Markets is now a global company and in 2008 carried out 21 million buying and selling trades – 98% of these trades were online.

Globalisation

> Globalisation affects the recipient country in a range of socio-economic and environmental ways – with some being advantageous, but many are not.

One of the consequences of de-industrialisation has been the global shift of production and services to lower cost locations around the world, but mostly to the newly industrialised countries (NICs), the rapidly industrialising country (RICs) and LEDCs. **Globalisation** is essentially the changing geography of economic activity and production brought about by the growth of new activities, patterns of investment by multinational companies and of course the receiving government's policies towards economic growth. Globalisation has had effects beyond the economic issues – it has changed people's political outlook and has had effects on cultural integration too. Globalisation in some form or another has been happening for hundreds of years, e.g. the UK's East India Company trading with colonies in the seventeenth century.

> Globalisation is seen as a stage in the development of MEDCs.

Why has globalisation leapt forward?

Five key factors have emerged.

- The growth and relative importance of **foreign direct investment** and multinational enterprises. Globalised companies bring wealth that is able to be invested in the recipient country creating jobs and new technology.
- The **internationalisation** of financial markets – with new markets come new stock exchanges which encourages trade and further speeds the globalisation process. Support from international organisations, such as the World Trade Organisation (WTO), also helps the globalising companies of the world.
- The continuing development of **communication** and **transport technology**, better communications and ever improving transport methods

It is globalisation that has also driven the so called economic and power groupings' of the world, such as the LEDCs and the NICs. There are groups that work together for mutual economic gain, such as G8 and OPEC and there are also groups that work together for mutual advantage such as trade blocs, like NAFTA.

It is these groups that really serve to highlight the differences in wealth and power between countries. The gap is widening between rich and poor.

(containerisation), mean that companies can easy locate out of their 'base' country.

- **Deregulation** and **liberalisation** – many countries have opened up in the last twenty years to outside influence, e.g. China, embracing both western culture and methods of working.
- **Privatisation** of public sector services. BT was released from UK government control in 1984 and since then it has globalised and become a multi-billion pound corporation and a leader in telecommunications technology and research and development. Re-branding set it up as a globally connected company.

Threat or opportunity?

Globalisation has the potential to generate wealth and improve living standards. For those countries with the products, skills and resources to take advantage of the opportunities provided by global markets, the benefits are evident. However, there are also significant drawbacks particularly for those nations that don't fall into this category.

Threat	Opportunity
• Increased flow of jobs from developed to developing nations as companies look for cheap labour, with people consequently losing their jobs.	• Increased free trade between nations.
	• Increased availability of capital allowing investors in developed nations to invest in developing nations.
• Economic disruptions in one nation affecting all.	• Companies have greater flexibility to operate across borders.
• Corporate influence on state decisions.	
• The control of world media by a handful of corporations will limit cultural expression.	• Global mass media and communications tie the world together.
• There is a chance that reactions to globalisation can be violent in an attempt to preserve cultural heritage (G8 riots).	• Increased flow of communications allows vital information to be shared between individuals and corporations around the world.
• A risk of diseases being transported unintentionally between nations.	• Greater ease and speed of transportation for goods and people.
• Spread of a materialistic lifestyle and attitude that sees consumption as the path to prosperity.	• Reduction of cultural barriers between nations.
	• Spread of democratic ideals to developing nations.
• Groups and bodies like the WTO infringe on national and individual sovereignty.	• Greater interdependence of nation-states.
• Increase in the chances of civil war within developing countries and open war between developing countries as they vie for resources.	• Reduction of likelihood of war between developed nations.
	• Increases in environmental protection in developed nations.
• Decreases in environmental management as polluting companies take advantage of weak regulatory rules in developing countries. Carbon emissions, waste and pollution increases.	• Working conditions might be better for the workers in recipient countries. Or they could be worse, e.g. sweatshops.

Solutions to the threats posed by globalisation include:
- fair trade should be developed.
- goods should be bought locally
- carbon trading should be extended.

11.2 Specific industrial examples

LEARNING SUMMARY	After studying this section, you should be able to understand:
	• the importance of the transnational corporation to globalisation
	• the importance and consequences of out-sourcing
	• the importance of a resilient ICT industry in the UK

The transnational corporation

AQA **A2 U3**
OCR **A2 U3**
Edexcel **A2 U3**
WJEC **A2 UG3**

> The most popular destinations for FDI are China, India and then Europe. China's share in 2009 was $90 billion.

Transnational corporations (TNCs), often referred to as **multinational companies**, are industries that control economic assets in other countries – usually at least a 10% share of such an asset. They are major forces in global economies. TNCs have enormous financial and technical resources and have an extensive global reach. They have the ability to link countries together through mergers and acquisitions and in terms of the **foreign direct investment (FDI)** that TNCs bring, developing countries received about one-sixth of available FDI.

Throughout the world, there are some 64 000 TNCs and they exist in all sectors of modern economies. Globalisation, particularly the dismantling of trade barriers, has allowed companies to spread widely in search of **cost efficiency** and to implement **integrated production strategies** across regions and even continents. This usually means investment in branch plants overseas and the creation of networks of firms supplying company's components.

Positives and negatives

> 29 of the 100 biggest world economies are TNCs!

TNCs and the FDI bring great benefit to developing countries. They have the potential to generate employment, raise productivity, transfer skills and technology, enhance exports and contribute to the long-term economic development of developing countries. There are though objectors to the activities of TNCs in developing countries.

- It is felt that TNCs don't always have the best interests of the host country at the fore. Much of the money made by TNCs never appears in the economy of the host country.
- Specific instances of serious problems with TNC practices, usually related to **environmental damage**, have created civil unrest against the presence of multinational companies, especially large and politically well-connected ones, e.g. the Shell Oil Company in Nigeria, Bechtel in Bolivia, Union Carbide in India and Chevron/Texaco in Ecuador.
- TNCs bring jobs, money and technology to areas that need them, but when they leave they often take jobs, money and technology with them, e.g. when Sony left West Java.
- TNCs may be discouraged from entering some countries because of issues with legal/regulatory systems, corruption, poor infrastructure or political issues. **Nationalisation** or **appropriation** may also be a concern to them if they stay in a country for a long period of time.
- Some developing country's governments fear that powerful TNCs will 'crowd out' domestic industry or damage 'infant industries' that they hope to move forward.
- Many feel that by admitting some TNCs into some countries they are gaining an unfair advantage in terms of avoiding the tariffs imposed by trade blocs.

Case study: Dell Inc., USA

Dell Inc. is a TNC computer/ICT company based in Texas, USA and globally employs 96 000 people. It became a world force within a year of its launch in 1984, grossing $73 million. In 1988 it started to expand into global computing, opening its first overseas factory in Limerick in Ireland. A period of acquisitions saw it purchase ConvergeNet Technologies, Alienware, Perot Systems and 3PAR. It moved into the global market and established factories in Malaysia, China, UK, the Philippines, Brazil and Poland. In these branch plants Dell configures computers to order, manufacturing them close to its customers, using just-in-time (JIT) technology.

Companies, like Dell, accelerate the process of globalisation linking their product to the people, to money and investment. Furthermore, Dell is able to establish clear and effective divisions of labour. The talented software and hardware designers in India provide the components for their near neighbours in China to construct and assemble the products.

Outsourcing

OCR   **A2 U3**
WJEC   **A2 UG3**

Outsourcing is when a company contracts with another firm or company to provide services that might otherwise be performed by in-house employees. The best known of the outsourced jobs is that undertaken by call centre services, e-mail services and payroll. These roles are taken on by different companies that have specialisms in them. They are often located abroad. To date India, the Philippines and Malaysia have been the countries of choice.

Clearly such advantages could apply almost equally to the Philippines and Malaysia.

There are then many reasons why companies might outsource various jobs, but the most obvious is that it saves money. An expanding company can use the new technology of the outsourcer; this is particularly cost effective when a company is setting up its operations.

The biggest disadvantage for a company with outsourcing is that it loses contact with its customers and clients. Communications to the parent company are often delayed and secrets and strategy are never secret for long!

The advantages of India as an outsourcing centre

Case study: HSBC, India

The financial services industry are leaders in outsourcing. Citigroup and Bank of America were pioneers, but banks across the world have now outsourced their tech support and maintenance, administrative offices and even customer service departments.

HSBC's software-development centre in Pune, India was opened in 2002 and is now one of its most important operations. It has moved to establish development centres in China and Brazil and it has opened back-office operations in five other Indian cities, but Pune remains the star centre.

The need to constantly improve technology was becoming difficult for HSBC, drawing attention away from its core financial services business. With banking services increasingly dependent on fast, efficient technology, HSBC needed to move to a new technology platform and become more efficient.

The Pune centre opened with just 30 workers. There are now over 2500 and though it is a 'captive centre' it operates like an independent contractor, getting paid at market rates for whatever it does for the bank. HSBC has developed a model that many have tried to copy, starting small, where assigned people move work forward, where customers are involved in software development. Staff manning the operations are young and eager to move the operation forward. For HSBC key to the success of the operation is that it is seen as a development wing of the parent company, not a solutions workshop.

The ICT industry in the UK

AQA	A2 U3
OCR	A2 U3
Edexcel	A2 U3
WJEC	A2 UG3

> Being footloose ensures that ICT in the UK is resilient. It responds to changes quickly and appropriately.

The high tech industry in the UK has characteristically been footloose (not constrained to a location by traditional location factors).

The main locating factor is the availability of a skilled, knowledgeable and innovative work force to cope with the rapid pace of research and development. Cheap labour and transport are also necessary to produce products and distribute them.

The industry in the UK produces medical goods, lasers, aerospace technology, pharmaceuticals, as well as computers and computer chips. The UK's experience of ICT has been one of decline and growth.

Examples are shown in the table below.

Decline in London and the south-east and Midlands	Growth in peripheral London and along the M4 corridor	Growth in Wales
There has been a general move from urban to more rural areas. Why? • Lack of space to expand. • Congestion. • Land price too high. • Unsuitable sites. • Planning issues.	• Good communications with proximity to London. • Links to universities (e.g. Oxford Science Park). • Skilled and talented workers are available. • Research and development firms have located in these peripheral areas. • Changes in technology (micro-tech/better internet access) mean that more rural sites are seen as desirable to operate from.	• Good communications network (motorways). • Help from the Welsh Development Agency. • High unemployment means workers available in the component construction industry.

11.3 Economic theories

LEARNING SUMMARY

After studying this section, you should be able to understand:

● how economic theory had changed over the years

Theories of economic growth

OCR **A2 U3**
WJEC **A2 UG3**

Cumulative causation

The concept of cumulative causation was developed by Swedish economist **Gunnar Myrdal** to explain the process that increases inequalities between regions. His model is displayed as a systems diagram. In Myrdal's original model, economic growth starts with new manufacturing industry. Of late it has been adapted to suit the new siting of service industries in the urban periphery.

The initial advantages of the new periphery site are likely to be strong infrastructural links and social acceptance of change. The expanding service industry attracts employment and wealth begins to grow. The increasing scale of activity on the new site has a **multiplier effect**, **agglomeration** occurs and the site becomes increasingly successful. In the peripheral regions backwash causes a degree of downward spiral until spread effects reduce this imbalance.

The model has many merits. It seeks to explain the growth of industry as a series of linked relationships on an undefined timescale. It works best at a national or international level, but can be applied at local levels too.

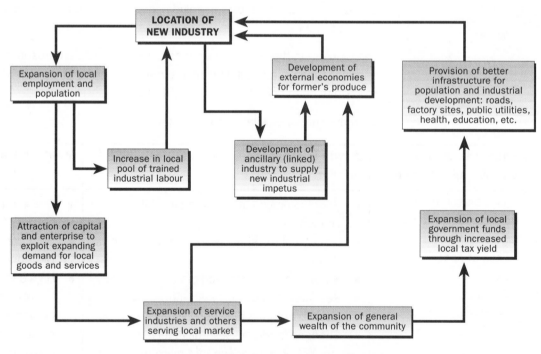

Myrdal's model of cumulative causation

Rostow's model of economic growth

The economist **Rostow** devised a model that attempted to explain the development of economies in five linear stages or sequences. Rostow's graph plots time against expanding wealth (see graph and table opposite).

Merits

- A useful starting point for understanding development.
- It can be used in conjunction with cumulative causation.

Limitations

- It is analogy based.
- It is based on Europe, USA and Japan.
- Growth in the economy can occur without development.

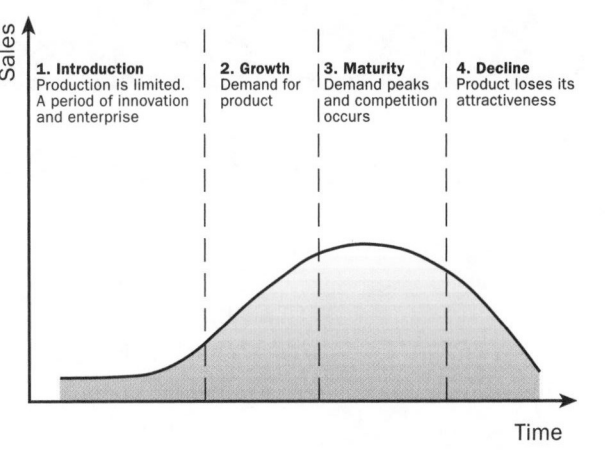

The Rostow model of economic growth

The product-cycle as a four-phase model

Stage 5: High production and mass consumption of goods

Stage 4: The drive to maturity – economic growth spreads to rest of economy

Stage 3: Take-off – manufacturing industries grow rapidly based on development in stage 2

Stage 2: Pre-conditions for take off – extractive industries develop, transport systems improve, agriculture develops

Stage 1: Traditional society – mainly agricultural, little science or technology

1. Introduction Production is limited. A period of innovation and enterprise

2. Growth Demand for product

3. Maturity Demand peaks and competition occurs

4. Decline Product loses its attractiveness

The product-cycle

Based on work by **Vernon**, the product-cycle (see above) is used to study industrial location in relation to regional development.

Advantages

- It is simple.
- Everyday objects can be exemplified.

Disadvantages

- Assumes the product is created in its final form and that it doesn't evolve.
- Improvements to a product are ignored.

PROGRESS CHECK

1. What are the features of a post-industrial society?
2. The advent of the small firm was spawned from de-industrialisation. Why did this phenomenon happen?
3. What do you understand by the multiplier effect in Myrdal's model?

Answers: 1. Increases in tertiary activity, secondary activity drops and quaternary industry increases. 2. Smaller more efficient sections of bigger companies always survive. By pooling the knowledge and experience of a few key workers drives such firms forward. Rapidly changing technology is better managed in small firms. 3. Ideas and wealth permeate through the area encouraging new industry and operations to develop.

Exam practice questions

1. **(a)** Define the terms (i) de-industrialisation and (ii) branch plant economy. **[4]**
   (b) For one named region where de-industrialisation has occurred, outline the effects of the process on the regional economy. **[6]**
   (c) Explain why a branch plant economy does not guarantee continued economic development. **[5]**

12 Agriculture and food supply

The following topics are covered in this chapter:

- **What is agriculture?**
- **Organised industrial agriculture (OIA)**
- **Future trends**

12.1 What is agriculture?

LEARNING SUMMARY

After studying this section, you should be able to understand:

- that agriculture is a system that farmers can manipulate to enable them to maximise output and profits
- that agriculture operates at a variety of scales and levels and can be classified in a number of ways
- that a variety of factors influence the decisions made by farmers

Defining agriculture

AQA **AS U1**
WJEC **A2 UG4**
Edexcel **A2 U3**
CCEA **A2 U1**

Importance of agriculture

- Agriculture employs millions around the world. In fact about 50% of the global economically active population are involved in agriculture. In some 90% are involved LEDCs. In MEDCs as few as 0.5% can are involved in agriculture!
- It makes a strong contribution to economic output.
- Being so important ensures the issues it raises are always in the news.

Classifying agriculture

Traditionally farms were classified by the dominant enterprise: be it dairying, arable production and so on. It is now recognised that a whole range of characteristic prefixes and suffixes can be used and applied to farming systems. Some common classification criteria are shown below.

> Traditionally questions on agriculture have focused on this aspect. However, most new specifications include this as a minor element in their agricultural 'orders'.

> Learn a case study for each classification.

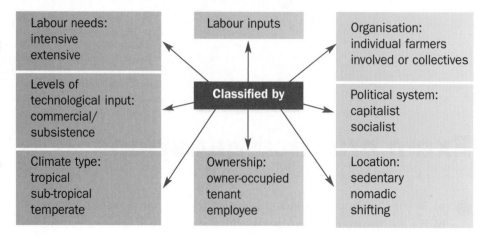

Labour needs:
intensive
extensive

Levels of technological input:
commercial/
subsistence

Climate type:
tropical
sub-tropical
temperate

Labour inputs

Classified by

Ownership:
owner-occupied
tenant
employee

Organisation:
individual farmers
involved or collectives

Political system:
capitalist
socialist

Location:
sedentary
nomadic
shifting

The agricultural system

AQA	AS U1
WJEC	A2 UG4
CCEA	A2 U1

Like any system there are inputs, processes and outputs linked to farming. At its most basic, various chains of people, groups and institutions labour to produce food.

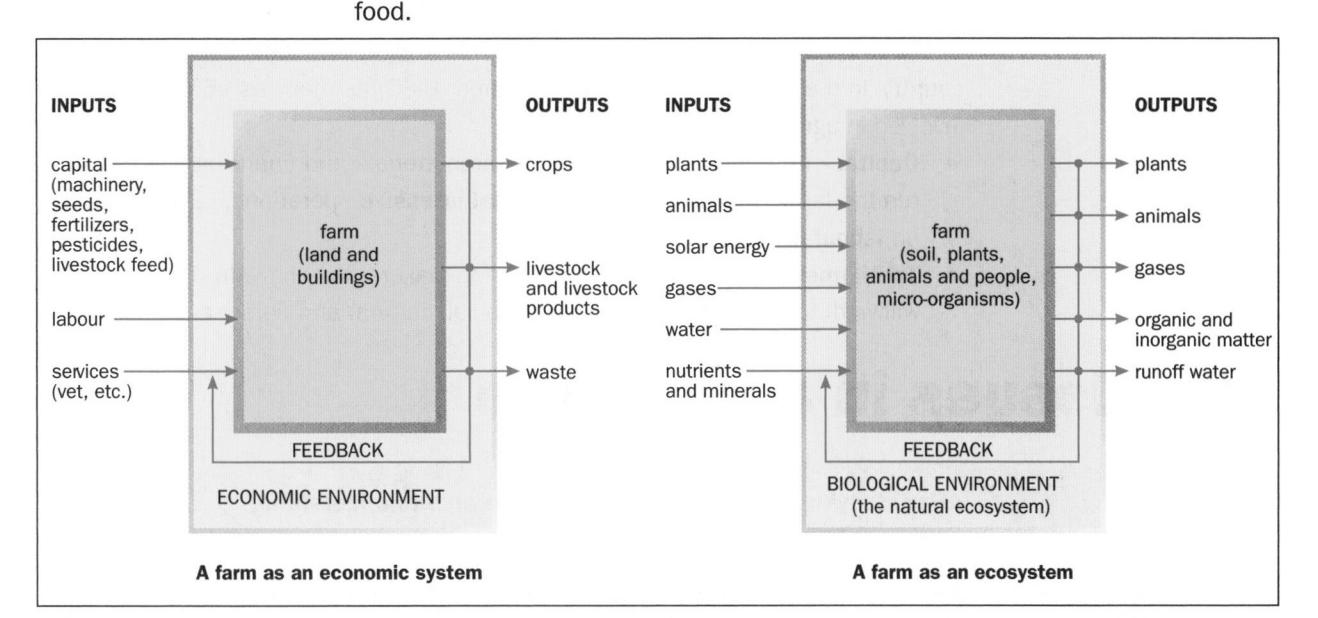

A farm as an economic system

A farm as an ecosystem

Influences, impacts and factors involved in agricultural production

> Worth remembering for the exam. If agrisystems are balanced in terms of inputs and outputs, such systems are said to be sustainable. Organic farming attempts to do this.

Influences

- **Climate** – precipitation is probably the most important climatic characteristic. The significant factors are: annual amounts; seasonal distributions; variability. Irrigation can overcome many of these problems.
- **Temperature** – varies with latitude. Influences tend to be at global, regional and local scales.
- **Soils** – these vary in their physical and chemical characteristics and this influences agriculture.

Physical factors	Chemical factors
• soil moisture budgets (relates to water holding capacities)	• soil acidity – the best soils are between pH 6.5 and 7.5
• soil texture (relates to drainage)	• soil nutrients – the most productive soils are high in nutrients
• soil structure (relates to the way humus and minerals stick together)	

Impacts

- **Environment** – farming has a distinct and profound effect on the environment.

Positive	Negative
• development of the agricultural landscapes of fields, hedgerows and trees	• land degradation (soil erosion, salinisation and desertification)
	• reduction in biodiversity
	• deforestation
	• water pollution

Factors

Like any economic activity agriculture depends upon:

- **Workforce** – in MEDCs the number involved in agriculture has declined. In some counties in the UK less than half of one percent are still involved in agriculture and the workforce is greying fast.

In LEDCs the number involved in agriculture is roughly the same country to country. In the highly indebted poor countries (HIPC) as many as 95% may be involved in agriculture.

- **Capital** – in farming this includes all the materials and financing needed to run the farm. MEDCs tend to be **capital intensive** operations. LEDCs tend to be **labour intensive** operations.
- **The farmers** – most farmers are decision-makers; they have to decide what will work to achieve their economic, social, cultural and personal goals.

12.2 Issues in agriculture

LEARNING SUMMARY	**After studying this section, you should be able to understand:** • the current issues facing agriculture • the issues relating to population increase and food supply • that ensuring food supply is a global challenge

The current situation

AQA **AS U1**
Edexcel **A2 U4**
WJEC **A2 UG4**
CCEA **A2 U1**

Issues are the main thrust of agricultural geography in the new specifications.

Agriculture has undergone many changes and revolutions over the last century, to ensure that food production has kept pace with the world's booming population, e.g. **high yielding varieties** (HYVs), fertiliser and pesticide usage, bio-technology in the form of **genetically modified (GM)** foods and so on. At the same time this has been going on there has also been an increase in the numbers of people that want 'chemical free' or **organic foods**. The other major change has been the concentration of power into the hands of small numbers of multi-national firms, concerned with the growing, marketing, processing and packaging of foodstuffs.

Population increase and food supply

AQA **AS U1**
Edexcel **A2 U4**
WJEC **A2 UG4**
CCEA **A2 U1**

The government of Haiti was toppled by food rioters.

Over the last 50 years population has more than doubled. Meeting the demand created by extra mouths is the responsibility of individual governments. The potential crisis of meeting demand has been coined **food security**. It is estimated that by 2050 just over 10 billion people will have to be fed in a sustainable way. In 2008 food prices rocketed, e.g. wheat prices rose by 130% (drought in Russia in 2010 further inflated prices) and rice by 74%. These inflated food prices saw the United Nations (UN) add $1.2 billion in food aid to help 75 million people in 60 poor nations.

Every year the UK throws away 8.3 million tonnes of food, worth £12 billion, or around £680 for the average family.

More people die every year from hunger and malnutrition than, for instance, AIDS. The World Bank has estimated that a 50% increase in cereal production and an 86% in meat production is needed by 2030 to meet world demand. In 2009 world food reserves were at their lowest level for many years and 32 countries were experiencing some sort of food crisis.

In July 2010 there were 18 million tonnes, 30% of India's grain reserves of wheat, rotting because of a lack of storage. This would feed 210 million poor Indians for one year! The figures for China (5–23%) and Vietnam (10–25%) are similar.

As well as increases in population many other factors are affecting global food production, including changes in weather patterns, a lack of water, desertification, soil erosion, fungal diseases, pesticide resistant insects, poor infrastructure causing losses through weak transport and storage capabilities.

Many of the issues indicated above are global problems. Ensuring the world's food security challenge will require a huge multi-national effort to integrate the best research from science, engineering and socio-economics so that technological advances can bring benefits where they are most needed.

Case study: Modern agriculture and food security – a brief history

Following the Second World War rising populations in many countries meant that governments had to try to increase the amount of food they could produce, thus avoiding future shortage problems experienced during the war. The decisions made led to increases in production using varieties of crops (HYVs) combined with the use of new pesticides, fertilisers and new machines. This period has been termed the **second agricultural revolution** in MEDCs (by 1990 the UK became 80% self-sufficient in food production) and the **Green Revolution** in LEDCs (10 years after the Second World War Mexico became an exporter of wheat). The period of greatest technology uptake was 1943–1973.

The Green Revolution has its origins in Mexico. The term was first used in 1968 by William Gaud, USAID Director.

There have been critics of the Green Revolution who say that the benefits have not been evenly distributed and that the methods employed required fertilisers, lots of water and pesticides. Farmers in developing countries could ill afford these chemicals and the equipment needed. Ecologists were also convinced that soil fertility, erosion and the effects of pests would increase. Economists felt that the green revolution had increased debt and forced poorer farmers from the land.

A second new and greener revolution has been called for, that includes lessons learned from the past, that spreads its benefits more widely and enables the landscape to be managed in a more sustainable and responsible fashion. It is also necessary to cope with the increase in global demand for food.

Ensuring food supply

AQA	**AS U1**
Edexcel	**A2 U4**
WJEC	**A2 UG4**
CCEA	**A2 U1**

Some examples: in 1989 Mexico's obesity rate was set at 10%, fifteen years later it hit 70%! In China in a similar time period the rates rose from 13% to 27%.

Two different perspectives

The fat world: globesity!

Alongside the issue of malnutrition is the escalating problem of obesity and, contrary to popular opinion, it is not just a problem in the developed world. It is estimated that there are in excess of 1.3 billion overweight people in the world today, which compares with 800 million who are underweight!

This is happening because people around the world (especially in the developing world) are eating more 'calories', oils and animal-source foods. The developing world is increasingly using technology that lowers physical activity. Changing lifestyles and affluence will continue to affect globesity!

The malnourished world

In 2009 there were approximately 200 million malnourished people in Sub-Saharan Africa.

It is unlikely that Sub-Saharan Africa will reach the 2015 Millennium Development Goal of reducing malnutrition. Economic growth, investment in agriculture, good governance, political stability, internal peace, rule of law, rural infrastructure, agricultural research, better education for children in rural areas and improving the situation of women are all essential for increasing agricultural production and reducing hunger and poverty. War and conflict (alongside unreliable weather over last 20 years) has conspired to actually reduce agricultural production by 12.5%.

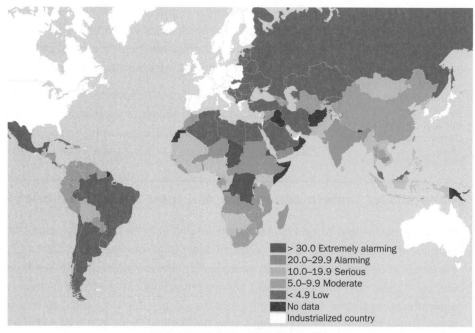

Global hunger index 2009 by severity

> 30.0 Extremely alarming
20.0–29.9 Alarming
10.0–19.9 Serious
5.0–9.9 Moderate
< 4.9 Low
No data
Industrialized country

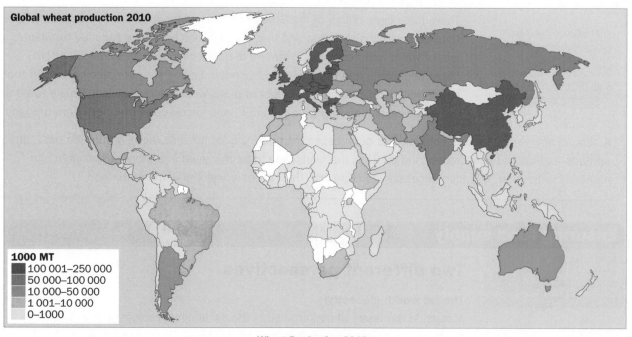

Global wheat production 2010

1000 MT
100 001–250 000
50 000–100 000
10 000–50 000
1 001–10 000
0–1000

Wheat Production 2010

It is unlikely that levels of malnutrition will drastically reduce principally because of the extreme environmental conditions, but also the overriding social and economic problems facing this area.

Case study: Failed harvests through drought and conflict, Darfur

There has been extreme tension, war and conflict in Darfur for many years relating to land and grazing rights. The consequence of this was that in 2009 very few farmers in Darfur had any seeds to plant. The seeds that were available would only enable less than one third of cultivable land to be utilised.

Darfur's population depends on locally produced grain for food. From the planting of 2009 only 15% of the populations needs were met! It is likely aid will have to supplement the rest of the country's needs for some time.

Governments and other groups have dealt with the unevenness of food production and consumption, in vastly different ways.

- **By importing food** – this has inbuilt problems, especially when the producer country has years of shortage itself and of course during years of conflict when food shortages are made worse.
- **Government policy** – after the Second World War it was thought that the UK would not be able to support its growing population and recognised the vulnerability of its supply chain. Wide-scale industrial farming was adopted which drove production up at some considerable environmental cost. However, the UK farming system is now extremely efficient.

> **KEY POINT**
>
> **US farming subsidies**
> Since the Great Depression USA has paid direct subsidies to farmers of around $20 billion/year, under farm income stabilisation bills. The notion is that farmers get extra money for their crops and a price guarantee.

- The maintenance of food supply was also behind the **adoption of a Common Agricultural Policy** by Germany and France in the 1950s. Its influence continues to affect the members of the EU, ensuring that the smaller farmer's production and prices they gain for products are maintained.

European Common Agricultural Policy

Always plenty in the press about CAP!

The **Common Agricultural Policy (CAP)** is the most controversial of all the EU policies. Originally designed to increase food production and to secure EU food supplies little has changed since its creation (in 1957 under the Treaty of Rome), though it also attempts to protect the rural landscape and lifestyle too. Attempts have been made to reform CAP and to allow overseas competition. The CAP is due to be renewed in 2013 and it is expected that monies will be spread from CAP to include supporting sustainable energy in Europe too.

At the World Trade Organization (WTO) meeting in Doha in 2006 it was agreed to phase out all agricultural export subsidies by the end of 2013.

A form of **protectionism**, CAP attempts to ensure that European producers are protected from cheaper products outside the EU by ensuring that subsidies were applied to agricultural produce. CAP is now achieved by discouraging imports from beyond the EU with a system of **import tariff and quotas** and by subsidising EU farmers through the **Single Farm Payment (SFP)** (removing the link between subsidies and production of specific crops, this system introduced in 2005 is thought to be more agriculturally sound).

The world agricultural market is constantly monitored – if surplus food is produced then the EU will intervene supporting with extra subsidies the export of the crop/product (generally to poor countries like Africa); storing or create '**food mountains**'; selling it later in the cycle, when prices are higher or by destroying it. The CAP also sets quotas on how much a farmer can produce or by paying them not to produce, in an attempt to control production.

The CAP budget for 2010 is €58 billion. The UK's contribution is about £7 billion per year. Under SFP the UK farmers receive around £230 per hectare, as long as they meet various environment and welfare standards.

Arguments for CAP	Arguments against CAP
• Looking after the countryside can be done by looking after the custodians of this landscape (farmers). • In a free market prices fluctuate. CAP ensures stability.	• CAP makes food much more expensive than in a free market. • Poverty in poor countries increases as they are unable to compete with the EU. • A small number of people benefit from massive amounts of money set aside for CAP. • CAP is an expensive scheme. • CAP displaces instability from the EU into the wider, world market. • CAP also conflicts with Article 110 of the Treaty of Rome, which states that member states should aim to contribute to the 'harmonious' development of world trade. • CAP threatens the USA agricultural system and leads to a trade wars. • CAP generates resentment among Eastern European countries. • Subsidies are controversial as money is redirected from the 'national purse'.

Polish farmers in favour of CAP!

Since joining the EU Poland's farmers have really benefited from CAP. After accession support for agriculture and rural development increased. 1.4 million Polish farmers, working 90% of the cropland in the country, benefitted. A by-product of CAP payments has been the increase in farmers who have bank accounts (up from 20% to 90% – this is because a bank account was necessary to receive CAP support.

Larger intensive producers have gained most on the scheme.

Reform of CAP is constantly called for. The 2010 EU agricultural budget is €58 billion.

Governments have actively encouraged changes in diet. These changes must be reflected in governments' approaches to food production. By midway through this coming century agricultural production is expected to have tripled or quadrupled to keep up with demand from the world's burgeoning population. Both MEDCs and LEDCs will have to maintain food supplies against growing demand, without substantially further damaging the environment or people's health.

KEY POINT

Sufficient food for the future?
- For over a hundred years the world has managed to increase food production faster than its population has grown.
- There is greater per capita food availability and decreasing real prices which equates to greater demand.
- Growth in agricultural production has slowed.
- Margins between supply and demand are smaller – arable land being lost to non-agricultural uses and extreme degradation.
- It is agreed that the world population will double in the next 40 years (possibly 7 billion in 2010 and beyond 8 billion come 2025).

So what of the future?
Most reports are optimistic, suggesting:
- small increases in cultivable land
- yields to increase 1.5–2%
- global food demand will continue to grow at a rate slightly behind supply
- food availability measured in calories per capita will increase
- prices will remain constant or fall slightly
- developing countries are expected to increase food imports
- the developing world will source its imports from the developed world.

Those reports that are less optimistic, assume:
- food insecurity will remain in Sub-Saharan Africa and South Asia, especially as markets, like those in China, destabilise markets
- weather and climate changes could adversely affect food production
- the demand for water could approach the hydrological limits the planet can support
- that plants are no longer responding to fertiliser and pesticide applications
- the loss of agricultural land to other uses.

So what must be done?
There should be fewer threats to the food system over the coming decades as long as:

- new and appropriate technology continues to be worked on and invested in
- the resources and environment are used effectively and sustainably
- barriers are removed to trade in food and agricultural commodities
- the world is flexible to any shocks in supply and demand occurring in vulnerable countries.

Poverty is the principle cause of food insecurity in poor countries. Poverty needs to be eliminated.

Issues related to food supply

AQA	AS U1
OCR	A2 U3
Edexcel	A2 U3/4
WJEC	A2 UG4
CCEA	A2 U1

About 20% of US produced corn is used to make ethanol.

The US government subsidises ethanol production as it contributes to America's targets for reducing greenhouse emissions.

With biodiesel you're left with the oilseed cake after the oil has been pressed out – depending on what seed is used, this is usually a highly nutritious, high-protein livestock feed.

The rising cost of tortillas, made from corn, has led to street demonstrations in Mexico!

The food vs. fuel trade-off – can you have your cake and eat it?
For a number of years now the claim has been that biofuel demand makes food expensive. The cost of corn in 2010 more than doubled. The reason for the increased price is demand from refineries buying up the corn to turn it into ethanol. The ethanol extracted is mixed with conventional oils for use in vehicles.

There is an effect being felt by corn users and consumers in the USA that basic foodstuffs costs will increase – cheap food in the USA is expected to become a thing of the past. Corn is also used as a feed for pigs and chickens and so higher costs at this point in the food chain will mean higher costs when it is brought to the table. Another consequence of this surge in biofuel use has been an increase in land prices across the USA as more farmers and companies jump on the biofuel bandwagon. In Asia and Brazil virgin rainforest has been removed (with all its consequent biodiversity loss) to enable hugely profitable oil palms to be grown.

The issue needs to be measured against the real situation of food supply and demand (half of all food in the USA is dumped). This was exemplified in Brazil during its 'ProAlcool programme' when, of 55 million hectares of land under production, only 1.7 million was used for ethanol production.

If it is not increased biofuel production, what is increasing food prices? Around the world it is thought that policies which are biased towards large acreage commodity export crops, hyper-inflation, currency devaluation and price controls, should be considered in discussions about reasons for increasing food prices. Furthermore, people starve because they are victims of an inequitable economic system, not because they're victims of scarcity and overpopulation.

Fair trade agriculture
Sales of fair trade goods in the UK have increased by about 40% in the period 2009–10. This is thanks to consumer awareness and a greater demand for such products. Demand drives supermarkets to stock a range of such items.

Coffee and bananas are produced under fair trade schemes.

GM crops are forbidden (and crops do not have to be organic!).

The trade in fair trade cotton

Cotton has been central to the textile industry for hundreds of years. The only difference between fair trade cotton and ordinary cotton is that the marginalised farmer at the bottom of the process chain is more fairly treated (7.5 million farmers in 60 countries, including India, Mali, Egypt and Peru; two-thirds of cotton is produced in LEDCs). Fair trade encourages careful sustainable stewardship of the environment. In real terms, even with fair trade, cotton farmers in developing countries earn less for their crop than they did in the early 1990s.

Problems faced by cotton farmers in the developing world	Benefits of fair trade for cotton growers
<ul><li>It is their only cash-crop.</li><li>No crop, or an unreasonable price for their crop, means families don't eat.</li><li>Unstable world prices.</li><li>The price of pesticides and fertiliser.</li><li>Cotton production in the developed world is subsidised.</li><li>When cotton farmers in the developed world over-produce, prices are driven down.</li><li>Cotton grown in most developing countries is at the hydrological boundary of production.</li><li>Lack of agricultural equipment/infrastructure.</li></ul>	<ul><li>A guaranteed price is paid.</li><li>A fair trade premium is paid to be used for community projects.</li><li>Access to global markets is achievable.</li><li>Contributes to general development.</li><li>Pre-finance of up to 60% of the value of the contract is available on request.</li><li>Fair trade registered buyers make a commitment to long-term, stable relationships with producers.</li></ul>

Case study: Mali's cotton industry

Cotton was introduced into Mali in 1996 when the peanut crop failed. About 600 000 tonnes are produced per year and 40% of rural dwellers on 200 000 farms are dependent upon the crop.

Dougourakoroni cotton producers co-operative

The Dougourakoroni co-operative was founded with the aim of improving living standards for its 169 members. Farms are about 7 hectares in size, with 1.3 hectares under cash-crop cotton, producing around 1000 kg of cotton a year. Farmers, in rotation with the cotton, grow millet, sorghum, maize and peanuts on a subsistence level. No irrigation is used for any of the farming undertaken.

The co-operative provides access to loans, technical advice and agricultural training. It supplies agricultural inputs and oversees commercial activities. Workshops are also organised where knowledge and experience are shared with non-members. They also collect the members' harvested cotton, weigh and classify it, store and guard it. The cotton is purchased by the government, at an agreed farm gate price and sold on the local and world market. The French company DAGRIS (Dévelopement des Agro-Industries du Sud) has helped Mali introduce fair trade cotton products into French markets.

The co-operative was inspected and certified in 2004. Its seed can now be exported and the quality of its cotton is such that it now exports part of its production under the conditions guaranteed by international fair trade standards. The co-operative has recently joined in union with 20 co-operatives in the Kita area and this will make the co-operative stronger and increase sustainability through improved agricultural methods, quality control and environmental protection.

The cultivation of marginal agricultural land

Most of the prime agricultural land in both developed and developing countries is now used for cash-cropping. It is to marginal land that most now look for agricultural land to further develop. Such areas characteristically exhibit degraded soil, poor land fertility, scarce water and improper irrigation methods. The unsuitability of such areas means that results are often disastrous, with such areas being susceptible to drought, landslides or flooding and salinisation. Farmers are coming up with a range of solutions for such areas.

- Planting biofuel crops.
- Forage crop cultivation.
- Innovative irrigation techniques, i.e. micro-irrigation.
- Innovative soil retention techniques, i.e. stone bunds.
- The cultivation of trees, such as white pine, that grow successfully and quickly.

Organic farming

Organic farming works in harmony with nature rather than against it. The aim is to achieve good crop yields without damaging the environment or the people 'working' it. The methods and materials that organic farmers use include keeping and building good soil structure and fertility through recycling, composting, rotating crops and mulching. Organic farmers control pests, diseases and weeds by using resistant crops, encouraging useful predators that eat pests, by increasing genetic diversity. Most organic farmers also use water carefully and rear animals using long established husbandry techniques.

Organic farming does not mean going 'back' to traditional methods – it takes the best from the past and combines it with modern scientific knowledge.

The market for organic chicken

In the USA the market for white meat is increasing. The production of organic meats is the fastest growing section of the organic market. In 2005–10 sales of organic meats have increased by over 70%. In the same time period, only a little over 1% of all laying hens in the USA were certified organic.

Food miles and carbon dioxide

Food miles (or food kilometres) describe the distance that food is transported as it travels from producer to consumer. In the UK, our food travels an amazing 30 billion kilometres each year. This includes imports by boat and air and transport by lorries and cars. Food transport is responsible for the UK adding nearly 19 million tonnes of carbon dioxide to the atmosphere each year. Over 2 million tonnes of this is produced simply by cars travelling to and from shops.

Choosing food that is local and in season means it does not have to travel so far. Reducing food miles can have a dramatic effect on reducing carbon dioxide emissions.

The concept of food miles has become overcomplicated! So the Peruvian avocados, Kenyan green beans, New Zealand lamb and all those other foreign foodstuffs that now fill the shelves of our supermarkets could in fact be greener than we thought. Experts are encouraging us to weigh up the fact that many products brought to our shores are grown using manual labour, nothing is mechanised, they use cow-muck for fertiliser (rather than oil-based fertiliser) and use low-tech irrigation. In fact driving 6.5 miles to buy your shopping emits more carbon than flying a pack of Kenyan green beans to the UK! In light of growing concern the Carbon Trust which, with the British Standards Institute, has been involved in calculating how a meaningful carbon inventory can be compiled for foodstuffs.

No organic farm is over-run with weeds or taken over by nature – they use all the knowledge, techniques and materials available to work with nature.

Nearly 75% of all meat bought in 2009 was chicken.

Locavores – the term coined by those that only eat locally produced products.

But is this all a big myth? Increasingly the bigger supermarkets, such as Tesco and the Carbon Trust think so!

1. Why is it that agricultural issues seem to dominate rural affairs?
2. Why is an understanding of agriculture and related systems important?
3. What is meant be food security?
4. Why is the issue of over-eating as resonant as issues related to starvation?

Answers: 1. Destructive nature of some agricultural practice/food security/sustainability issues/subsidies/… lots of areas to explore. 2. Employment issues – especially in LEDC's. Contribution to wealth. 3. Making sufficient food available to meet the demands of the people. 4. More people are over-weight in the world than are starving, including in some LEDCs. This is the result of changed diet and just eating too many calories.

12.3 Organised industrial agriculture

LEARNING SUMMARY

After studying this section you should be able to understand:

- the reasons for and the effect of the rise in OIA approached on both arable and livestock production
- that the use of toxic chemicals to fertilise and to ward off pests can be harmful to the environment
- that biotechnology has an increasing effect on crop and livestock production
- that OIA can affect both the environment and our health
- the global food system
- features and cost of OIA

The global food system

AQA	AS U1
OCR	A2 U3
WJEC	A2 UG4
CCEA	A2 U1

The current global food system is highly productive and means that out-of-season foods can be available all year round, often in plentiful supply. Due to trading methods and preservation techniques, food from all around the world is available – to those who can afford it. The food industry has extended from a local level to a global one. Those involved want to control costs and production in such a way as to ensure that costs are minimised whilst profits are maximised. The global food system of today is optimised for those who can afford to buy food – mainly the populations of MEDCs (only a quarter of the world's population).

Features of organised industrial agriculture

AQA	AS U1
Edexcel	A2 U3
WJEC	A2 UG4
CCEA	A2 U1

There is little doubt organised industrial agriculture (OIA) could support the biggest proportion of the world's population.

Specialisation

Most money is made in the processing portion of agriculture.

OIA has brought an end to peasant agriculture in the MEDCs. In the UK, for instance, less than 1% of the working population are maintained by agricultural activity (with a further 7% involved in food processing activities).

Tea, coffee, cotton.

OIA has also brought an end in MEDCs to the self-sufficient community. It has also meant, and enforced for many countries, a rise in monoculture. **Colonial monoculture** was imposed by many of Europe's colonisers on LEDCs well before the 1900s. On the whole these plantation crops have very little effect on the host GDP and provide no 'food' support. In some LEDCs monoculture has been put in place by the countries themselves to ensure that their population can be fed, but also to ensure that yields can be sustained year on year.

> Colonial monoculture is especially vulnerable to pests and diseases.

Issues in agricultural employment in the EU

- A fall in agricultural employment and north–south differences – on average agriculture accounts for less than 10% of jobs. The low percentages (3%) are found on the arable and 'animal' farms of the northern EU countries and highest (near 10%) in the labour intensive southern areas of the EU around the Mediterranean.
- In relative terms, employment in the agri-food sector is highest in Denmark and Ireland. The number of farmers per hundred inhabitants has tended to fall while the number of agri-food jobs per hundred inhabitants has remained static.
- Family labour is the keystone of agriculture – family labour predominates in agriculture, accounting for four out of every five jobs.
- Part-time work predominates in southern Europe.
- Agriculture remains a driving force for economic and social cohesion – even as a minority in the countryside, farmers are still the main managers of the land and agricultural work largely determines the degree of attractiveness of these regions, particularly where the landscape is concerned.
- In the UK agri-environmental measures are leading to a small increase in agricultural work and a substantial increase in work by specialist firms.

> Farmers are becoming more educated; in the UK 11% have undergone higher education.

Mechanisation

Changes brought on by two world wars meant the use and adoption of machinery and new technology was inevitable to ensure food supply and have ensured that the USA and Canada have become effectively the 'bread basket' of the world. The Green Revolution has mirrored the productivity of the intensified MEDCs' agriculture. Gains have been made, not by mass mechanisation, but through the use of fertiliser, pesticides and above all else the use of HYVs of seed.

> Despite year-on-year use, soil still sustains yields with no special additives, i.e. GM crops ensures yields.

Case study: GM rice – the rice that could feed the world

Genetically modified (GM) rice could boost yields by up to 35%, solving the world's impending food shortage. Rice stocks are dwindling and there are fears of famine as the world's population increases. Half the world's population is dependent on rice. Genetic material from maize can be inserted into rice, boosting the rate of photosynthesis so the plant is able to produce more sugar and increase grain yields. Farmers must consistently produce an extra 6.7 million tonnes of rice a year, using less land and less water, just to maintain current nutrition levels.

In India and Bangladesh the idea of having to buy seed each year from giant companies has brought serious resistance to GM technology, and GM companies have been heavily criticised for not helping the developing world more. Supporters of genetically modified crops pointed to the potential benefits of another GM rice announced in 2000, called golden rice because of the colour caused by the modification. This adds vitamin A to the grain and could cure the vitamin A deficiency of 124 million children worldwide.

All the crops commercially developed so far, such as soya, maize and oil and seed rape have been aimed at rich farmers in the USA, Europe and Brazil, adding to surplus stocks rather than feeding the world's hungry.

Intensification

On the open prairies of North America **intensification** has led to a consolidation of farms into larger and larger units, using less labour. The levels of mechanisation needed to run and manage these 'estate' farms means that increasingly financial groups and institutions have a considerable stake in them and are a driving force for farmers, ensuring continued high production and profitability.

Commercialisation

A further consequence of OIA has been the rise of the controlling large company – the rise of **agri-business**. These companies not only influence what is grown, but also what seeds and fertiliser are used! Undoubtedly the industrial approach is better suited to the more developed world, with their big fields, equitable climates and big machines. But even in LEDCs the influence of industrialised agriculture is increasing. This, against a background of steep terrain, sometimes arid and tropical climates and environments that on the whole are unsuitable for such techniques.

Organised industrial livestock farming

The industrialisation of agriculture has had particularly dramatic effects on the farming of livestock and impacts on meat supplies.

Mass production methods

Many farmers now consider the rearing of animals for slaughter in individual small units too inefficient. Many treat animals as units of production (ignoring basic needs such as exercise, fresh air and wholesome food) and have designed systems for turning huge numbers of animals into vast quantities of cheap meat extremely quickly, such as battery units and broiler sheds for chickens, sow stalls and farrowing crates and poultry sheds.

Abuse of drugs

New technology is further extending livestock productivity, e.g. injections of bovine somatotrophin can increase milk production in cows and clenobuterol is used to boost the rates of growth gain in cattle for beef. Factory farms attempt to counter the effects of intensive confinement, and its associated disease ridden conditions, by administering high doses of antibiotics and other drugs to animals. The most commonly used are penicillin and tetracycline. The squandering of these important drugs in livestock production is wreaking havoc for physicians in the treatment of human illness. As well as suffering from anaemia, influenza, intestinal diseases, mastitis, metritis, orthostasis, pneumonia and scours, the animal's behaviour is also affected!

The upshot of growing awareness of the conditions in which livestock are kept means that many thousands are turning to vegetarianism as a matter of conscience. Other concerns to do with livestock are the massive amounts of waste they produce and its effect on the climate.

Intensification has ensured a quadrupling of production in MEDCs.

Most farmers in MEDCs have a heavy debt burden, brought about by the need for massive machinery to farm their vast estates. Profit, necessarily, has to be at the forefront of their operation.

Plenty of press coverage available for this aspect.

600 million broiler chickens are reared per year in the UK. At slaughter they are 6–7 weeks old.

2000 human deaths per year are related back to antibiotic misuse. As a result more virulent bacteria harmful to health appear.

In Norfolk 'chicken poo' power stations have been built to deal with the waste produced!

Case study: Battery hens

There are 300 million egg laying hens in the USA (30 million in the UK) confined in battery cages – small wire cages stacked in tiers and lined up in rows in huge warehouses. The United States Department of Agriculture (USDA) recommends giving each hen 4 inches of 'feeder space', which means the agency would advise packing 4 hens in a cage just 16 inches wide. The birds cannot stretch their wings or legs, cannot fulfil normal behavioural patterns or social needs, suffer from severe feather loss and their bodies are covered with bruises and abrasions. They have part of their beaks cut off in order to reduce injuries resulting from excessive pecking, behaviour which occurs when the confined hens frustrated. De-beaking is a painful procedure. Laying more than 250 eggs per year the hens' body's are severely taxed. They suffer from 'fatty liver syndrome' when their liver cells, which work overtime to produce the fat and protein for egg yolks, accumulate extra fat. They also suffer from what the industry calls 'cage layer fatigue' and many die of egg binding when their bodies are too weak to pass another egg. A hen will use a quantity of calcium for yearly egg production that is greater than her entire skeleton by 30-fold or more. Inadequate calcium contributes to broken bones, paralysis and death. After one year in egg production, the birds are classified as 'spent hens'. They end up in soups, pot pies or similar low grade chicken meat products where their bodies can be shredded to hide the bruises. Male chicks are killed at the hatchery. They are of no economic value.

The cost of OIA

AQA	AS U1
Edexcel	A2 U4
WJEC	A2 UG4
CCEA	A2 U1

Nitrogen phosphorus and potassium are very necessary – 113 million tonnes of fertiliser is applied annually to fields around the world.

Many countries have signed the Convention on Biodiversity which enshrines the precautionary principles in relation to transboundary movements of GM crops.

The march of OIA has been just about unstoppable. Over the last 40–50 years grain production has doubled, even trebled, in some countries and resulted in a range of environmental and health problems. Below are some of the problems and changes that result from the increasing grain harvest.

- **Fertilisers** – these effectively extend the life cycle of the soil by artificially adding nitrogen. The application of fertiliser is often overdone, the excess being washed into watercourses causing **eutrophication** and contaminating water more generally with nitrates. Nitrates are linked with a range of disorders including 'blue baby syndrome'.
- **Seeds and plants** – there was a time when all countries and areas grew their own crops which were suitable for the areas and region in which they lived. However, in today's 'industrial economy' large economies of scale and production could not be achieved using such a system. Today, less than half a dozen high yielding, pest and disease resistant varieties are grown and they supply nearly three-quarters of the world's calorie intake. To enable these varieties to 'provide' as they do, considerable **genetic engineering** has been undertaken. Some consider that this **genetic modification (GM)** is intrinsically wrong or 'unnatural'. Others are uneasy about what GM means for our relationship with the natural world. For many it is a step too far, concerned as they are with the consequences, rather than the rightness or wrongness of manipulating genes.

The potential opportunities and risks of GM crops in developing countries

Socio-economic:
- Higher agricultural yields from reduced labour.
- Reduced costs for producers, e.g. from reduced dependency on fertilisers and pesticides and increased effectiveness of herbicides and cheaper products for consumers.
- Help for poorer people, e.g. by enabling cultivation on land that was previously considered useless.
- Cheaper staple foods for importing to developing countries.
- Increased corporate control over seed and agro-chemical markets, e.g. through monopoly production, at the expense of poor farmers.
- The range of crops that can be produced in northern temperate zones is increased, which compete with developing country exports.

- Potential opportunities bypass smaller farmers due to a lack of investment in research on crops/applications.
- Biotech companies bear little liability for harm caused to the environment/public health.
- Large-scale GM crop farming makes GM-free and organic farming impossible.
- GM crops and technology reduce production costs, but are only adopted by larger farmers. This increases problems for small-scale farmers.

Environmental:
- Reduced use of pesticides and herbicides.
- GM plants can be developed to remove toxic chemicals from soil.
- Biodegradable plastics can be produced.
- May lead to loss of biodiversity from monocultures.
- Alien genes transfer from GM crops to other varieties and other species with unknown effects.
- Increased resistance of weeds and pests to agrochemicals, leading to increased use.
- Reduced natural soil fertility (due to reduced activity of nitrogen-fixing bacteria).

Health and consumer choice:
- Eradication of allergens and toxic substances in crops.
- Production of vaccines.
- Increased allergens result in antibiotics losing their effect so viruses spread between species, e.g. from plants to bacteria in the human gut.
- Improved quality and shelf-life of fruit and vegetables.
- Better flavour, texture and nutritional content of food.
- Lower prices for food products.
- Choice to make buying decisions based on social, ethical, religious and dietary preferences may be threatened.
- **Water irrigation** – much of the water used by man finds its way onto the world's crops: irrigation is carried out on 25% of the world's cropland. Agriculture must in the future balance its needs against the needs of growing populations.
- **Fuel consumption** – farmers are big users of fossil fuels, especially in MEDCs where high mechanisation uses agricultural diesel.
- **Pesticides, herbicides, fungicides and insecticides (biocides)** – in China 150 000 hectares of land have been planted with genetically modified cotton and the country is spending some $80 million a year on research and development of GM crops. Similarly, the Department of Biotechnology in India urged the Indian government to provide nearly $3.5 billion for a 10-year plan to develop biosciences.

> In LEDCs it has led to salination and water logging.

> 1 tonne of fertiliser = 1600 litres of fuel to make it! 10 units of fuel = 1 unit of food energy!

> Approximately 20–30% of harvested food is spoilt by pests and disease. A bigger proportion never makes it to harvest!

Problems with biocides are mostly confined to LEDC usage:
- Dichlorodiphenyltrichloroethane (DDT) (an organochloride) and organochlorines are still used that will persist in the soil. The link between accumulations in food chains and deformities, deaths and illness is well documented.

> The world market for pesticides is worth about $35 billion.

- Poor education ensures that biocides are not being properly applied or effectively used in many LEDCs.
- Chemicals banned in the West are often found as toxin residues in/on imported LEDC products, e.g. bananas. Within most agricultural circles there is recognition that biocide use needs to be minimised, as chemicals used are soon superceded by a naturally produced 'counter chemicals'. In MEDCs strict regulations apply to the use and application of biocides with resulting increased use of biotoxins or disease resistance being built into the genetic structure of plants.

Amounts of biocide
used/year (in tonnes):

USA	380 000
Italy	174 000
Brazil	68 000
Spain	45 000
Mexico	34 000
UK	30 000

12.4 Future trends

LEARNING SUMMARY

After studying this section, you should be able to understand:

- that concerns about the environment and animal welfare have led to an increased interest in more sustainable agriculture.

Sustainable agriculture

AQA	**AS U1**
OCR	**A2 U3**
Edexcel	**A2 U4**
WJEC	**A2 UG4**
CCEA	**A2 U1**

Sustainable farming conserves nutrients, soil and water, whilst negating the need for lots of fertilisers and biocides. Sustainable farming is unable to compete with OIA in its intensity of yield, but competes instead through its appeal to its customers. Such individuals are concerned about nutritional quality, health and taste. Furthermore, farmers are increasingly concerned about what they are doing to the environment, causing soil erosion, water contamination and so on.

Sustainability in agriculture can range in the MEDC from using a low tillage plough disc, which affects only the top layer of soil, to fully fledged and accredited small-scale organic farms that may well use no-tillage agriculture or **perennial polyculture**.

In the LEDCs sustainable agriculture usually involves appropriate or transfer technologies and invariably returns the country back into subsistence management of farms.

The essential attributes for our agricultural future are summarised below.

- Farms should be very productive, producing safe, high quality crops and animals.
- Farming should be sustainable, i.e. using physical resources to allow long-term development.
- Farms must be biologically sustainable avoiding biological agents and pesticides.
- All farms should satisfy agreed standards for human and animal welfare.
- Pollution must be avoided.
- Products must be grown/reared that are wanted, i.e. biofuel vs. fuel debate.

Exam practice questions

1 The figures below show the growth of organic agriculture in the UK.

UK Sales of Organic Products 1995-2009.

Year	Sales
2009	£1,840 million
2008	£2,113 million
2007	£2,078 million
2006	£1,900 million
2005	£1,600 million
2004	£1,200 million
2003	£1,100 million
2002	£1,000 million
2001	£920 million
2000	£802 million
1999	£605 million
1998	£390 million
1997	£260 million
1996	£200 million
1995	£140 million

(a) Assess the relationship between area under organic production and numbers of organic farms in the UK. [3]

(b) Suggest how consumer pressure and environmental concerns may have caused the pattern of growth that appears on the graph. [7]

(c) Discuss the idea that the advantages of organic farming do not outweigh the problems for either the farmer or the consumer in MEDCs. [10]

2 Assess the role of agricultural change in meeting the needs of a growing global population. [25]

13 Global challenges and issues

The following topics are covered in this chapter:

- Energy – how secure is its future?
- Superpowers
- Water conflict
- The geography of technology and the technological fix
- Pollution
- Trade alliances
- Health and welfare issues

13.1 Energy – how secure is its future?

LEARNING SUMMARY	After studying this section you should be able to understand:
	• energy and its security into the future.

Energy supplies

AQA	AS U1
OCR	AS U2
Edexcel	A2 U3
WJEC	A2 UG4
CCEA	A2 U1

On our contested planet there are **players** (the decision makers – organisations, groups and individuals). All the players hold different opinions, values and perspectives. The response to any issue warrants an **action**. In geographical terms such actions are commercial and market led, perhaps focusing on national planning and are **top-down** in nature or are sustainable, often community led and **bottom-up** in nature. Based on the players and their actions our futures are decided. The routes taken are either 'radical', where attempts are made to return the environment to how it was, or are 'sustainable' where some stabilisation is attempted. Alternatively, they just focus on carrying on as we've always done, the so called 'business as usual' route.

Energy supplied is either primary or secondary in nature. Energy could come directly from an oil fired source or it could secondarily heat water, producing steam and then turning turbines. The UK's primary energy mix is shown opposite.

Other 2%
Coal 33%
Renewable 5%
Nuclear 16%
Gas 43%

The UK's primary energy use

Demand, supply and security

Candidates might be expected to know how the energy in the UK is produced.

Non-renewable	Renewable	Recyclable
These resources will run-out (they are finite).	Some 'infinity', can be replaced – continual flow.	Management needed – can be re-used.
Examples include coal, gas and oil.	Wind, tidal, solar and hydroelectric power.	Biomass and nuclear
Environmental impacts are high.	Medium impact – NIMBY ('Not In My Back Yard') issues.	Low impact – is nuclear 'green'?

NON-RENEWABLE ➡️ RENEWABLE

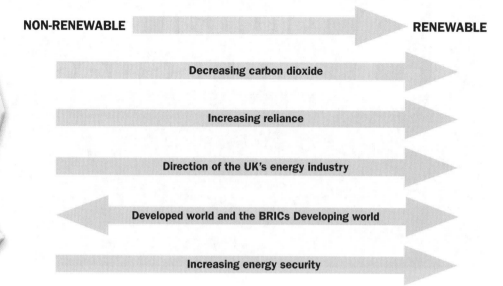

Decreasing carbon dioxide

Increasing reliance

Direction of the UK's energy industry

Developed world and the BRICs Developing world

Increasing energy security

Energy issues and different countries: The UK imports lots of gas from overseas. France is dependent on the nuclear industry. Singapore would be unable to produce renewable energy.

BRICs = Brazil, Russia, India and China.

Global energy distributions

Some countries produce lots of energy, while others produce very little energy. Production is relative to the amount of resource available and the money to exploit it!

Know examples of countries with high and low energy consumption and how their needs are supplied.

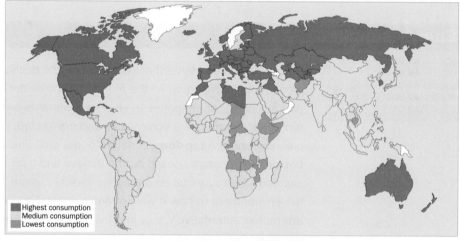

World Consumption of Energy

Highest consumption
Medium consumption
Lowest consumption

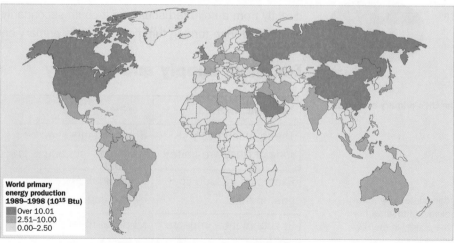

World Energy Production

World primary energy production 1989–1998 (10^{15} Btu)
Over 10.01
2.51–10.00
0.00–2.50

There is a very strong link between gross domestic product (GDP) and energy consumption. The rich countries generally use more energy than poor countries.

Problems with fossils fuels

Many pollution issues have been identified relating to fossil fuels.

- **Acid rain:** Burning fossil fuels releases gases that make precipitation acidic.
- **Oil spillage:** Causes environmental damage.
- **Global warming:** Fossil fuels burned during energy production release carbon dioxide – this is a **greenhouse gas** which is linked to climate change.
- **Mining:** Open cast mining destroys huge areas of land.

The politicisation of energy

Shortages of energy resources ensure that energy resourcing and production is top of many countries' agendas.

Issue	Outcome
Middle East tensions	• Oil and gas reserves might be withdrawn. • Countries are willing to go to war to secure supplies. • Prices increase.
Superpower energy	• Superpowers establish strong links to ensure continuation of supply. • Rich countries pay over the odds for resources. • The Gulf States and Russia hold the bulk of resources.
Cost of oil	• The cost of crude oil is very volatile reaching highs of over $120/barrel in early 2010. • Stockpiling and price fixing cause massive problems for countries.
Arctic and South Atlantic supply	• Arguments over territorial water mean problems ahead for oil and gas exploitation.
Going nuclear	• Waste disposal issues. • Terrorist issues.

Alternative and sustainable energy sources

With fossil fuel resources being depleted at an increasing rate, many countries have looked to other methods of energy production.

Accident can happen – Chernobyl, Russia (1986) and Three Mile Island, US (1979) are two examples. There have been 97 other accidents around the world since 1970 – two-thirds of them in the USA.

In 2010 a total of 264 wind farms contributed 4580 MW to the National Grid – 40% of Europe's wind passes over the UK!

Energy source	Advantages:	Disadvantages:
Nuclear power (6% of world power). Heat energy from uranium or plutonium heats water, steam turns turbines and electricity is produced.	• Less waste • Less toxic material! • Lots of cheap electricity • Low levels of carbon dioxide emitted	• Radioactive material is difficult to manage, store and decommission • Accidents can happen
Wind power (1% of world power). Turbines turned by the wind to produce mechanical energy that turns a generator/turbine: located in exposed areas with regular wind.	• No fuel • No pollution	• Not always windy • Large turbines needed • Noisy and unsightly

La Florida plant, Alvarado, Spain covers 550 000 m² and produces 50MW. Spain produces 20 000MW of solar power, 53% of its needs!

Energy source	Advantages:	Disadvantages:
Solar power (>1% of world power). Works by either concentrating or storing energy from the Sun: photovoltaic cells convert energy into electricity.	• Clean • Free • No carbon dioxide produced	• High installation costs • Sun needed! • Large land requirement
Biomass power (4% of worlds power). Could be wood, plants and animal waste that are burned to produce energy.	• Renewable if properly managed	• Polluting • Waste has to be disposed of • Noxious gases released
Hydroelectric power (20% of world power) HEP projects trap water and release it to turn turbines	• Renewable • No carbon dioxide released • It's 'free'	• Habitat destruction • The production "plant" is expensive

The Three Gorges Dam Project, on China's Yangtse river cost $26 billion to build, capable of producing 22 500 MW of power.

Case study: Two different biomass power plants, UK

Sembcorp Biomass Station is a £60 million biomass plant built in the Tees Valley, UK. It uses sustainably cropped wood to produce 30 MW of electricity.

Thetford Biomass Power Station is a £69 million biomass plant built in Norfolk. It is the largest power station of its type in the UK. Burning poultry litter it produces 385 MW of electricity. The by-product of burning is a high quality fertiliser which is used locally.

Energy can be conserved at home and in industry. This contributes to sustainable supply and production.

Other sustainable sources are wave and tidal power – both rely on expensive and largely unreliable equipment. They are both at the mercy of the elements. A potential advantage of wave and tidal power is that there is a seemingly limitless potential supply of energy from the sea and tides.

Case study: Energy targets and harnessing the waves

The UK could generate all the electricity it needed from around its long coastline. However, there are enormous hurdles to overcome, not least being able to construct a viable generator. By the end of 2010, 10% of electricity provided by the electricity supplier must come from such sources. It is thought that this requirement will help many rural communities to initiate and run their own renewable power plants.

Islay in the Inner Hebrides may be one of the first communities to benefit as they now have what is believed to be the world's first commercial wave-powered generator.

Part of the reason for the 10% renewable energy requirement is bound up in the government's pledge in its report *Energy – the changing climate*, that is by 2010 carbon dioxide levels will be cut by 20%.

The government offered four ways in which these targets could be reached. The UK is embarking on a pathway that leads to a sustainable energy policy that protects the interests of future generations.

Recklessly causing large-scale disruption to the climate by burning fossil fuels will affect all countries. It is the poorest countries that would suffer most.

The government recognised the value of nuclear power in providing carbon-free energy, but did not believe it was indispensable. The report confirms that fossil fuel economies, such as the UK's, are on the wrong path, but it also shows that wind and solar power can break our addiction to oil, coal and gas. It makes clear that tinkering around the edges, which is what all governments are doing now, won't stop climate change.

Future challenges for energy

For the future there will be a need to reduce our fossil fuel dependency as it becomes expensive and output drops. Fossil fuels also contribute to gas emissions and their contribution to global warming is well documented, as is governmental commitment to their reduction. So how will we cope?

- Through generating energy for ourselves – ground source, or local wind and solar power.
- By encouraging renewable energy – offering a greener future.
- Energy production must come from a mix of sources to ensure stability.
- Those that use the dirty fossil fuels must be heavily taxed – **green taxes**. The receipts from these taxes will pay for alternative energy sources.

13.2 Superpowers

LEARNING SUMMARY	After studying this section, you should be able to understand:
	• that superpowers seek to influence our future
	• that the superpowers are changing

Superpowers

Edexcel **A2 U3**
WJEC **A2 UG4**

Superpowers are globally important economic, political, cultural and military powers. Who are they? The two superpowers are the USA and the European Union (EU), emerging superpowers are Brazil, Russia, India and China (the BRICs) and important regional superpowers are Japan, Mexico and South Africa.

> Superpower theory is sometimes linked to Rostow's work and Frank's dependency theory and on occasions even Kondratiev theory – see Chapter 10.

KEY POINT

Over time power 'shifts'... until 1920 the superpower was the British Empire; from 1945–90 the Cold War dominated the world and from 1990 to the present the USA is the dominant power. As to the future, complex groupings of established superpowers exist alongside the BRICs and others.

Superpower roles

Edexcel **A2 U3**
WJEC **A2 UG4**

During the period of the British Empire direct and close control was exerted over the countries that were part of the Empire. Today, superpowers still take control over troubled and troublesome countries, such as Iraq, Bosnia, Afghanistan and Kuwait, but rarely do they stay very long!

What is obvious these days is that superpowers are able to use many methods, other than invasion and war, to maintain power and to manipulate countries. This is in effect a form of **neo-colonialism**. Aid, trade and resources are invariably distributed to former colonies in preference to other countries and invariably the countries that are over-looked really need such help. But nothing is for free. Aid and trade that flows into poorer countries is invariably tied with huge **reciprocal debt payments** being demanded.

> The North Atlantic Treaty Organisation (NATO), G8 (a group of rich nation's including UK, France, Germany and USA) and the World Bank.

The role of superpowers in international trade, particularly the USA and the EU, is significant (particularly their influence in the World Trade Organisation). Furthermore, the EU and the US's transnational companies (TNCs) also exert huge influence over less economically developed countries (LEDCs), along with **international governmental organisations (IGOs)**.

In today's world the USA undoubtedly is the most influential country. Creeping **Americanisation** brings global brands, global media, TNCs and well-developed transport and communications to every part of the world.

The influence of the USA will decline as the BRICs challenge their supremacy. What is more important is that old allegiances will change. China, for instance, has developed a massive foothold in Africa and in a number of other countries that have not traditionally been served by the developed world.

The superpower future is bound up in a number of agendas – strategically speaking into the next decade the **conflict boundary** will continue to be the Middle East and **economic** growth will occur in those areas with resources, such as Russia, China and Africa. But one agenda will continue to be given only lip service by the new order, that of **climate**.

> It is likely that tensions and conflict will increase with such changes.

13.3 Water conflict

LEARNING SUMMARY	After studying this section, you should be able to understand:
	• that there are conflicts and issues over the future availability of water.

Water: demand and supply

AQA **AS U1**
OCR **AS U1/A2 U3**
Edexcel **A2 U3**
WJEC **A2 UG4**
CCEA **A2 U1**

The conflict over water relates to the gap between demand and supply. Add in climate change, the geopolitics of water transfers and pressures from population increase and the problem becomes only too obvious. By comparing the two diagrams below what should become obvious is that by 2025 the countries growing most rapidly will experience the very lowest water availability.

> Global water consumption rose six times greater than population increase between 1990–95

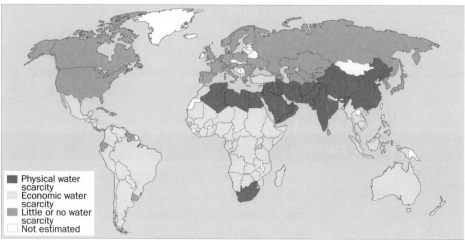

Physical water scarcity
Economic water scarcity
Little or no water scarcity
Not estimated

Projected water scarcity in 2025

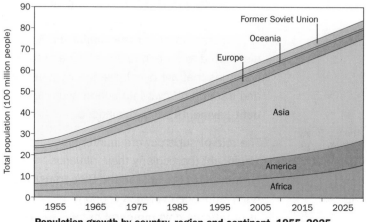

Population growth by country, region and continent, 1955–2025

World distribution of potable accessible water

Background

Only about 3% of the Earth's water is potable (useful to drink or for agriculture and industry). Three factors determine supply.

- **Climate** is critical in determining water supply.
- **Rivers** move water around. But they experience seasonal flow and water balance variation (see Chapter 4); both factors determine water supply possibilities.
- **Geology** a country with little permeable rock (sedimentary chalks and sandstone for instance) will have no storage of groundwater.

Shortage, scarcity and stress (SSS)

> Sahel, South America, China, India and Pakistan all experience SSS.

Shortage

Shortages of water have become common place in many areas of the world – too many areas receive low levels of water supply relative to basic needs.

Case study: Sucking India dry

Water from rainfall is only available from surface supply from July to September in many parts of India. Competing demands for water causes scarcity. In Kishangarh, Rajasthan in India the operations in a marble factory is sucking the ground dry (over extracting water causing shortages of drinking water), but in the fields crops die through lack of water and people die through starvation.

> Poor water quality increases disease. Stored water harbours mosquitoes – two-thirds of countries by 2025 will live in water stress.

Scarcity and stress

When water supply falls below 1000 m^3/year a huge imbalance between supply and demand can occur. The world is said to be 'water bankrupt' and that the water scarcity crisis is a bigger threat to the world than the financial crisis. Scarcity of water and hydrological shock (where suddenly there is no rain and no replenishing of groundwater and river water) brought on by climate change, will increase the risk of major international and national security threats – the **water wars**.

Israel – lots of competition for water in the Middle East. Israel shares the River Jordan with Jordan and draws water from the Sea of Galilee. Groundwater is polluted and depleted – it is shared with Palestine. Israel is going down the desalinisation route. It also buys water from Turkey.

Ogalla Aquifer – stretching from Texas to Dakota has a maximum of 250 years' worth of water left. Irrigation has caused a massive decline.

Mexico City – is sinking as water is pumped from the ground; 80% of Mexico City's water is from the ground and 27% of water leaks from pipes in Mexico City.

Chad – Lake Chad has 5% of its original surface area left. Local weather changes and over abstraction being the causes. Population of nine million are affected by its demise.

Turkey – sells water from Manvgat river system. Turkey has spent billions of $'s building dams; 22 on the Tigris and Euphrates have provoked angry criticism from Iraq and Syria.

Nile – population of 160 million rely on the Nile. The tensions are high in the other seven countries the Nile flows through, as Egypt sticks to the 1929 treaty, that 'no work be done on it to reduce the volume of flow'. Cairo has said it will use force to protect the river and treaty.

North China – 30 km^3 of water is pumped to the surface every year by farmers which is never replaced by rain. For 200 days/year the yellow river is dry in its lower course.

Ganges – is depleted (the source, the Gangotri glacier is retreating fast), polluted (with arsenic and sewage), and disputed (the Farakka barrage controls flow and Bangladesh suffers!).

Australia – the continent with the least rainfall. The flow of the Murray and Darling (Australia's two main rivers) has been reduced by three-quarters to provide for dams, irrigation and drinking water.

Water scarcity hotspots

Case study: Two stressed drainage basins

The Tigris-Euphrates Basin

A situation with serious international implications is the demand for the waters of the Euphrates by Turkey, Syria and Iraq. The Euphrates is the primary water source for millions of people who depend on it for power generation and irrigation in an extremely arid climate. Conflict over water in this area is decades old. It has intensified in recent years as a result of a massive Turkish dam building programme known as the Greater Anatolia Project (GAP), designed to provide a supply of water and power adequate to fuel the development needs of Turkey's population, which is growing at 1.6% annually. It provides Turkey with a generating capacity of 7500 MW of electricity, which is nearly four times the capacity of Hoover Dam, and will open up at least 1.5 million hectares of land to irrigated cultivation.

Full implementation of the GAP system of dams could result in a 40% reduction of the Euphrates' flow into Syria and an 80% reduction of flow into Iraq. This will reduce the electrical output of Syria's Tabqa Dam by up to 12%, while Iraq could lose irrigation water to 1 million hectares. The levels of salinity will increase, as well as the amounts of agricultural and industrial pollution, in the remaining water. Syria and Iraq have already threatened war over their access to the Euphrates. As the populations of these nations continue to expand, driven by fertility rates well above the global average, the competition for fresh water between agriculture and development could endanger stability in the region.

The Nile

In Egypt it has been recognised for some time that the headwaters of the Nile might be developed. Egypt has threatened war, to preserve its access to fresh water. Ethiopia has already built some 200 small dams on the Nile since emerging from civil war and famine.

Conflict and the future of water supply

In the future water will undoubtedly be in short supply. There are five important points to consider.

1. Demand outstrips supply. There are three possible scenarios for water usage in the world (see diagram below).

Blue water: clear and unpolluted.
Green water: released through the hydrological system. **Grey water:** polluted water.

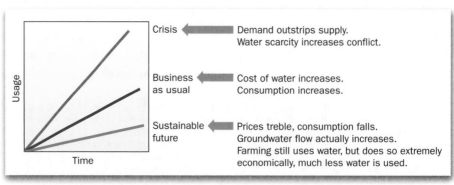

2. United Nations Millennium Development Goals – by 2015 'to halve (1.4 billion) numbers without access to surface water and sanitation'.

Remember – water is needed to generate energy. Energy is needed to deliver water. Both are running out fast.

3. In the future **political** (superpowers, G20, World Bank, UN, WTO TNCs); **social** (NGOs); **economic** (TNCs); **environmental** (WWF, Friends of the Earth) 'players' and decision makers will determine how water is divided up amongst the 6.9 billion consumers.

4. Water conflicts will have to be managed.

5. Low-tech and high-tech fixes will help achieve sustainable water supply long into the future.

These points suggest five alternative futures for water supply.

Low-tech fixes include: stone dams, tub wheels, water harvesting, micro dams, drip irrigation and compost latrines.

High-tech fixes include: dams, diversions, pipelines, tankers, recharging, desalination, **fertigation** and water meters.

13.4 The geography of technology and the technological fix

LEARNING SUMMARY	After studying this section, you should be able to understand:
	• that technological fixes help many of our present day problems

The technological fix

Edexcel **A2 U3**

The technological fix is where humans solve issues and problems facing them using technology – machines, tools or computer systems.

Why do we need technology?

We are short of energy for the continuing development of the world and this means getting to the remaining fossil fuels. For the developing world the ability to buy into the new technology, such as solar power, will be the biggest problem due to financial constraints.

- To deal with world poverty and social inequality.
- To cope with climate change.
- To deal with the scarcity of water.
- To effectively feed the world.
- To cope with rising population.
- To tackle pollution.

Types of technological fix

Compare the cost of a variety of technologies and technological fixes.

All technology has a life–cycle. The technology costs are less as sales grow. All technologies are superseded by better technology.

Low-tech is intermediate technology (easy to master by locals)

In 2010 one-ninth of the world was on-line!

Labour intensive and low-tech

Increasing cost

Capital intensive

Increasing cost

Technology (appropriate to skills, income and knowledge)

Civil engineering (dams)

Micro technology (ICT)

Bio technology (Green Revolution)

Geographical engineering (Space)

KEY POINT

Attitudes to technology can be tempered by **environmental determinism**. This is the belief that the environment determines human culture and societal development.

Access to technology

Various factors control the access to technology:
- levels of development
- physical factors
- political and historical reasons
- religious and military reasons.

> Those who have been fortunate to embrace technology are developing at the fastest rate.

The **knowledge economies** (well-developed educational systems) in the East Asia, the USA and UK lead the way in technology and as a result are best placed to gain wealth and prosper. Furthermore, their access to such technologies 'shrinks' global distances.

Technology and the future

The use of technology has long been seen as determining population and resource relationships. Two futures are predicted.

- A **divergent** world with a technological core (the developed world) and a technologically backward periphery (the developing world).
- A **convergent** world with technology for all.

13.5 Pollution

LEARNING SUMMARY	After studying this section you should be able to understand:
	• that pollution and waste are out of control

What is pollution?

AQA	A2 U3
OCR	A2 U3
CCEA	A2 U1

> Three things determine the severity of pollution:
> - The chemical nature of the pollutants.
> - The concentration of the pollutants.
> - The persistence of the pollutants.

Pollution occurs when contaminants are introduced into the environment – they harm or destabilise ecosystems. Annually the Blacksmith Institute in the USA issues its world's 'most polluted listings'. In 2009 they identified Azerbaijan, China, India, Peru, Russia, Ukraine and Zambia as the most polluted countries.

Pollution may be categorised as:

- water
- noise
- soil.
- air
- visual

> Where long-term health is affected the poverty of such areas deepens.

Toxic pollution is a health and economic catastrophe – mostly the poorest in developing countries are affected. The pollution of air, water and soil contribute to 40% of deaths annually around the world. In some instances on-going toxic activities are to blame. In other places the legacy of now defunct activities causes problems.

Modern awareness of pollution dates from the Second World War. In the UK the Great Smog of 1952 in London prompted the Clean Air Act (1956). In the USA various noise, air, water and environmental Acts; dioxin, PCB and Chromium-6 releases and various radioactive releases, Bhopal and land and sea oil disasters have drawn attention to pollution and encouraged clean-ups.

Case study: Bhopal, India (1984)

The world's biggest industrial disaster occurred in Bhopal in India, 1984. Following an explosion 42 tons of methyl isocyanante was released into the atmosphere. Approximately 8000 people died and a similar figure have died since. At the time of the accident 558 125 people were injured, 38 475 were disabled and 3900 permanently disabled. The American owners Union Carbide paid $470million in compensation.

For the future

It is likely pollution will be controlled via one or more of the four 'P's!

- **Proximity** – pollution is dealt with at the source, e.g. waste computers will be decommissioned in the UK not sent to West Africa for them to deal with!
- **Polluter pays** – cost of pollution incidents are paid for by the polluter. This is part of the Kyoto Protocol (1998) and Copenhagen Conference (2009) detail.
- **Prevention** – stop at source, i.e. pollution 'scrubbers' in industrial chimneys, smokeless zones and zero emissions areas.
- **Precautionary** – action taken whenever risk is proven, i.e. chemical bans (DDT for instance), chlorofluorocarbon (CFC) release acted on to reduce ozone depletion.

> Scrubbers are chemicals installed in the chimneys to scrub gases out as they pass over them.

> **KEY POINT**
>
> **Carbon trading** targets carbon dioxide (calculated in tonnes of carbon dioxide equivalent) and is used by countries in order to meet obligations outlined in the Kyoto Protocol. It is an attempt to mitigate future climate change.

Pollution types and remedies

Examples of pollution incidents and how they were dealt with are detailed below.

Water pollution	Air pollution
Location: Shanghai, China. **Pollutant:** Residential and industrial waste. **Cause:** Dumping of raw sewage in waterway, compounded by flooding. **Population affected:** 13.4 million. **Health impact:** Untreated sewage spreads cholera, typhoid and dysentery among urban populations. **What was done:** The Shanghai Government and Asian Development Bank teamed up to design and implement a 12 year project which included sewage treatment, injection of oxygen into the waterway and flood controls to bring water quality to acceptable levels for household use. **Outcome:** The first phase of the project improved water quality. The second phase aimed at maintaining water quality while incorporating sustainable urban design elements in the rehabilitation system. **Result:** This rehabilitation project demonstrated success in restoring contaminated water bodies in an urban setting, as well as generating additional benefits such as increased green space and higher property values in the area. **Remaining challenges:** Despite ongoing billion dollar investments and relocation/shutdown of industrial facilities, upstream pollution continue to threaten the Yangtze River.	**Location:** Delhi, India. **Pollutant:** Sulphur dioxide, particulates, carbon monoxide and other urban air pollution. **Cause:** Emissions from automobiles, power plants, as well as other local industrial and urban sources. **Health impact:** Long-term exposure to relatively low levels of pollutants can cause health problems, e.g. cancer, respiratory and cardiovascular diseases, asthma. The health effects of outdoor air pollution fall disproportionately on infants, children and the elderly. **What was done:** India's Ministry of Environment and Forests undertook extensive measures to reduce vehicle emissions in the late 1990s, including the introduction of a fleet of buses powered by compressed natural gas, mandatory inspection and maintenance requirements, emission norms, more stringent clean fuel requirements and a pollution tax. **Outcome:** 1993–2000 – ambient carbon monoxide (CO) decreased by almost half and lead concentrations fell by 75%. 2000–03 as buses were converted to compressed natural gas (CNG), sulphur dioxide levels (SO_2) decreased by 34.8%, and particulate matter (PM10) levels, by 7%. **Implications:** Use of alternative fuel vehicles can yield significant urban air quality improvements; however, early gains have been neutralised due to the increase in the number of personal vehicles.

Pollution from mine tailings	Indoor air pollution
Location: Candelaria, Chile. **Pollutant:** Copper mine tailings. **Cause:** Open pit mining and mineral processing could contaminate limited water sources in the desert. **Health impact:** Tailing slurries with toxic components (such as cyanide) would contaminate groundwater from limited existing sources and cause nervous and immune system damages. **What was done:** A tailing impoundment/cut-off wall system was constructed to dispose of tailings and conserve scarce water in the desert region; 80% of the water bound up in the tailings is treated and re-circulated into the supply system. **Result:** The multi-stage treatment process ensures trapping of tailings with toxic chemical content, e.g. cyanide. The system, as a whole, collects around 365 million tons of tailings and treats the water content found in them. **Implications:** Successful design and implementation of this comprehensive system serves as a prime example for tailings management in developing countries, which is one of the most severe and underestimated environmental and health issues. **Remaining challenges:** Extract and treat the last 20% of water bound up in the tailings.	**Location:** Accra, Ghana, West Africa. **Pollutant:** Indoor air pollution (carbon dioxide, particulate matter). **Cause:** Burning of biomass (wood, charcoal, dung, crop residue) for cooking. **Health impact:** Cooking fumes lead to an estimated 1.5–2 million deaths worldwide according to a report published in 2000. **What was done:** To address the health effects of indoor air pollution due to cooking with biomass, the Gyapa stoves project has created an innovative commercialisation scheme for producing and distributing an improved cooking stove, coupled with a campaign to drive demand by raising awareness of the health risks posed by indoor air pollution and teaching women how to cook traditional foods with the new stoves. **Result:** In 2008, 68 000 stoves were sold in Accra and Kumasi, potentially providing cleaner indoor air for approximately 400 000 people, including approximately 160 000 children. There was a 45% improvement in air quality with the new stoves. **Remaining challenges:** The Gyapa stove confronts issues of pricing, manufacturing, and quality control.

13.6 Trade alliances

LEARNING SUMMARY	After studying this section, you should be able to understand: • that trade groupings influence and affect global trade

Trade agreements

AQA **A2 U3**
Edexcel **A2 U3**
WJEC **A2 UG4**

The 2009 Doha Trade negotiations had the effect of lowering barriers and increasing global trade. It also effectively broke the stranglehold of agricultural import rules, though the EU amongst others refused to sign up to this part of the negotiations!

Towards the end of the 1990s a number of regional trade agreements came into existence. Today there are in excess of one hundred. It is thought that just about every country in the world is involved in at least one agreement. These agreements strengthen the political base of a country, ensure security and can even ensure food security (and fair prices) between neighbours/trading partners through the use of price fixing, quotas and tariffs. Some agreements have even worked to allow countries into the global economy, such as Vietnam whose present day success is due almost exclusively to its entry into the Asian Free Trade Area (ASEAN). For the future it is thought that the EU (with lots of extra Eastern Bloc countries) and the North America Free Trade Agreement (NAFTA), with all of the Americas, will dominate world trade.

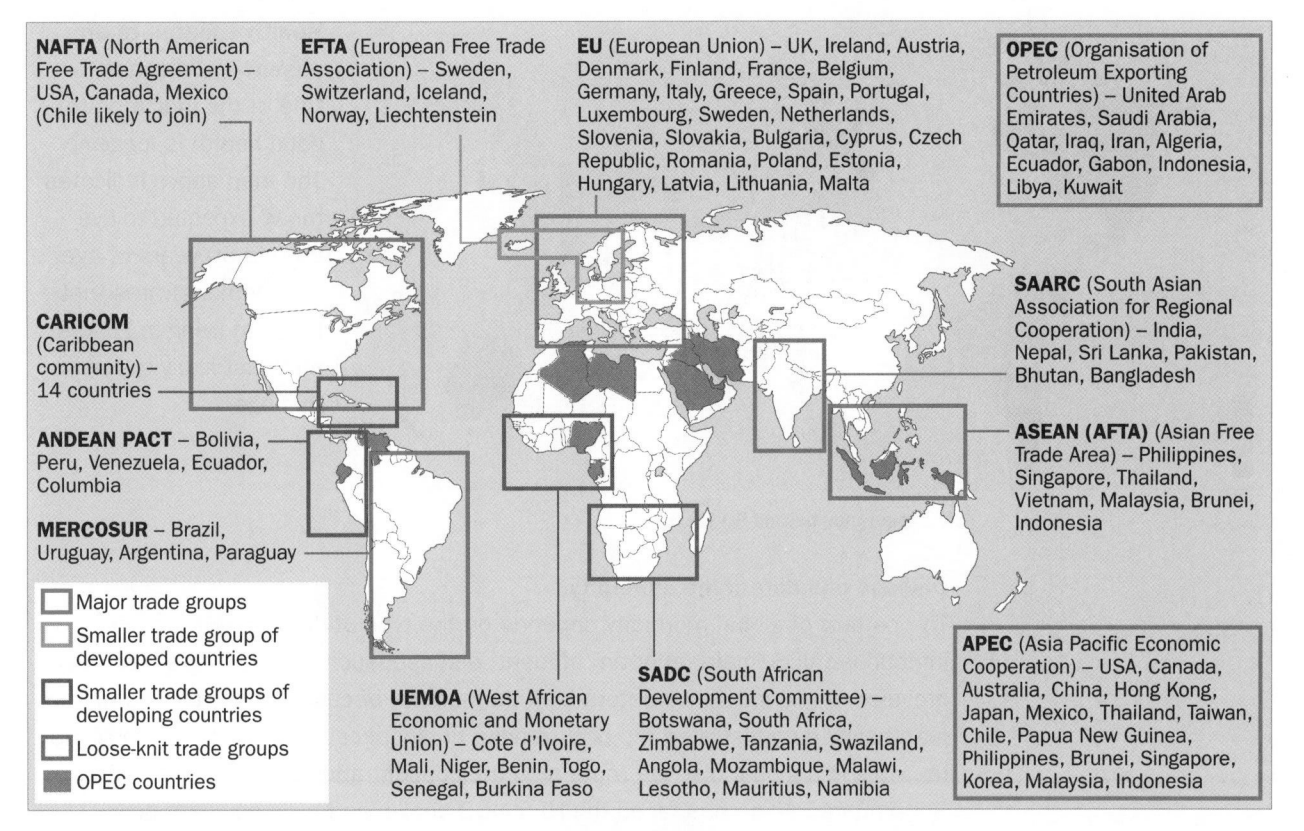

NAFTA (North American Free Trade Agreement) – USA, Canada, Mexico (Chile likely to join)

EFTA (European Free Trade Association) – Sweden, Switzerland, Iceland, Norway, Liechtenstein

EU (European Union) – UK, Ireland, Austria, Denmark, Finland, France, Belgium, Germany, Italy, Greece, Spain, Portugal, Luxembourg, Sweden, Netherlands, Slovenia, Slovakia, Bulgaria, Cyprus, Czech Republic, Romania, Poland, Estonia, Hungary, Latvia, Lithuania, Malta

OPEC (Organisation of Petroleum Exporting Countries) – United Arab Emirates, Saudi Arabia, Qatar, Iraq, Iran, Algeria, Ecuador, Gabon, Indonesia, Libya, Kuwait

CARICOM (Caribbean community) – 14 countries

ANDEAN PACT – Bolivia, Peru, Venezuela, Ecuador, Columbia

MERCOSUR – Brazil, Uruguay, Argentina, Paraguay

SAARC (South Asian Association for Regional Cooperation) – India, Nepal, Sri Lanka, Pakistan, Bhutan, Bangladesh

ASEAN (AFTA) (Asian Free Trade Area) – Philippines, Singapore, Thailand, Vietnam, Malaysia, Brunei, Indonesia

☐ Major trade groups
☐ Smaller trade group of developed countries
☐ Smaller trade groups of developing countries
☐ Loose-knit trade groups
■ OPEC countries

UEMOA (West African Economic and Monetary Union) – Cote d'Ivoire, Mali, Niger, Benin, Togo, Senegal, Burkina Faso

SADC (South African Development Committee) – Botswana, South Africa, Zimbabwe, Tanzania, Swaziland, Angola, Mozambique, Malawi, Lesotho, Mauritius, Namibia

APEC (Asia Pacific Economic Cooperation) – USA, Canada, Australia, China, Hong Kong, Japan, Mexico, Thailand, Taiwan, Chile, Papua New Guinea, Philippines, Brunei, Singapore, Korea, Malaysia, Indonesia

Trade groupings

The future for world trade

Likely scenarios relating to future world trade are listed below.

- TNCs will continue to dominate trade.
- Older developed nations will face increasing competition from the NICs, RICs and BRICs.
- The disproportion between imports and exports between the LEDCs and MEDCs will remain.
- Raw material trading will increase, as countries run out of resources.
- Trade will grow between the less developed nations.
- It is likely that the WTO will aid developing nations by helping with trade reform, to settle disputes and by promoting clear trade regimes and reforms.

13.7 Health and welfare

LEARNING SUMMARY	After studying this section, you should be able to understand:
	• the changing health and welfare of the people of the world

Global health and welfare

AQA   AS U1

Health and welfare relates a population's standard of living and quality of life to economic, social and environmental factors. Generally, better health equates to high levels of wealth.

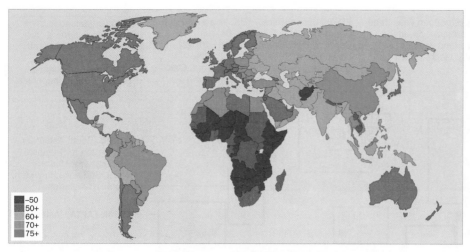

Life expectancy beyond 50 years

Health equates to an absence of disease
One of the indicators of good health is longevity. The map above indicates those expected to live beyond 50 years of age. This demonstrates that those in living in LEDCs are least likely to reach this age.

Disease can determine morbidity

The pattern of global morbidity depends on the type of disease – it can be infectious (AIDS/malaria) or an 'affluent' disease, such as heart disease. The prevalence and incidence determines the number of cases at any one time. The map below demonstrates the occurrences of global cases of AIDS. It is obvious that this is a global disease (note also the concentrations of AIDS in LEDCs). Sub-Saharan Africa has faced the HIV/AIDS challenge for some years now.

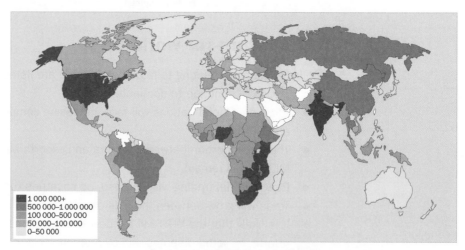

Occurrences of global cases of AIDS

Case study: AIDS in Sub-Saharan Africa

Sub-Saharan Africa is heavily affected by HIV and AIDS. An estimated 22.4 million people are living with HIV in the region – around two-thirds of the global total. In 2008 around 1.4 million people died from AIDS in Sub-Saharan Africa and 1.9 million people became infected with HIV. Since the beginning of the epidemic, more than 14 million children have lost one or both parents to HIV/AIDS.

In the absence of massively expanded prevention, treatment and care efforts, it is expected that the AIDS death toll in Sub-Saharan Africa will continue to rise. The social and economic consequences of AIDS are widely felt, not only in the health sector, but also in education, industry, agriculture, transport, human resources and the economy in general.

Sub-Saharan Africa faces a triple challenge.
- Providing health care, anti-retroviral treatment, and support to a growing population of people with HIV.
- Reducing the annual toll of new HIV infections by enabling individuals to protect themselves and others.
- Coping with the impact of over 20 million AIDS deaths, on orphans and other survivors, communities and national development.

Diseases of affluence

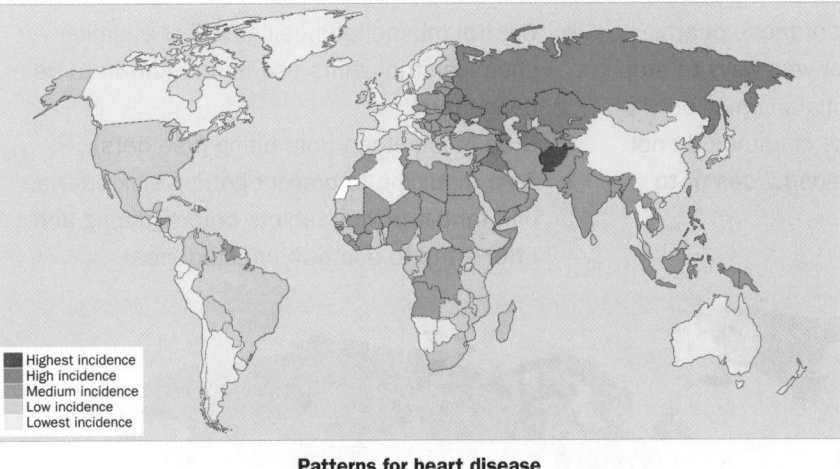

Patterns for heart disease

The map (left) shows morbidity patterns for heart disease around the world. Affluent countries on the whole dominate this map, though as the populations of developing countries eat more processed food their totals are expected to climb rapidly. It is also worth comparing this map with longevity as heart disease is mostly a disease of old age, but also relates to poor health style.

Legend on map:
- Highest incidence
- High incidence
- Medium incidence
- Low incidence
- Lowest incidence

Impact of disease

Generally high morbidity causes high mortality, but MEDCs have the resources to reduce the incidence of death, whereas LEDCs don't. Examples of resources in MEDCs include:

- education and training
- medical advice and its availability
- better housing, sewage and cleaner water systems
- vaccination to protect against disease
- other measures to reduce disease, such as spraying mosquito infested areas
- improved nutrition for all.

Health in the UK

Some general points about the state of health of the population in the UK.

- Women live longer than men, but can die of chronic disease.
- With people living longer there is less tax base from the working population to support a stressed National Health Service (NHS).
- Income can affect health! Wealth = better health care = south-east UK is the healthiest region.
- There is a north *vs.* south divide in general healthiness in the UK.

Case study: Malaria

The malaria life cycle is as follows.

1. Infected mosquito bites human.
2. Parasite goes to the liver within 30 minutes.
3. The parasite starts reproducing in the liver, some lie dormant in the liver, to reactivate and cause diseases often long after the initial infection.
4. The parasite gets into the blood stream, attaches to and enters red blood cells.
5. Infected red blood cells burst, infecting other blood cells.
6. This repetitive cycle causes fever and depletes the body of oxygen-carrying red blood cells.

Infected red blood cells clog up the circulation in vital organs such as the brain and kidney.

7. As infection progresses, sexual forms of the parasite are released into the blood stream. When a mosquito bites, it takes up these parasites and the cycle of infection is perpetuated placing others at risk.

Every year 300–700 million people get malaria. It kills 1–2 million people every year. The biggest problem is in Africa where 90% of the people who contract malaria die. Most of the people who die from malaria are children. In Africa, 20% of children under five die from malaria. Most of these deaths could be stopped with medicine or with ways to stop mosquitoes. Many of the places malaria may be found are in poor countries. These countries do not have enough money to stop the mosquitoes or to give people medicine.

There are three ways to prevent malaria.

- Control mosquitoes using DDT or a similar chemical, though its use is now known to be dangerous!
- Stop mosquitoes from biting (use nets).
- Take medicine to prevent getting sick after a bite (anti-retroviral tablets before, during and after a trip to a known infected area).

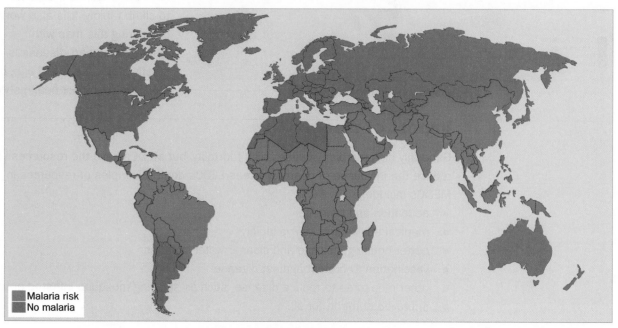

Malaria risk
No malaria

Global malaria risk

PROGRESS CHECK

1. Why is there a decreasing amount of carbon dioxide released as countries move from non-renewable to renewable resources?
2. What is the difference between biomass and biofuel?
3. Why is there a gap between demand and supply of water around the world?
4. Why are trade alliances not always beneficial to LEDCs?

Answers: 1. Less carbon dioxide is being released from fossil fuels. 2. Biomass is a waste product and biofuel is grown. 3. Growth in population is exceeding amounts of water available. 4. They deny, through the use of tariffs and quotas etc…the ability of LEDC's to trade fairly in the world markets.

Sample questions and model answers

1. Study the diagram, which shows the direction of change of international trade of the member states of the European Union.

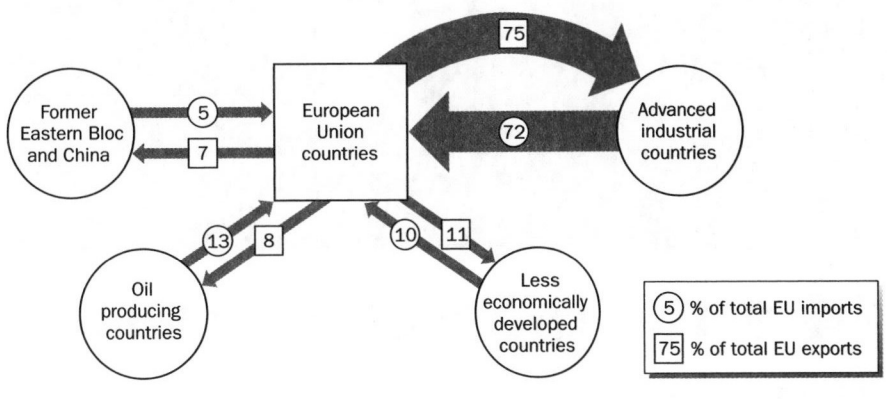

(a) Describe the pattern of trade shown in the diagram. **[4]**

Trade from Eastern Bloc, OPEC countries and LEDCs is far less than that between Advanced Industrial Countries and EUC countries. For instance exports to LEDCs are 11% whereas to Advanced industrial countries they are 75%. Imports are 10% from LEDCs and 72% from Advanced industrial countries.

Makes correct observation in first line.

Recognises the imbalance between LEDCs and MEDCs.

(b) Account for the relative importance of the trade with advanced industrial countries compared with less economically developed countries. **[4]**

Trade with LEDCs is important as very often LEDCs possess large amounts of raw materials that can be acquired cheaper than from anywhere else. However, advanced industrial countries, e.g. USA, Japan, have much better technology and money, their investment and trade is good for Europe e.g. UK-NE and Nissan.

This answer has rather lost its way. There is no reference to the lack of trade with the eastern bloc. Europe as a trading bloc is also not covered. What do LEDCs make that is needed in great quantity?

(c) Outline the benefits to a less economically developed country of increasing its trade with EU countries. **[5]**

LEDCs have in general many natural resources but by comparison their manufacturing industry and their ability to create high quality, high technology goods is very poor. By trading with EU countries (UK, Germany, France) they can obtain higher technology goods and machinery to aid them with their development and raise their standards of living. It is also important as foreign investment (often by private companies and multi-nationals, e.g. Ford) provides money and labour for the country. Paradoxically it has the effect often of draining money out of the country to multi-national HQs.

Simple comment to the relationship between levels of development and what's produced.

Changing dependency.

Multiplier effect.

Idea of inward investment. A better answer but still lacks specifics.

(d) Comment on the view that economic development always leads to an increase in human welfare. **[7]**

Economic development has its benefits; mainly increased amounts of money leading to better education, healthcare, amenities, luxuries, and better public services. This has been especially true in the development of western countries. On the other hand, development by multi-nationals in Africa has often been in their own self-interest rather than the welfare of the indigenous population, e.g. oil companies such as BP have interests in Nigeria, but although the economy has developed much of the money is simply drained back to Europe and some of Nigeria's poorest people live in the oil-producing area of the coast.

Use 'indicators' to elaborate the process of development.

Material consumption increases.

The following topics are covered in this chapter:

- The theoretical framework
- Trends in tourism in the UK
- International tourism
- Sustainable ecotourism

14.1 The theoretical framework

LEARNING SUMMARY

After studying this section you should be able to understand:

- that the term leisure includes recreation, sport and tourism
- that leisure participation is influenced by economic, social, cultural and political factors
- that leisure exists on a demand-supply circle and makes demands on resources
- that the demand for tourism is constantly growing
- how tourism can be classified using a range of terminology
- how growth, stagnation and decline of tourism can be modelled
- the features and qualities of tourists that can be modelled

The leisure industry

Edexcel **A2 U4**
CCEA **A2 U1**

Leisure can be defined as time away from work, (paid or unpaid); time when you are free from other obligations and where you choose to be involved in an activity very different from the usual things you do. Like all human activities leisure impacts on our resource base, both locally and globally, and our environment. Leisure is demand driven and as such others profit by supplying various 'needs'.

With more leisure and recreation time available it is important you know and understand its dynamics.

As we enter the new millennium we are now spending more on leisure than on housing and food. One-sixth of household expenditure goes on leisure. The age group 65–74 spends the most on leisure and services.

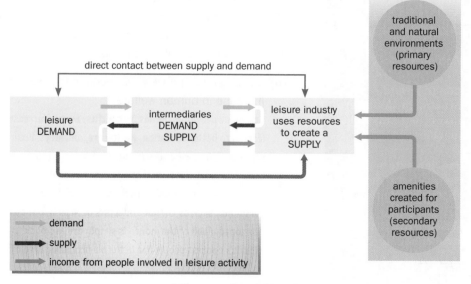

Leisure supply and demand

> Generating 10% of the world's GNP and it employs nearly 10% of people in employment.

The economic potential of leisure is enormous, up to 20% of total income is spent on leisure. The leisure industry is now the biggest world business. Globally the industry is larger than the world trade in many raw materials e.g. iron and steel.

Leisure is socially and culturally linked and is also influenced by fashions and fads, e.g. mountain biking in country parks. However, the basic factor underlying our leisure involvement is **time**.

> High Lodge, part of Thetford Forest; has tens of miles of bike trails and has been successful in attracting a large number of mountain bikers.

On the whole, people with limited leisure time tend to seek out and be involved in activities close to home, whilst those with more time travel much further afield.

short break to swim $\longrightarrow$ medium break to play a round of golf $\longrightarrow$ long break to go on holiday

The leisure/tourist continuum

Edexcel **A2 U4**
CCEA **A2 U1**

> Leisure can be defined in time terms, is voluntary and can be enjoyed at home or close to home. Tourism involves staying away from home for at least one night or more.

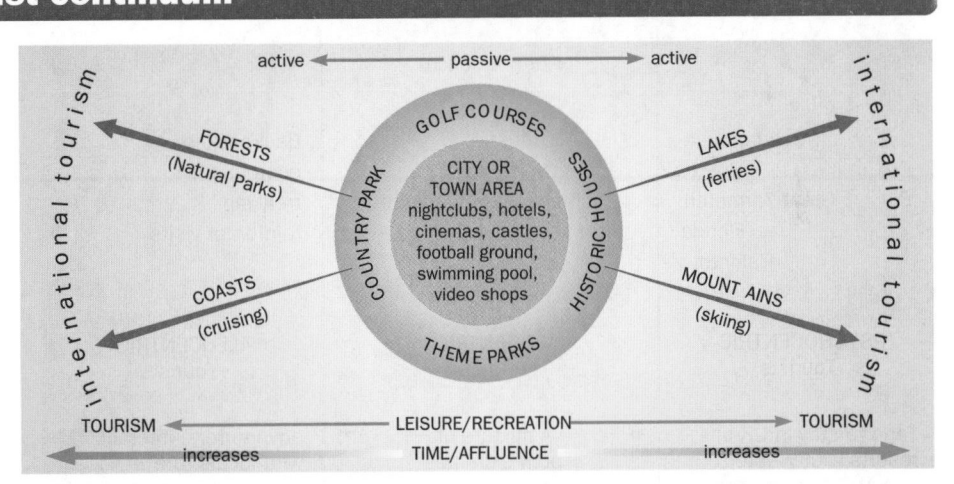

The leisure/tourist continuum

Analysing tourism

Edexcel **A2 U4**
CCEA **A2 U1**

> We all travel; know how it is classified.

KEY POINT

Classification

- **Domestic tourism** – visits within own country.
- **Inbound tourism** – visitors to UK.
- **Outbound tourism** – destinations abroad.

Why travel?

- Touring and sightseeing culture.
- To relax.
- To do recreational activities.
- To visit friends and relatives.
- Business and specialist trips.

Where do tourists go?

Destinations that:

- are attractive, e.g. beaches
- offer the right facilities (amenities) i.e. hotels
- are accessible, e.g. by air.

How do tourists get to their destinations?

This is changing all the time. In the last 30 years there has been a 24% increase in air travel, a 16% drop in boat traffic. Transport is 'evolutionary'; methods change.

How long do tourists take for their holidays?

This depends on:

- age and family commitments
- entitlement to holiday
- disposable income.

When do people travel?

- During closures of schools and universities, e.g. in the summer holidays.
- Over public holidays.
- Traditionally in the warmer summer months and in the winter for skiing.
- Factories close to coincide with school closures.

Models of tourism

Edexcel **A2 U4**
CCEA **A2 U1**

There are a number of models that attempt to classify tourism.

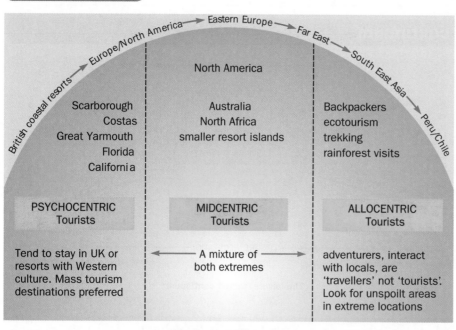

Plog's personality model

- Plog's model (opposite, top) looks at personality types of tourists.
- Butler's model (below, bottom) looks at the way tourism has evolved.
- Doxey's Index (page 280, left) shows how attitudes to tourism change as the tourist industry develops.
- The Enclave model (page 280, right) shows how socio-economic factors influence tourism.

Exploration
- small numbers of allocentrics or explorers
- little or no tourist infrastructure
- natural or cultural attractions

Involvement
- local investment in tourism
- pronounced tourist season
- destination advertised
- emerging market area
- public investment in infrastructure

Development
- rapid growth in visitation
- visitors outnumber residents
- well-defined market area
- heavy advertising
- external investment leads to loss of local control
- artificial attractions emerge to replace natural or cultural
- mid centrics replace explorers and allocentrics

Consolidation
- slowing growth rates
- extensive advertising to overcome seasonality and develop new markets
- psychocentrics attracted
- residents appreciate the importance of tourism

Stagnation
- peak visitor numbers reached
- capacity limits reached
- resort image divorced from the environment
- area no longer fashionable
- heavy reliance on repeat trade
- low occupancy rates
- frequent ownership changes
- development peripheral to original developments

Decline
- spatial and numerical decrease in markets
- a move out of tourism
- local investment might replace abandonment by outsiders
- tourism infrastructure is run-down and might be replaced by other users

Rejuvenation
- completely new attractions replace original lures, or new natural resources used

Butler's tourist area life cycle, 1980

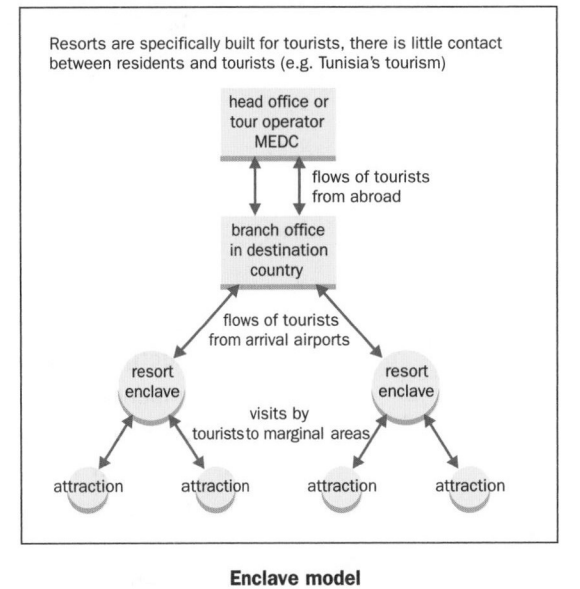

Doxey's Index e.g. Bali

EUPHORIA — investment welcomed by host country (Bali – late 1960s)

visitors taken for granted (traditional dances adapted to suit tourists) — APATHY

ANNOYANCE — hosts have misgivings about visitors (Bali 1970's Costa del Fosters)

visitors seen as a cause of problems in resorts (e.g. growth of sex tourism on Bali) — ANTAGONISM

Enclave model

Resorts are specifically built for tourists, there is little contact between residents and tourists (e.g. Tunisia's tourism)

head office or tour operator MEDC

flows of tourists from abroad

branch office in destination country

flows of tourists from arrival airports

resort enclave — visits by tourists to marginal areas — resort enclave

attraction attraction attraction attraction

> It is possible to model tourism. This allows more accurate descriptive and analytical work. This sort of information can be incorporated into essays.

PROGRESS CHECK

1. In the leisure system what are the traditional and natural environments called?
2. Give examples of passive leisure.
3. What 'parts' make up the structure of the tourist industry?
4. Why does stagnation occur in tourist areas?

Answers: 1. Primary resources. 2. Nightclubs, hotels, cinemas, castles and football supporting. 3. Transport, travel agents, accommodation, tour operators, attracting custom, development and promotion. 4. As peak numbers are not reached, capacity limits are reached, too much reliance on repeat trade and low occupancy rates.

14.2 Trends in tourism in the UK

LEARNING SUMMARY

After studying this section, you should be able to understand:

- that UK tourism has changed greatly, with domestic holidays giving way to holidays abroad
- that resorts have a distinctive character in the UK and have been forced to 'change' to survive
- that the tourism product in the UK is diverse
- how foreign visitors are important to the UK tourist industry

Why tourism has increased into and from the UK

Edexcel **A2 U4**
CCEA **A2 U1**

Since the early 1960s the type and amount of tourism into, and out, of the UK has changed.

- **Society** – in our stress filled world the annual holiday, day trips and visits are considered absolutely essential.
- **Income** – with greater disposable income we are able to spend more on holidays.

> Cheap air fares and package holidays ease many into the annual holiday.

- **Development** – in the UK in 2009 tourism revenue was £21 billion. Tourism is seen as a major revenue earner in developed and developing countries. Revenue from tourism had dropped by 12% and the East Anglian Development Agency has had to work hard to put tourism 'first' in the area. Its efforts are beginning to bear fruit by promoting cultural heritage, the regional food and drink market, the area as a gateway to the UK and has encouraged the building of hotels. In Botswana, Wilderness Safaris uses profits from tourists to support the environment and the animal and human populations.

> The 45–65 age group is the group that tops all tourism tables in the world.

- **Elderly population** – the pensioner pound/dollar/yen has a very important influence on the growth of tourism.
- **Communication** – IT makes finding holidays and last minute deals very easy.
- **A smaller world** – people can fly quickly and cheaply all over the world. Infrastructure has also improved journey times, e.g. motorways (in the UK and on the continent) and access through the Channel Tunnel have helped to speed journeys. Ease of movement encourages people to take holidays.

> Use mnemonics to allow you to remember the points opposite:
> **S**ociety
> **I**ncome
> **D**evelopment
> **E**lderly population
> **C**ommunication
> **A** smaller world
> **M**ore holiday time

- **More holiday time** – in the developing world there have been an increase in the number of holidays a person receives in employment according to the law. Because of unions, strong economies, higher incomes, societal changes and people receiving a minimum four holiday weeks' paid holiday, there is a greater chance they will go on holiday.

UK tourism predicted to increase

Over 30 million tourists visited the UK in 2010. Britain's pound is weak and is expected to drive an increase in tourism with visitors to the UK predicted to spend more than £21 billion.

> It is believed there is a growing demand for business travel and attractive sport and leisure events held in the UK.

VisitBritain, the group that promotes tourism in the UK, has established a trio of three-year global campaigns to support the promotion of tourism, adopting the themes of 'Classic', 'Dynamic' and 'Luxury'. They are designed to promote all that is traditionally British and are aimed at the youthful and high values markets.

The traditional UK holiday resort

Edexcel **A2 U4**
CCEA **A2 U1**

Over the last 25 years progressively less has been spent on UK holidays. At present the British people spend almost £25 billion on foreign holidays. Almost £19 billion is spent on holidays in the UK.

> Tourism has been described as the industry without chimneys.

However, UK seaside resorts still provide for domestic holidays. Each resort has its own unique location, morphology and degree of dependence upon its resort function. Resorts are a distinctive type of urban settlement in the UK with characteristic land use patterns, explainable using bid-rent theory. In a resort highest land values are found near the beach/sea with a peak where the frontage meets commercial property.

To attract visitors and investment to survive, resorts have to have quality attractions, character and quality accommodation. This is known as **critical mass** in the tourist industry.

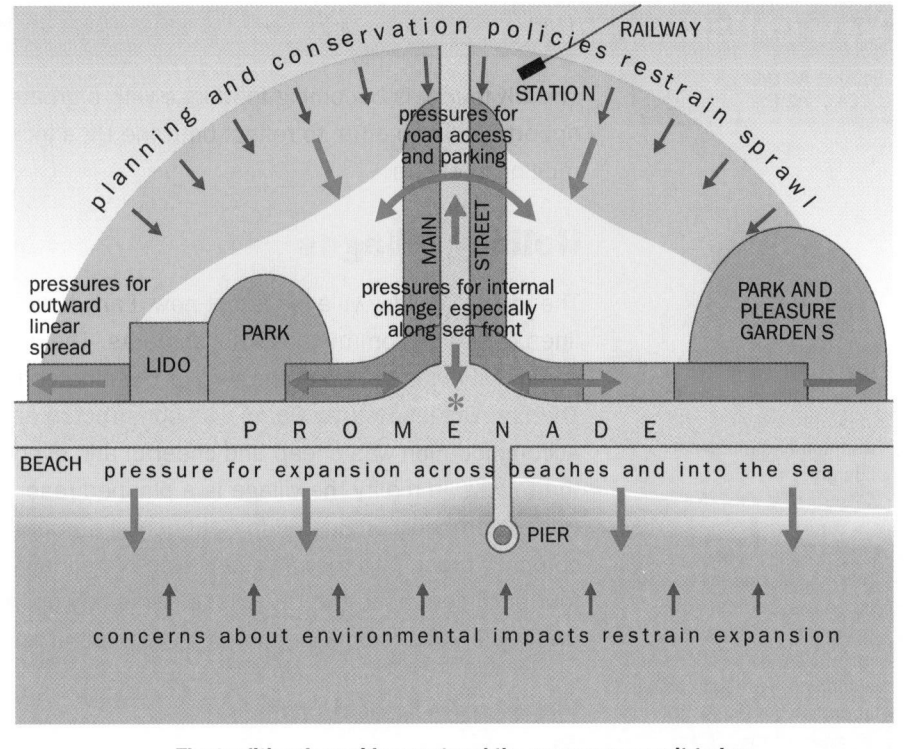

The traditional seaside resort and the pressures upon it today

Legend:
- large hotels
- small hotels
- entertainments
- tourist business district
- guest house/bed and breakfast
- residential
- * peak land value intersection (PLVI)

planning and conservation policies restrain sprawl

RAILWAY

STATION

pressures for road access and parking

MAIN STREET

pressures for outward linear spread

LIDO

PARK

pressures for internal change, especially along sea front

PARK AND PLEASURE GARDENS

P R O M E N A D E

BEACH pressure for expansion across beaches and into the sea

PIER

concerns about environmental impacts restrain expansion

Case study: Holidays in the UK

Holiday camps in the UK have seen a 25% increase in bookings, that they partly put down to upgrading of their accommodation and food, and the addition of new facilities, like health spas and entertainment improvements. They have also benefitted from 'recessional lag' and the British move towards the so-called **staycation**. In the UK, the growth of the staycation has had a major impact on the domestic tourist industry. Holiday park owners have reported a 40% rise in sales of static caravans. Even the relatively remote Scilly Isles are reporting capacity bookings in every holiday period. UK tourism contributes £114 billion every year to Britain's economy, or nearly 8% of its GDP.

Many resorts have succeeded in reinventing themselves and spruced themselves up as **short-break** destinations or as being ideal for the conference trade.

Another factor that has drawn people to the traditional seasides are festivals which have all aided the staycation trend.

'Bucket and spade', Scarborough

Scarborough is found on the East Coast of Yorkshire. It claims to be the first seaside spa resort. It was fashionable for the Victorian aristocracy who used its sea bathing facilities. The railway brought the middle classes to the town in 1846 and by the late nineteenth century Scarborough was well established.

Over the last few years Scarborough has seen a switch from **high-season** tourism to year round visits. Tourism impacts on Scarborough in all the usual ways. On the positive side it generates significant amounts of income (in 2006, £381 million) and nearly 20% of all employment. The permanent population also enjoys many modern facilities. On the negative side, congestion, housing and pollution problems are all too common. For the future, there are challenges to face. But strategic planning should ensure that the town is able to compete. Recent developments have tagged Scarborough the 'National Seaside Resort for the twenty-first century'!

New initiatives and diversity

| Edexcel | A2 U4 |
| CCEA | A2 U1 |

The UK tourist is becoming footloose with a greater, broader range of holiday opportunities on offer, to reflect both the UK's increased mobility and wide ranging interests.

Holiday villages

Heritage in its broadest context allows visitors to look at nostalgia, tradition and historical customs.

The idea of holiday villages is not new. Early holiday villages were built along the lines of similar communal establishments. The first in the UK, a 'tented' community, was set up at the turn of the nineteenth century at Caister–on-Sea. The first Butlins Holiday Camp was constructed at Barry, South Wales. Accommodation was cheap and cheerful and everything was provided on site for the tourist. In reality the village is a planned resort, or tourist enclave. Such villages probably originated in Continental Europe. In the late 1980s a Dutch company brought to the UK a new type of holiday village. Centre Parcs now has four such centres in the UK. These parks rely upon a mobile public wanting to be involved in leisure activities in an all-year-round environment.

Heritage tourism

80% of inbound arrivals to cities are tourists. MICE travellers (meetings, incentives, conventions and exhibitions) are increasingly important for cities, not only because of the higher per capita expenditure of MICE travellers, but also because of its promotional impact.

Nearly 55% of all domestic holidays now make use of the countryside and heritage attractions. All city and county areas have to have structure plans with tourism as a prime concern.

Case study: Re-imaging destinations

London well ahead of the pack

London attracts about 16 million international visitors per year. Bangkok, Paris and Singapore follow London each with about ten million tourists. But can London keep the tourists coming?

Airports are definitely the gateway to twenty-first century cities. Heathrow's Terminal 5 opened in March 2008, which allows London's airports to handle a further 30 million passengers per year, providing much needed relief to the overcrowded older terminals and many more visitors for London's burgeoning tourist industry. Furthermore, cheap flights, booked on the internet, mean that millions of tourists from all across Europe can access London within a 2–3 hour flight. What remains to be seen is if tourists that have discovered London will repeat their experience again and again as the number of routes within continental Europe expands. There are other cities that are reshaping their skylines in an attempt to usurp London into top destination. London has undoubtedly worked hard to ensure what tourists want. The 2012 Olympic Games are also expected to give the capital a 'ten year boost' after the games.

New cities 'on the block'

In Asia cities such as Kuala Lumpur, Taipei and Shanghai have all worked hard to re-image themselves. Dubai has opened the tallest hotel in the world called the Burj Al Arab Hotel. Dubai has also created the largest artificial islands in the world. Called The World, they provide the city with luxury properties, hotels and leisure centres all close to the beach! Branches of the Louvre and Guggenheim museums are expected to open soon in the Middle East. The USA is also working to plan and construct landmark buildings such as libraries, airports and art galleries. New and innovative buildings have attracted tourists to cities for years and will continue to do so.

The Spanish urban transformation

Barcelona's 1992 Olympic Games saw the Catalan city transform itself with heavy infrastructure investment and develop a whole new city on the back of the Olympics. Millions of tourists have visited since the Olympics and hosting the Universal Expo in Seville also raised the country's profile.

Theme parks

It is said a successful theme park works by closing off the real world. The British theme park is vastly different to its American counterpart. Although thrill rides are in evidence there is much more emphasis on the educational aspects of the theme parks. So in the UK, theme parks range from the purely educational Iron Bridge Gorge Centre to the thrill seeker's paradise at Alton Towers. One feature of all these theme parks is that people are willing, and prepared, to travel long distances to reach them. The parks are highly accessible and are within easy reach of the densely populated parts of the country. In what is a very volatile market continual adaptive change is a must as visitors' tastes change. Windsor Safari Park is a good example. It was a leisure casualty and no longer exists – Legoland took its place!

> Other attractions include farm visits, zoo attractions and animal sanctuaries.

Sport and leisure

> Demand has grown by 50–60% through the 1990s, for sailing, water-skiing, windsurfing, rowing and canoeing.

As disposable income has increased so has people's involvement in **medium-to-high** cost sports. This has led to a plethora of developments related to **water-based** activity. Second generation sports centres have also started to appear in the urban landscape. These are a level above the 'original' sports hall, offering swimming, courts, fitness suites, climbing walls and other activities. They reflect both changing tastes, and our increasing affluence, and the fact we have more time available for such activities.

Foreign tourists in the UK

Edexcel **A2 U4**
CCEA **A2 U1**

Some 30 million tourists visited the UK in 2010. Most were from Europe, though tourists from the USA continue to make up a significant proportion of our visitors. Annually, foreign tourists to the UK spend some £16.7 billion. Nearly 50% of all visits are focused on and within Greater London. The other half of visitors congregate around tourist honeypots such as Stratford-on-Avon, Oxford and Warwick.

PROGRESS CHECK

1. According to the mnemonic SIDECAM the A stands for 'a much smaller world'. What does this mean?
2. Why has the so called staycation become so important?
3. What is meant by critical mass in the tourist industry?

Answers: 1. Flights, motorways and tunnels all speed all journeys. 2. Accommodation upgrading, resort improvements and recessional lag have all made the staycation more appealing. 3. The numbers and investment in tourism needed to survive.

14.3 International tourism

LEARNING SUMMARY

After studying this section, you should be able to understand:
- why people choose certain destinations over others around the world
- that all tourist destinations are affected by a range of impacts

Choice of destination

Edexcel **A2 U4**
CCEA **A2 U1**

Tourism is not evenly dispersed throughout the world – both physical and human factors affect the patterns of distribution. Significant factors include:

Opportunities to travel have increased. It is important you know why and what effect we have on the environment.

- **resource distribution** (both cultural and natural)
- a country's ability to provide the **infrastructure** for growth
- **weather and climate** – temperature and seasonal variation of weather have an effect and hotter areas in particular are advantaged!
- **government controls** – some allow tourism to grow in an ad hoc way, others regulate and plan, e.g. Kenya
- changing **tastes**
- **war** and **political problems** can deter visitors
- the state of the **economy**: business tourism is an important facet of tourism.

Despite the unevenness of its distribution, international tourism has grown due to:
- awareness of other places, education and the media have created a demand on providers to allow entry – countries have become consumer packages.
- the advent of the internet has revolutionised the way we purchase holidays
- cheap charter travel, makes travel possible for a wider audience
- relative affluence, increased standards of living, paid holiday and early retirements give people more time to travel.

The impacts of tourism

Edexcel **A2 U4**
CCEA **A2 U1**

Costs and benefits

Tourism development can be closely linked to a series of environmental and socio-economic impacts.

Benefits of tourism

As much as 90 per cent of revenue leaks from some Caribbean tourist destinations.

- Tourists have huge **spending power**. Tourism stimulates economies and helps the balance of payments.
- **Investment** in host countries is encouraged. Increased levels of foreign currency circulate in the countries. GDP increases.
- **Jobs** are created. This job creation has a multiplier effect through the economy.

TIM (Tourist Income Multiplier).

- Local food and drink is in **demand** and farmers **benefit**.
- Local crafts and goods are also in demand. This is both a money earner and a way of ensuring the craft type is preserved and revitalised.
- As the local infrastructure firms up to cope with the tourists the local **standard of living** is boosted.

The Gambia, Seychelles and Tunisia rely heavily on foreign exchange.

- The host's **culture** is fortified by interest from the tourists and visitors.
- **Healthcare**, **education** and **social security** develops.
- The **role of women** is enhanced.
- Horizons and experiences are **broadened**.

Costs of tourism

- Tourism is affected by events in a country. Israel has a high reliance on foreign currency from visitors; conflict between Jews and Muslims discourages visitors to the Holy Land!
- Development of core and peripheral areas. Seaside resorts (core) usually do very well economically but areas on the periphery do less well and often demonstrate deprivation and are run-down.
- Economic benefits are selective, only a small amount of revenue generated may reach those that need it. Much is returned to foreign investors. Jobs are seasonal and part time.

In Luxor Egypt local 'squatters' are being relocated from Al Gourna so that tourists have an uninterrupted view of the Nile!

- Western values and demands are often at odds with those of the locals.
- Tourism may force locals out of their traditional homes or off their ancestral lands.
- Locals are driven out by rich retirees moving in.
- House, clothes, and food prices are inflated.
- Tourism literally drains countries dry of their water resources.
- Diet changes can cause problems, e.g. increased heart disease in Hawaii.
- Honeypot sites suffer chronic congestion problems, e.g. the volcanic Timanfaya National Park, Lanzarote.
- Crime increases.
- Health risks increase.
- Antagonism – references to dress, behaviour and religion.

Case study: Has Majorca lost out to tourism?

Majorca has been a favourite of European tourists since the 1960s.

The island of Majorca is the largest of the Balearic chain, which lies off the eastern Mediterranean coast of Spain. It has a rugged landscape, balmy climate, and old world architectural charm.

The advent of tourism in Majorca saw the agricultural dominance all but fade away. The industry that typified the island has been lost. Where it survives today high value crops are cultivated such as vines and fresh vegetables (using intensive, technical, irrigation-based cultivation methods). Where farming and farms were abandoned speculation has increased land and property values. At the time this happened many properties and land were taken on by Northern Europeans, especially Germans.

Tourists initially stayed at the coast, however, over time they have spread into rural areas, staying in second homes, farmhouses in the agro-tourism sector, or in small rural hotels. This type of tourism actually supports rural economies and is being encouraged.

The benefits of tourism in Majorca are heavily outweighed by the negatives. The benefits of tourism in Majorca include:
- bringing employment
- more money is put into improving the island.

The problems with tourism in Majorca include:
- pollution, noise and litter
- drunkenness

- the effects on Spanish culture and tradition
- the jobs provided are seasonal, unskilled and poorly paid
- locals cannot afford to use tourist facilities
- most of the money from tourism goes out of the area
- discos, bars and other tourist attractions spoil the local way of life
- tourism raises prices so locals cannot afford goods in shops.
- the better tourist jobs rarely go to the local people.

Case study: The Kleenex Trail, Nepal

Nepal is among the poorest and least developed countries in the world with almost one-third of its population living below the poverty line. Agriculture is the mainstay of the economy, providing a livelihood for 75% of the population and accounting for 38% of GDP. Nepal is a popular tourist destination for mountain climbers and hikers. The tourism industry, however, has been threatened by political instability and a sluggish global economy. Opening the doors to the world and tourists has brought not only advantages, but also disadvantages to Nepal. Tourism and foreign aid have brought money, but also environmental pollution.

Oxygen cylinders, PVC waste and synthetic wrappings and other refuse lie strewn along the expedition routes – many refer to it as the Kleenex Trail; the detritus of latrines guiding the way to the foothills of the Himalayas! Up to 500 kgs of rubbish are left by each expedition.

Namche and Phortse have forests that have been destroyed by endless streams of trekkers and climbers, causing bare mountain tracts, landslides and "rock-streams". Moreover the shrubbery has to be sheltered from the sherpa's yaks, sheep and mountain goats. With little work available many become part of the environmental problem in Nepal; acting as guides and porters for trekkers. Others leave to find a place in the armies of Nepal, India and Britain. There is also a regular seasonal migration from Nepal to the paddy fields of the Terai Arc and even into India. Furthermore, many feel that Nepal has sold out to the tourists.

Environmental impacts and management strategies

Area	Concern	Management approach
Spain	• Coastal, high-rise development.	• Strict planning controls.
Caribbean	• Coral destruction, Mangrove destruction. • Silting of seabed/algal blooms.	• Marine Park designation, coastal development prohibited. • Buffer zones replanted along coast land.
Himalayas	• Depression/rubbish disposal.	• Many areas run strict permit entry only.
French Alps	• Damage to delicate flora on piste slopes.	• Zoning and no new development allowed.
Greece	• Athens – traffic congestion leads to air pollution. • Zante/Loggerhead turtle's breeding ground is being destroyed.	• Limit, restrict car usage. • Limit access to beach concerned.
UK	• In the Peak District, ecological damage through footpath erosion. • The Hadrians Wall National Trail, a conservation *vs* recreation issue.	• Reinforce paths and rebuild. • Issued ticketing, clear tourist management with guide posts.

PROGRESS CHECK

1 Why has international tourism increased?

2 List some examples of environmental damage caused by tourists.

Answers: 1. Education and awareness has increased, cheap holiday trade and affluence has increased. 2. Traffic congestion, pollution, damage to vegetation, coral destruction and path destruction.

14.4 Sustainable ecotourism

LEARNING SUMMARY	**After studying this section, you should be able to understand:**
	• that non-sustainable tourism has economic, socio-cultural and environmental dimensions
	• that eco/green tourism is becoming increasingly popular
	• why countries of destination have to accept stewardship of their environments

Non–sustainable tourism

Edexcel **A2 U4**
CCEA **A2 U1**

As tourism grew, the protection of the environment was a low priority. Landscapes were destroyed and congestion and pollution became an increasing problem. Profit became the driving force behind most developments. Nowadays, investment, and financing, of tourist development projects usually hinges on the sensible and sustainable development of resources. The World Bank, the EU and so on will no longer be linked with environmentally destructive projects. Areas worthy of protection are not just 'natural'. The cultural and built environment is also given due consideration. Consider the information below.

Tourism can be said to help conservation efforts by financing them and by encouraging countries to preserve the resources that they want tourists to visit. But travelling abroad is certainly a destructive activity: air travel produces a high level of pollution due to the huge quantities of carbon dioxide it produces.

Regardless of where they travel to, tourists have expectations, and tour companies try to meet these expectations – basic expectations such as showers and flushing toilets must be provided, often at the expense of the local environment. However, there are some companies that have tried to lessen the damaging effect on the places tourists visit. This **low-impact tourism** takes place with boats, trains and bicycles, and the money spent by tourists goes towards encouraging initiatives such as tourists staying in local people's homes, and investing in local projects. This kind of tourism can be classed as a much more ethical activity and must surely be the way forward for the tourist industry.

Sustainable tourism

Edexcel **A2 U4**
CCEA **A2 U1**

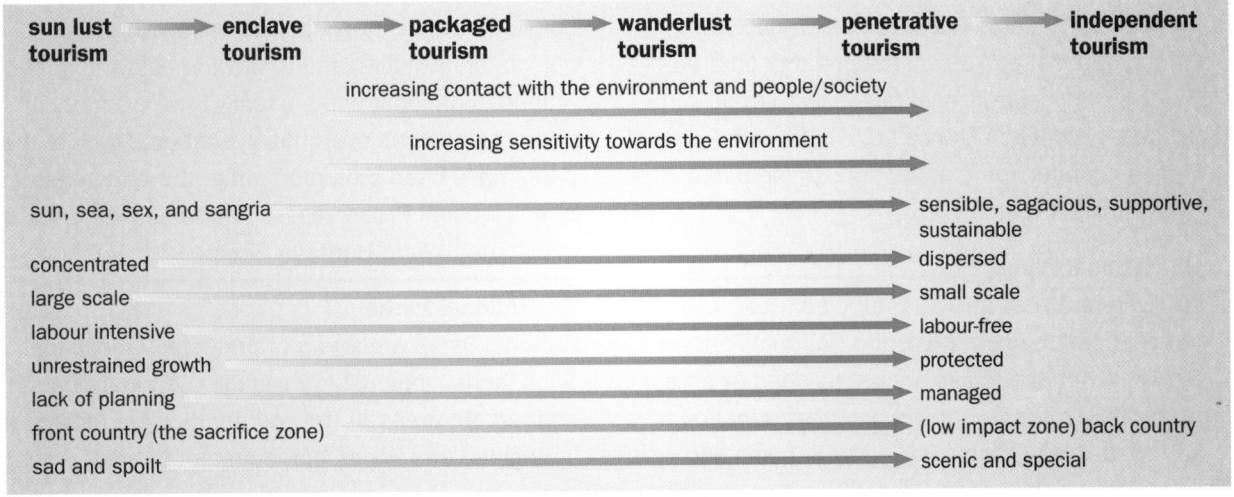

sun lust tourism	→ enclave tourism	→ packaged tourism	→ wanderlust tourism	→ penetrative tourism	→ independent tourism

increasing contact with the environment and people/society →

increasing sensitivity towards the environment →

sun, sea, sex, and sangria	→ sensible, sagacious, supportive, sustainable
concentrated	→ dispersed
large scale	→ small scale
labour intensive	→ labour-free
unrestrained growth	→ protected
lack of planning	→ managed
front country (the sacrifice zone)	→ (low impact zone) back country
sad and spoilt	→ scenic and special

The extremes of tourism

Tourism can be seen as an exploitive industry, depleting the resources at one location before moving on to the next. To avoid this scenario when tourism in a country goes into decline (see Butler's tourist life cycle page 280), efforts must be made to resurrect the situation. But laudable as many efforts are to adapt to sustainable regimes, the huge size, scale, diversity and the swiftness of change in the tourism industry means that 'strategies' are slow to take-off and become adopted.

Case study: The Sarawak experience – ecotourism in action, Malaysia

Substantial resources have been committed to build a viable tourism industry around Sarawak's natural landscape, mostly through the Sarawak Economic Development Corporation (SEDC). The SEDC views tourism and leisure as a 'strategic business'.

The SEDC made an initial investment of over 550 million Malaysian ringgits (equivalent to approximately $154 million) to build five international standard hotels and other tourism properties, plus downtown shopping complexes in Sarawak's capital city, Kuching. The cultural village (a 6.9 hectare tourist attraction 35 km from Kuching in the Damai Beach resort district) is a living museum that enables visitors to experience, the authentic dwelling styles, arts, crafts, games, foods, music and dance of seven of Sarawak's major resident cultures.

Two of the SEDC resorts, Bukit Saban and Royal Mulu, are located deep within Borneo's interior rainforest and serve as a comfortable base for ecotouristic adventures. Sarawak is working hard to attract and please the visitor – its approach is sensible and sustainable.

Case study: Green tourism and the Alpine environment

Switzerland's first national law on the environment was passed in 1902 to protect the forests, although local by-laws had existed long before that to provide avalanche protection. Since 1980, tough laws have been passed on air pollution for industry and motor vehicles, water pollution and waste disposal. Cars and motor cycles have to be fitted with catalytic converters, speed limits are lower than in many other European countries and domestic heating systems using oil have to be monitored for sulphur output. Switzerland revised its waste disposal laws in favour of the environment, instead of least-cost solutions. Laws include regulations on non-returnable waste, especially in the drinks industry. Such laws are clearly important to the tourist industry. All construction projects have to be assessed to ensure that they conform to all legislation.

Case study: Protected landscapes

Parcs Nationaux de France
Nine national parks exist in France and all are maintained by *Parcs Nationaux de France*. The French national parks scheme protects an area of 3710 km² in central zones and 9162 km² in secondary zones in France – 2% of the total area of France is under some protection. These parks draw over 7 million visitors/year.

US National Parks
The United States has 58 protected areas known as national parks (operated by the National Park Service). National parks are established by an act of the United States Congress. The first national park was Yellowstone followed by Sequoia and Yosemite. The America National Parks have a very similar reason for being; they were established to conserve the scenery, the natural and historic objects and wildlife, and to provide enjoyment. The American national parks usually have a variety of natural resources over massive areas. Many of the parks have been protected under the Antiquities Act or as National Preserves. A number of the parks are designated World Heritage Sites.

UK National Parks
National Parks are areas of protected landscape. Such landscapes usually include the most beautiful and remote areas of the country they are located in. In England and Wales these areas are additionally

used for forestry and agriculture, for residential areas and even in some areas for industry, e.g. limestone quarries and military training. They also contain over a quarter of a million people! The National Parks in the UK were set up by an Act of Parliament in 1949. The first parks to be established were the Peak District, the Lake District, Snowdonia and Dartmoor in 1951. These areas are part of a global network of National Parks that cover 149 million square kilometres or 6% of the Earth's surface (8% of the UK)!

Contribution to the local economy of National Parks in the UK

Tourism contributes a significant revenue to the local economies of the National Parks. It is an important part of the economy of the regions which contain national parks. There are a range of economic opportunities created by the millions that visit national parks, e.g. in 2008 there were 2800 businesses in the Peak District National Park and 75% of these businesses were micro-businesses employing less than five people. In total 14 000 were employed (25% of the total employment in this area having a direct tourism link) and from 12.8 million visitors some £135 million was spent.

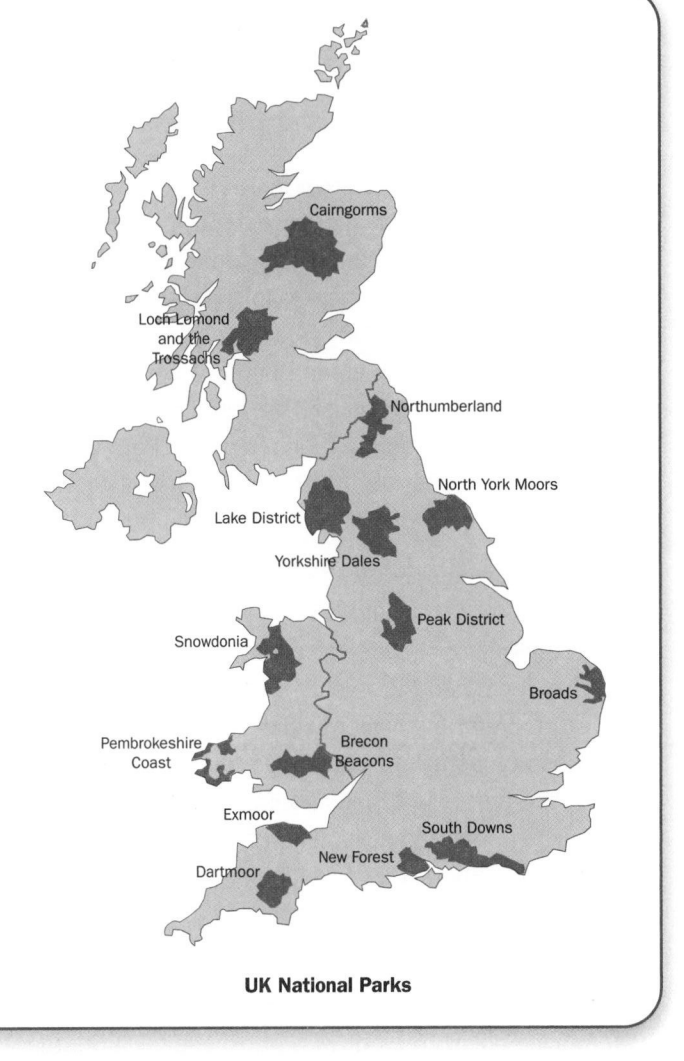

UK National Parks

Conflicts in National Parks in the UK

The two objectives of the UK's National Parks cause frequent conflict between different groups of people.

- To **conserve** and **enhance** the natural beauty, wildlife and cultural heritage of the area.
- To **promote opportunities** for the understanding and enjoyment of the park's special qualities by the public.

Tourism brings undoubted benefits to an area but it also brings a number of problems.

- **Congestion** of villages and beauty spots – these so-called 'honeypot areas' attract masses of visitors especially on Sundays and over holiday periods which leads to busy car parks, blocked roads and local facilities unable to cope, e.g. Bakewell in The Peak District.
- **Erosion** – most of the popular paths in the National Parks are severely eroded.

> The advent of the mountain-biking culture has further accelerated damage, e.g. Dovedale.

- **Damage** and **disturbance** to wildlife – the busiest parks have noted drops in all animal types. Nesting birds in particular are easily disturbed.
- **Litter** of all kinds is left in the National Parks.
- Damage to farmland – crops get trampled and sheep are frightened by dogs not under proper control.
- Local community **displacement** – the costs of accommodation for locals has rocketed. Communities feel pushed out by the tourists.

- **Conflict** between recreational users, e.g. those involved in noisy water sports and the bird spotting community.

Case study: The Last Shangri-La – 'high value-low volume', Bhutan

Tourism in Bhutan started in 1974. The goal for the government was to raise revenue by promoting the isolated nature of the country to the outside world. Since 1974 the number of visiting tourists has increased to about 30 000. The most important centres for tourism are in Bhutan's capital Thimphu and in the western city of Paro, near India. Taktshang, a cliff side monastery overlooking the Paro Valley, is one of the country's attractions. Druk Air is currently the only airline operating flights in Bhutan.

Bhutan is acutely aware of the environmental impact tourists could have on its unique and virtually unspoiled landscape and culture. Therefore, they restricted the level of tourist activity from the start, preferring higher quality tourism rather than mass tourism. The Bhutanese tourism industry is based on the principle of sustainability. Furthermore, the government's view is that modernisation and development should be guided by the 'Gross National Happiness' of the Bhutanese people, rather than by the Gross National Product!

The Bhutan government controls tourism by using a pricing policy. The only way of experiencing Bhutan is by using a government tour operator. The whole itinerary must be a cultural tour, festival tour, trekking or a mixture of both. An entry tariff is also levied. Holidays in Bhutan are not cheap – the phase used in Bhutan is 'that there is a favourable income to arrival ratio'! Bhutan's visa formalities are processed in advance of arrival and must be gained before entry to the country. The revenue from tourism gets to the local communities in several ways.

- Park managers awarding contracts to local people for the maintenance of footbridges and trails.
- Rest and community houses along trekking routes have been handed over to local community residents. The revenue generated from tourism goes directly to the local community and they have full control over these shelters.
- Tourists are given a local diet rather than imported foodstuffs by local communities on their trekking route.
- All transport is provided by the local people.
- Tourists are encouraged to buy handicrafts from local communities.

Ecotourism

Edexcel **A2 U4**
CCEA **A2 U1**

Well planned **ecotourism** conserves fragile ecosystems, whilst at the same time providing income for local companies.

It is possible for tourism to balance its activities with the natural environment.

A topical subject and a favourite with examiners.

KEY POINT

Key principles of ecotourism
1. Maintain the quality of the environmental resource. Develop it in an environmentally sound way.
2. Produce first-hand participation and inspiring experiences.
3. Educate all parties before, during and after the trip.
4. Encourage recognition of the value of the resource by everyone involved.
5. Accept the resource as it exists and accept that this may limit the number of visits involved.
6. Encourage understanding and partnerships between governments, non-governmental groups, industry, science and local people.
7. Promote moral and ethical responsibilities and behaviours.
8. Develop long-term benefits to the resources, community and industry.

Economic advantages of ecotourism include:

- a greater variety of economic activities in rural and non-industrial regions
- long-term economic stability
- tourists involved in ecotourism tend to spend more and stay longer
- demand for local goods and services
- development of infrastructure such as roads, airports and bridges
- increase in foreign exchange earnings.

Benefits of links with conservation organisations include:

- donation of portion of tour fees to local groups
- education about the value of the resource
- opportunities to observe and take part in scientific activity
- involvement of locals in providing support services and products
- involvement of locals in explaining cultural activities or relationship with natural resources
- promotion of a tourist and/or operator code of ethics for responsible travel.

Case study: 'Thai style' tourism

Tourism is very important to the economy in Thailand. Umphang is one of the remotest spots in Thailand and as such is a pristine destination, an area of rich forest and animal resources, and serves as one of the country's major sources of water. It is an area with many diverse customs and traditions. Road building has opened up the area. It was soon realised that the many tourists visiting the area were actually destroying it, leading to many problems including overcrowding, rubbish accumulations and environmental deterioration.

It was decided that strict adherence to the principles of ecotourism were necessary to ensure that tour activities continued, but preserved the environment, and which allowed other adventure activities to be initiated to take pressure off the traditional tourist activities. The Umphang Tourism Promotion and Preservation Club, have imposed various measures to help solve the problems they were encountering. They are the familiar ecotourism principles.

- To limit tourist numbers in accordance with the area's carrying capacity.
- Visitors have to gain permission to enter into the preserve.
- Litter and rubbish reduction and cleanliness enhancement.
- A requirement to make a donation.
- A coordination centre from tourist activities.
- Management of the activities to ensure a high quality service.

- Zoning of tourist areas and information signs help guide and direct tourists around sites.

The Thai Government has also laid down best practice in developing ecotourism.

- By enabling locals to be involved in planning and decision making.
- By teaching locals how to be good hosts.
- To ensure that locals produce good quality goods for the tourists and that they are sold exclusively to tourists. Also that money is returned to the local economy from all sales.
- That the identity of the local Umphang indigenous peoples is preserved.
- That energy saving and environmental sound structures are to be built to house tourists.
- That locals are taught how to market their products.
- All forms of media have been invited to visit Umphang and to publicise the area.

A number of prestigious honours have been won and serve to highlight the success of Umphang as a model for south-east Asia's efforts to seriously move ecotourism activities on apace, but in a sustainable way. Thailand clearly intends to maintain the kingdom's uniqueness, in terms of tourist attractions, high quality services and environmental protection, all of which will have a direct impact on the country's social and economic development.

Ethical tourism

Edexcel **A2 U4**
CCEA **A2 U1**

Three questions are central to ethical tourism.

- Socially: Do we have a right to behave as we wish in our destination resorts?
- Economically: Should the big operators be able to squeeze rentals so low that owners can barely survive?
- Environmentally: Do we have the right to intrude?

As tourists we must:

- protect and preserve natural and cultural heritages
- use energy and water resources efficiently
- help to preserve traditions, customs and local regulations
- avoid damaging the environment
- only buy products that demonstrate social, cultural and environmental sensitivity.

The tourist industry must:

- strive for excellence in all aspects
- encourage an appreciation of the environment
- contribute to community identity, pride and quality of life for residents
- protect cultural and aesthetic heritage
- manage waste and pollution appropriately
- support the tourists' quest for greater understanding.

> Changes in many areas, political, social and economic, will affect tourism in the future. Ensure you keep in touch with all the developments as they crop up.

Tourism will remain a growth area for the foreseeable future. Economic development brings more disposable wealth and travel. But what is the future of tourism?

Case study: What future tourism?

A recent study found that profitable tourist destinations could be turned into 'holiday horror stories'.

By 2020, visitors to the Costa del Sol could risk contracting malaria as global warming brings more frequent **heat-waves**, making the country a suitable habitat for **malaria-bearing** mosquitoes. Increases in summer temperatures to more than 40°C could make parts of Turkey and Greece **no-go** areas in July and August. The study suggests that operators and countries which rely on tourists for revenue will need to take account of the changing climate when planning new resorts or upgrading facilities.

The tourist industry is not just a potential victim of global warming – it also contributes to the causes of climate change itself. Air travel is the fastest growing source of greenhouse gas emission, and contributes to continued global warming. The study reports that:

- a decline in cloud cover over Australia will increase exposure to the sun's harmful rays, increasing the risk of sunburn and skin cancer
- winter tourism will be affected in the Alps and other European skiing destinations from the impact of less snowfall and shorter skiing seasons.
- the south-east coastline of the USA, including parts of Florida, may be threatened by rising sea levels
- islands in the Maldives could disappear as they are submerged by rising sea levels.

PROGRESS CHECK

1 Why are micro-businesses important in many National Park areas?
2 What are the three key principals of ethical tourism?

Answers: 1. They provide employment in a range of small industrial set-ups – they are revenue earners. 2. Social – we should behave as we wish in resorts? Economic – should tourist firms be allowed to squeeze small hotels? Environmental – should we be intruding?

Sample questions and model answers

1. Parkland Resorts is a holiday company with four purpose-built leisure complexes based on the concept of a recreational theme park in a woodland environment. The parks are all car-free and the accommodation has been built around a central climate-controlled dome. Study the extract below from a Parkland Resort press release.

> Parkland Resorts has won the Landscape Preservation Organisation's award for Landscape Management, in recognition of Parkland Resorts' ongoing commitment to developing exceptional natural environments at each of its three holiday parks. The Parkland Resorts Landscape Management Scheme is held up as a 'dynamic, co-ordinated and fully effective plan.' Each holiday park has a range of waterways, glades and woodlands, which vary from ancient deciduous woodland to mature coniferous plantations and replanted coniferous woodland areas. The Parkland Resorts Landscape Management Scheme has a 10-year plan for each holiday park, aimed at achieving landscape variety, diversity of habitats and spatial character. This is supported by yearly ecological studies to monitor the environments. Each holiday park also has its own Biodiversity Plan, which indentifies goals for each park, according to local targets for species recovery.

(a) Evaluate the costs and benefits that arise from the management scheme outlined in the extract.　**[7]**

(b) Account for the increasing number of visitors attracted to such theme parks in MEDCs.　**[13]**

(a) The clear outcome of the management scheme for Parkland Resorts has to be based upon sound financial considerations as they are clearly a commercial company and need to make a profit. The environmental benefits are clear in that they are doing much work to plant and reintroduce rare species into the woodland environment, so encouraging biodiversity and preserving different natural environments. The management scheme also includes monitoring and goal setting, all of which can only be beneficial for the environment. The costs come from the money that needs to be invested by the company to carry out and maintain this extensive management programme, and any damage that the construction of the actual building might cause to the natural environment. This all creates a very positive environmental image for the company in our increasingly environmentally aware society, and might help to reduce the costs of advertising by creating its own publicity. The benefits clearly outweigh the costs in both environmental and economic terms, as long as the plan is kept going.

(b) In most MEDCs there is a growing proportion of middle class people who are looking for more regular, short-term breaks and days out. Such places provide them with this opportunity and the owners of such parks know what people want when they are there. Alton Towers has a wide range of family-friendly rides and attractions to keep the whole family entertained for a whole day, and although many of the parks are in themselves expensive, in the new millennium people are beginning to spend more of their disposable income on leisure and recreation.

Also, increasing car ownership and the reduction in the relative price of public transport has increased people's personal mobility in the UK and enabled them to get to such parks over a weekend. The most successful example of this is Disneyland, Paris, which after a poor opening, is now one of the top tourist destinations in the whole of Europe. It also offers that little piece of the USA on the doorsteps of most Europeans, with the high standards of hygiene and safety that all responsible families would look for in a weekend break. Ultimately there is a demand for such parks as well as an increasing number of them being built, and while the standards remain high we will continue to visit them in increasing numbers.

Probably too much detail for a 7 mark question. Excellent coverage, but too long in the time given.

Excellent understanding of processes.

Some statistics would be good to include here, i.e. UK car ownership statistics.

Excellent clear conclusion which relates back to the question.

15 Fieldwork, investigative work, statistics and skills

The following topics are covered in this chapter:

- Fieldwork and investigative requirements
- Investigative and cartographic skills
- Statistical requirements

This chapter provides an overview of skills and references for all geography students.

15.1 Fieldwork and investigative requirements

LEARNING SUMMARY

After studying this section, you should be able to understand:

- the requirements of the examination board that you follow
- the statistical requirements at AS and A2
- the cartographic skills requirement at AS and A2

Investigative and research expectations

The following information shows what the different types of questions require you to do and the skills they want you to demonstrate.

AQA	AS Unit 2: Geographical Skills	Application of knowledge and skills to unseen information and reference to own fieldwork.	Q1. Tests skills and uses resources relating to Unit 1 topics. Q2. Tests the candidates' actual experience of fieldwork.
	A2 Unit 4A: Geographical Fieldwork Investigation	Preparatory work should be fieldwork relating to a geographical topic.	Section A tests knowledge and understanding of the fieldwork undertaken. Section B tests fieldwork skills in unfamiliar circumstances.
	A2 Unit 4B: Geographical Issue Evaluation	No fieldwork, but pre-release information issued.	Allows candidates to develop a range of skills and knowledge and understanding. The exam tests knowledge and understanding related to the investigative fieldwork related to the pre-release information.

OCR	AS Unit 1: F761 – Managing Physical Environments; Unit 2: F762 – Managing Human Environments	There is no explicitly required fieldwork at this level. The inclusion of river, coasts and urban geography do make fieldwork possible.	At this level it is suggested that candidates become familiar with fieldwork skills.
	A2 Unit 3: F763 – Global Issues; Unit 4: F764 – Geographical Skills	For Unit 4 (F764) individual investigative and research work is undertaken, related to the Global issues section of the specification.	Section A is a data response exercise that tests the skills of fieldwork. In Section B the question focus is on geographic skills.
Edexcel	AS Unit 2: Geographical Investigations	As part of Unit 2 fieldwork, research and practical work are intrinsic – all are tested in the examination.	Resource booklet issued. Fieldwork is seen as part of the competence of a candidate at AS.
	A2 Unit 3: Contested Planet; Unit 4: Geography Research	There is no fieldwork explicitly identified, though the paper is tagged as being geographical research.	Six research topics can be found in the specification. Candidates study one; pre–released information develops the candidate's choice. Candidates display their knowledge and understanding in the exam.
WJEC	AS Unit G1: Changing Physical Environments; Unit G2: Changing Human Environments	At AS, both the physical and human papers (G1 and G2) contain assessments of fieldwork experiences. Half the marks are available for the fieldwork element of Section B in each paper.	Fieldwork has to be undertaken in both areas.
	At A2, a topic area is set by WJEC in the year before the exam. A decision-making exam is set.	A two-part question is set in Section B of the exam, based on pre-released information relating to research into one of ten topics.	Fieldwork is recommended in the preparation of this paper.
CCEA	AS 1 and 2 papers include an assessment of fieldwork	Involves some data collection. A summary and data table can be taken into the examination.	Section A on both papers tests fieldwork skills. Section B gives candidates the chance to use GIS and the Internet.
	A2 Unit 1: Human Geography and Global Issues; Unit 2: Physical Geography and Decision-Making	No explicit fieldwork as such.	A decision making exercise is undertaken in A2 based on pre-released information.

KEY POINT

How you approach fieldwork and the route you choose is a decision for you and your school or college. The detail that follows covers the common methods used and those that cause consternation and confusion in candidates.

15.2 Investigative and cartographic skills

LEARNING SUMMARY	After studying this section, you should be able to understand:
	• investigation and cartographic skills

Investigation and cartographic skills

How to choose a suitable geographical question or hypothesis for investigation

Make sure you learn the aims, hypotheses and location of your geographical investigation and are able to justify them.

Before any fieldwork or data collection is carried out you must make sure that you are clear about the **aim** of the investigation. The aim should be a broad statement of the purpose and area of the investigation and be based on relevant **geographical theory** that you have studied, e.g. river or coastal processes for a physical geography investigation or urban land use models for a human investigation. Once the aim has been decided it can then be developed into more **specific hypotheses** (statements of what you would expect to find) or questions to be answered. As well as being based on relevant geographical theory, the aim/hypotheses/questions must be capable of being researched at a suitable scale.

Do remember that the title or hypothesis may well be chosen for you by the examination board!

Developing a plan and strategy for conducting the investigation

A successful investigation requires careful planning which balances the requirement for accuracy against restrictions imposed by time, resources or safety considerations.

Once the aims, hypotheses and/or location have been decided on the next stage is to plan exactly how the investigation will be carried out. You need to think about what data you need in order to test the hypotheses, or answer the questions, and exactly how this data will be collected.

Types of data

Primary data

Primary data is information collected by the student that has not been subjected to any manipulation or processing. It will usually include data collected during fieldwork, e.g. vegetation surveys along a sand dune transect or questionnaires conducted in a town centre. It could also include 'raw' census data or historical maps where the student has to process the data into a useable format.

It is important to include secondary data in a geographical investigation to give background information or to help in the explanation of results.

Secondary data

Secondary data refers to information that has already been collected and published by someone else. Sources of secondary data include:
- maps, e.g. town centre plans, geology maps
- newspaper and magazine articles
- reports or information from action groups
- council reports/surveys
- the internet.

Make sure you can distinguish between the primary and secondary data collected for your fieldwork investigation/ research project.

When using secondary data it is important to evaluate the source for reliability, accuracy and bias.

Be able to distinguish between quantitative and qualitative data with reference to your investigation/research project.

Quantitative data

Quantitative data can be measured or counted and expressed numerically, e.g. river velocity, particle size, temperature, environmental quality scores, traffic counts or indices of deprivation. Such data can be represented using graphs and charts and analysed using descriptive and inferential statistics.

Qualitative data

Qualitative data refers to non-numerical or 'descriptive data'. This could include responses to questionnaires, e.g. *What are your views about the proposal to build a new shopping centre?* or observations about sites in a rivers investigation.

Data collection strategies

Sampling

Very often it will be impossible to measure or survey the entire 'population' (the data set or area from which information is being obtained) due to time and resource constraints. Therefore it is necessary to select a **sample** – that is a portion of the population that should be **representative** of the total or parent population. There are three types of sampling generally used by geographers: random, systematic and stratified.

Make sure you understand the characteristics of each of these sampling strategies, as well as their strengths and weaknesses. You should also make sure you understand, and are able to explain and justify, the sampling strategies used in your investigation, as well as being able to suggest a sampling strategy for an unfamiliar investigation scenario.

- **Random sampling** – a random sample is one in which every member of the population has an equal chance of being selected for the sample. This is normally achieved by using random number tables or random number generators on calculators or the internet. In reality, it is very difficult to avoid bias in a random sample and in actual fieldwork situations subjectivity is often a problem, e.g. when choosing people to stop and interview in a town.
- **Systematic sampling** – in a systematic sample measurements are taken, or data is collected, at fixed intervals, e.g. every tenth person passing in an urban area for a questionnaire or every 100 metres along a river for channel characteristic measurements. The advantage of systematic sampling is that it is quicker and easier to use than random sampling and ensures even coverage of the population. However, it is possible to miss important points. There may also be practical constraints, e.g. safe access problems for river sites. In this case a **pragmatic sampling** strategy can be used to choose a site as close as possible to the one selected by systematic sampling, but which can be accessed more easily.
- **Stratified sampling** – stratified sampling is based on advance information about the population or area in question, e.g. the age distribution of a town for a human geography investigation. A stratified sample will contain the same proportion in each age group as the overall population therefore ensuring that the sample is representative of the whole. When the 'strata' have been identified random or systematic strategies are then used to select the required number from within each.

Risk assessment

Specific risks vary depending on the nature of the fieldwork.

Before any fieldwork is done it is essential to carry out a risk assessment to identify any hazards, who could be harmed, the likelihood of the hazard occurring and the specific measures that need to be taken in order to reduce the risk. All participants in the fieldwork must be made aware of the risk assessment and follow the procedures at all times. An extract from a risk assessment for a rivers investigation is shown on the following page.

Hazard	Who is at risk	Likelihood of incident	Measures to reduce risk
Drowning	All	Low	Students to be supervised by a member of staff at all times and only allowed to enter the water if river is below knee height. When in water face upstream at all times in order to be aware of any large debris being carried by the river.
Injury	All	Low	Wear sturdy footwear and be careful when entering or leaving the river. Check carefully for any dangerous objects in the river, e.g. barbed wire or broken glass.
Water borne diseases	All	Low	Wash hands or use antiseptic gel before eating after leaving the river.

Be able to describe and justify the risk assessments that you carried out for your geographical investigation and suggest the risks that might be associated with an unfamiliar situation and how to reduce them.

KEY POINT

Once you have collected your data/information you have to decide on methods that are suitable to use to present your findings to the examiner.

Data presentation

Learn the data presentation techniques that you used in your investigation, their strengths, weaknesses and alternative methods. You should also be ready to comment on the appropriateness or quality of a given piece of representation and be able to suggest methods that could be used to present unfamiliar data.

As the 'raw' data on its own is often difficult to interpret, making it difficult to identify patterns or any conclusions, an important part of any geographical investigation is the presentation of data. There is a wide range of data presentation techniques that can be used and it is essential to choose methods that are appropriate to the data collected and then to present them to the highest standard. Very often ICT is used for data presentation although many students find some techniques, e.g. sketch maps or field sketches, easier to hand draw.

Cartographic skills

Ordnance Survey maps

You should be able to read and interpret **Ordnance Survey (OS)** maps at a range of scales, especially the most common (1:25,000 and 1:50,000) and understand how to choose a map of an appropriate scale to locate or give information about the area studied. Remember that a 1:50,000 map will cover a larger area, but does not show as much detail as a 1:25,000 map, e.g. field boundaries are shown on a 1:25,000 map which can be very useful for rural land use surveys.

Maps with proportional symbols

Maps with proportional symbols use symbols, e.g. squares, circles, semi-circles or bars, that are proportional in area to the value they represent, e.g. population density or river particle size. When constructing such a map it is important to choose a scale for the symbols that allows them to be clearly located, but avoids overlap. This can be difficult to achieve, making the maps time consuming to construct – a disadvantage of this technique.

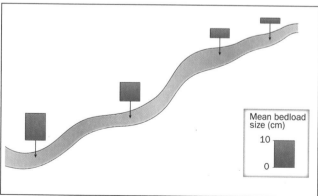

Mean bedload size (cm)

10

0

Located proportional rectangles map

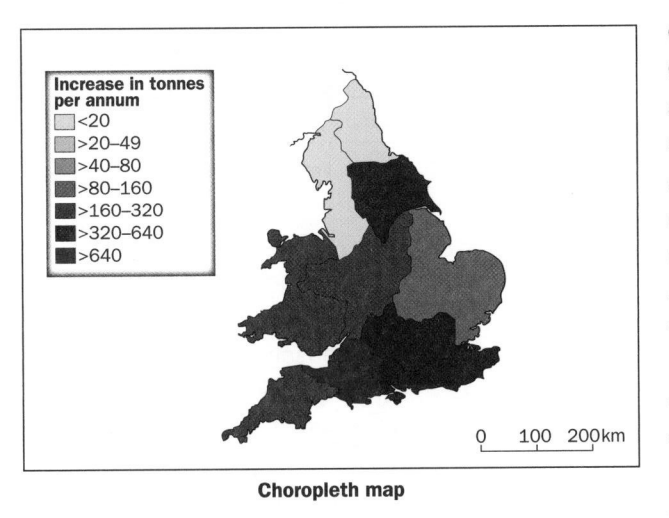

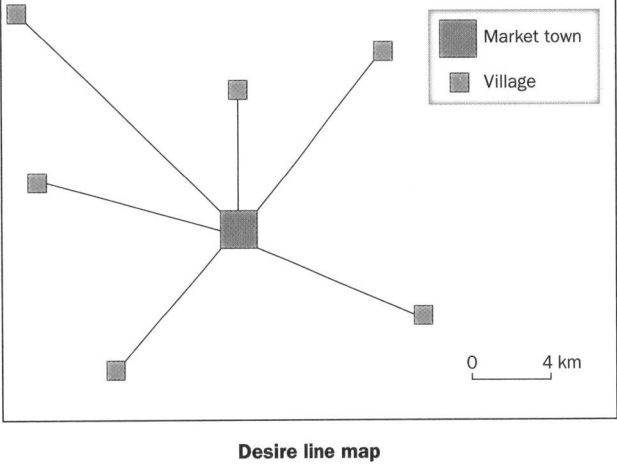

Flow line map to show river discharge

Desire line map

Flow line and desire line maps

Flow line maps are used to show the volume of movement along a certain route, e.g. traffic flows in urban areas or discharge along a river and its tributaries. The width of the line is proportional to the amount of movement being represented.

Desire line maps can be used in catchment area surveys to show the origin of shoppers at a particular destination to allow its catchment area to be determined.

Increase in tonnes per annum
- <20
- >20–49
- >40–80
- >80–160
- >160–320
- >320–640
- >640

0   100   200km

Choropleth map

Choropleth maps

Choropleth maps use density of shading within an area to represent data values, e.g. population density per square kilometre. In order to produce a choropleth map the data must first be grouped into categories or classes. A range of shadings is then used to cover the range of data, with the lightest shade used for the lowest values and the darkest for the higher values. Choropleth maps are quite straightforward to construct and give a very clear visual representation of patterns in an area. However, the method hides any variation within an area and makes it appear that there is an abrupt change at the boundary which is often not the case in reality.

Sketch maps

Hand drawn sketch maps are very useful to show the location and key features of an area of investigation, e.g. a river channel. Make sure that clear labels point out the key features and that the map includes a simple scale and orientation.

Isoline (isopleth maps)

On an isoline map points of equal value are joined. A good example is the use of contour lines on OS maps where the contours join points of equal height. Other examples include lines joining places of equal temperature (**isotherms**) and pressure (**isobars**) on weather maps. A common use of isoline maps in geographical investigation is to join areas of equal pedestrian densities in urban areas to investigate how this changes with distance from the central business district (CBD).

Dot maps

Dot maps are used to show the spatial distribution of a geographical phenomenon where the location and value has been measured, e.g. showing population distribution in a region where each dot on the map represents a given number of people. They are very useful in giving a simple visual representation. However, they are less useful in showing precise values due to the difficulty of counting large numbers of dots.

Graphical Skills

Line Graphs

Line graphs are used to show changes in data through time or with distance, e.g. population, temperature or river discharge. Time (or distance) should be on the x axis. Line graphs can be **simple**, **comparative**, **compound** or **divergent** so it is important to check the axes and keys very carefully when reading off data.

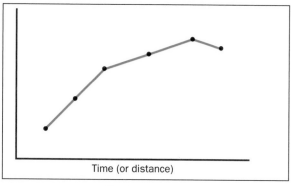

Simple line graph

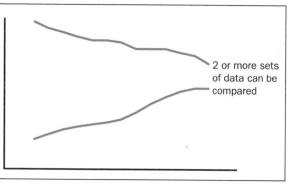

Bar graphs

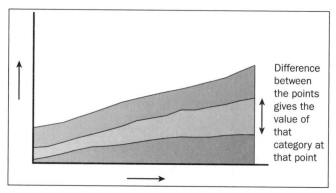

Compound line graph

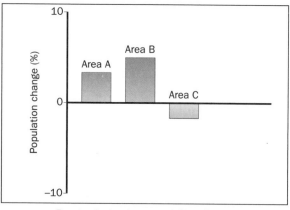

Bar graphs to show population change

Bar graphs

Bar graphs are very simple to read and can be used to represent a wide range of data. They use vertical columns to represent either absolute values or percentage figures which can be read off the y axis. As with line graphs they can be simple, comparative, compound or divergent, so it is important to check the axes labels and keys carefully when reading the data.

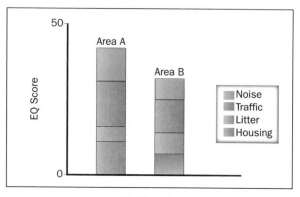

Divided bar charts

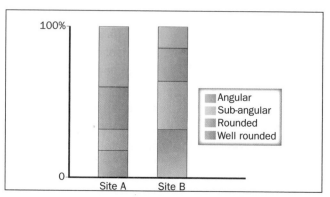

Compound bar charts

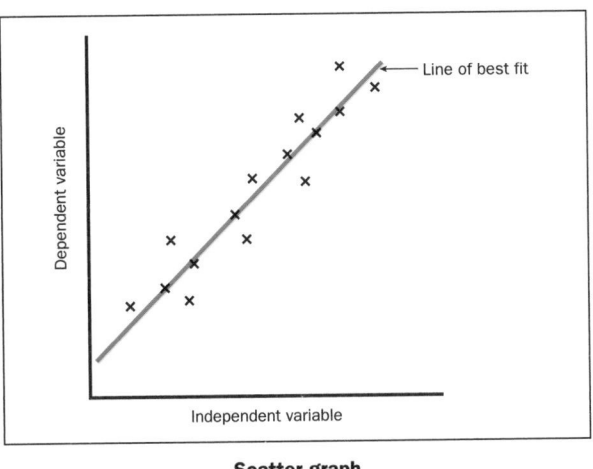

Scatter graph

Scatter graphs

Scatter graphs are used to show the relationship between two variables and are very useful in identifying patterns or trends in the data when a **best fit line** is added to the graph. When constructing a scatter graph you must ensure that the **independent variable** (the one under your control, e.g. distance downstream) is plotted on the x axis (horizontal axis) and the **dependent variable** (the one that you expect will be influenced by the independent variable, e.g. river velocity) is plotted on the y axis (vertical axis).

Pie charts and proportional divided circles

Pie charts show the proportion of the whole represented by each category. The angle for each category is calculated by dividing the value or number in each category by the total, then multiplying by 360 to give the angle in degrees. They give a simple visual representation and are relatively simple to construct. However, it can be difficult to see exact figures when there are a large number of categories and to make comparisons between large numbers of pie charts. When a number of pie charts are drawn on a map, e.g. to show the percentage in each roundness category for clasts in a river, it is possible to make the radius of each pie chart proportional to the value it represents in total, for example mean particle size at each point. The radius can be calculated using the formula:

Where V is the total value to be represented, r is the radius of the pie chart and $\pi = 3.142$.

$$r = \sqrt{\frac{V}{\pi}}$$

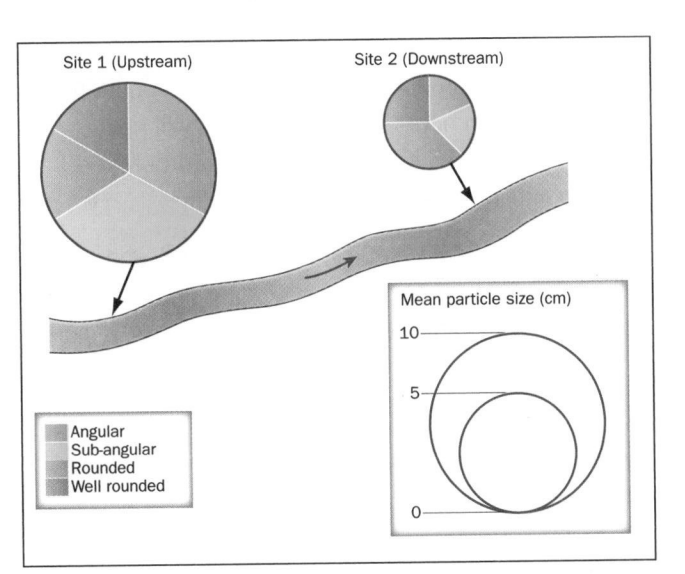

Located proportional circles to show clast size and roundness at two different sites

Triangular graph

Triangular graphs

Triangular graphs can be used when the data can be broken down into three components expressed as percentages, e.g. the percentages employed in primary, secondary and tertiary industries in a range of countries. This does mean that the technique is only suited to a limited range of data, but it can be helpful in identifying 'clusters' of points to help with further classification.

Kite diagrams

Kite diagrams show the change in percentage, e.g percentage cover of a particular plant species over distance. The width of the 'kite' shows the percentage cover of that species at a particular point and enables a simple visual comparison of the distribution of each species to be made at each point.

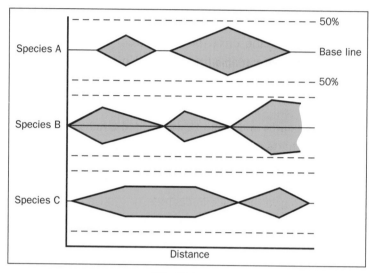

Kite diagram

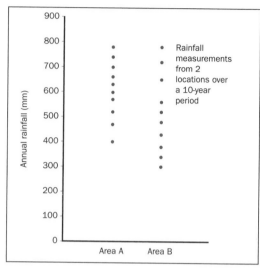

Dispersion graph

Dispersion graphs

Dispersion graphs clearly show the range of data and can be very helpful when calculating the **median** and **interquartile range**. Each value collected is simply plotted against a vertical scale (see pages 306–307).

ICT skills

The use of ICT in both data collection and processing is increasingly common in geographical investigation and the exam boards expect that you will have made use of ICT at some stage. This could include remotely sensed data, e.g. photographs and digital images, including those collected by satellite.

It is expected that you will be familiar with a common word processing and spreadsheet package to produce the write up of the investigation and some of the graphical/cartographic techniques mentioned above.

Geographical Information Systems

Geographical Information Systems (GIS) is a rapidly growing and increasingly essential area of geographical study.

If you have used a route planner site, or consulted an online weather forecast, then you have made use of a GIS. Essentially a GIS integrates hardware (computers and global positioning systems (GPS) enabled mobile devices for data collection), software and data for capturing, managing, analysing and displaying all forms of geographical data. A GIS allows you to view, understand, question, interpret and visualise data in many ways that reveal relationships, patterns, and trends in the form of maps, globes, reports, and charts. They can be extremely useful in an A-Level investigation.

In simple terms. a number of 'layers' are put into the system in the form of digitised base maps to which further data is added. When data has been

More detailed information can be found at www.esri.com/what-is-gis/index.html

collected this can be added to the base map as both 'locational' and 'attribute' data, e.g. in a rivers investigation you could add layers for the course of the river channel, relief, geology and land use. When channel characteristics have been measured the locational data would be the location of each site and the attribute data would be the velocity, clast size/roundness, etc. It would then be very simple to investigate relationships between land use, geology, distance downstream and channel characteristics and then display the results using maps, charts or tables.

Examples of GIS software that could be used for this are ArcView and AEGIS. However, simpler examples of GIS are useful in fieldwork such as Google Earth.

Don't worry if you have not used a sophisticated GIS in your investigation. The exam boards recognise that not all students have access to these and that some require training to use effectively. Don't forget that obtaining location maps from multi-map and images from Google Earth counts as making use of GIS. Be prepared to talk about how word processing and manipulation of the data in a spreadsheet was used as well.

KEY POINT

Once you have demonstrated the information and data collected in diagrams and other techniques you have to demonstrate your analysis of them in a written format.

Presenting a summary of the findings and evaluating the investigation

At this final stage of a geographical investigation, you should use the evidence presented in the earlier sections, and any conclusions derived from statistical analysis, to state a clear conclusion that relates to the hypotheses being tested or questions being answered. You should evaluate how far your results correspond with the underlying theory and refer back to any anomalous results with explanations, or areas, for further investigation if possible.

The examiner will be looking for you to show a critical awareness of the process of geographical enquiry and how the study has furthered your geographical understanding. In fact, an examiner would be rather suspicious of an investigation where every single result fits the hypotheses perfectly!

You should also show an awareness of the limitations of the study in terms of the data selected or data collection methods and be able to suggest improvements. Expect to be questioned about this area in the exam and remember that a successful geographical investigation does not have to prove every hypothesis or answer every question perfectly.

The benefit of fieldwork is that it helps students to appreciate why theoretical models do not necessarily apply perfectly in the 'real-world' and how human influence can often lead to unexpected results, e.g. building debris in a river leading to larger than expected and unusually shaped clasts. In answering these questions make sure that you can refer in detail to the results you obtained and are able to recall details of specific sites and unusual results.

15.3 Statistical requirements

LEARNING SUMMARY

After studying this section, you should be able to understand:
- why geographers use statistics
- statistical tests

Why do geographers use statistics?

Geographical investigations often involve the collection of large amounts of numerical data. Without further processing this can be difficult to interpret. Therefore, geographers often use statistics to simplify the data or to look for patterns or differences. There are two types of statistics that you should be aware of.

Descriptive statistics are used to give an idea of the 'central point' and 'spread' of a set of data and are generally very straightforward to calculate. **Statistical tests** (sometimes known as inferential statistics), whilst initially appearing more complex to calculate, can be very useful when looking for relationships between data, testing for differences or looking for patterns. It is important to remember that in the exam you will be assessed on your appreciation of the usefulness and application of statistics as a tool for geographers, rather than your understanding of the maths behind them (and examination boards would not expect you to remember the formulae!).

> A-level Geography students, especially those who do not also study Maths, can find the statistics section of the specifications daunting!

Descriptive statistics

Measures of central tendency

The **mean**, **median** and **mode** are all examples of measures of central tendency. These are used to indicate the average, or 'middle point', of a set of data, e.g. velocity measurements at a certain point on a river.

> You should be able to calculate each of these and understand their strengths and weaknesses.

Mean

To calculate the mean total, take the values in the set of data then divide by the number of values (n) – sometimes known as the arithmetic mean or $\bar{x}$.

$$\bar{x} = \frac{\Sigma x}{n}$$

The mean is easy to calculate and can be used in further statistical analysis. However, it is distorted when there are extreme values in the data set.

Median

The median is the middle value in the data set when the items of data are arranged in rank order. The median value will be at the $\frac{n+1}{2}$ position in the ranked data set, e.g. if there are 15 values (odd number) in the data set the median will be at the (15 + 1)/2 or 8th position. If there are 16 values (even number) in the data set the median will be at the (16 + 1)/2 or 8.5th position. In this case the mean of the 8th and 9th values gives the median value. It can be more complex to calculate than the mean, as the data has to be ranked first, the median is a useful measure as it is not distorted by extreme values.

Mode

> Many data sets have more than one modal value and this is the main weakness when using the mode as a measure of central tendency.

The mode is the value in the data set that occurs most frequently. If there are two modal values the data set is 'bi-modal' and if there are more than two modes then the data set is called 'multi-modal'.

Measures of dispersion or variability

Measures of dispersion or variability tell us how 'spread' the data is around the central point.

Range

The range is the difference between the highest and lowest values of a data set. It gives a simple indication of the spread of data, but as it is based on just two values, it is a very crude measurement and affected by extreme values.

Inter-quartile range

The **inter-quartile range (IQR)** is a more sophisticated measure of dispersion and is found by calculating the difference between the **upper quartile** and the **lower quartile** when the data has been placed in rank order.

- The upper quartile (UQ) is the $(n + 1)/4$th item in the data set (when ranked from highest to lowest).
- The lower quartile (LQ) is the $3(n + 1)/4$th item in the data set.
- The difference between the upper quartile and the lower quartile is the inter-quartile range.

IQR = UQ – LQ

The IQR measures the spread of the middle 50% of the data set about the median value and is therefore not influenced by extreme or 'outlying' values in the data set.

Standard deviation

Standard deviation measures how dispersed, or 'spread', around the mean value a data set is. The smaller the standard deviation the more closely grouped around the mean the data is. A high standard deviation indicates that the data is widely spread around the mean and that dispersion is large. The standard deviation is calculated by the following method:

- calculate the difference between each value in the data set and the mean value
- square each difference (eliminates negative values)
- total the squared differences
- divide this total by the number of values in the data set (n) – this figure is known as the **variance** of the data
- calculate the square root of the variance.

$$\text{Standard deviation} = \sqrt{\frac{\Sigma(\bar{x} - x)^2}{n}}$$

The standard deviation is a useful statistical measure as it links the data to the **normal distribution**. In a normal distribution approximately 68% of the population are within one standard deviation either side of the mean and 95% of the population are within two standard deviations either side of the mean. Any value outside of this range may require further investigation as an **outlier** or **extreme value**.

Statistical tests

The following statistical tests are used when testing for relationships (correlation), differences between data sets and patterns within a distribution. Whilst the detailed procedure and calculations in each are different, it is important to remember the basic process involved in carrying out any statistical test.

- State a '**null hypothesis**' – a simple statement that the test aims to reject, e.g. 'There is no relationship between velocity and distance downstream.' or 'There is no difference between particle size at upstream and downstream sites.'
- Carry out the test. This may involve ranking the data (as in the **Spearman's Rank Correlation Coefficient (SRCC)** and **Mann-Whitney U Tests**) or calculating 'expected' values as in the chi–squared test. The result of the test, however, is always a statistical value.
- Test this result for **significance**. This decides how confident you can be in rejecting the null hypothesis and accepting an alternative hypothesis, e.g. there is a relationship between particle size and distance downstream.

Spearman's Rank Correlation Coefficient

The SRCC is used to test for **correlation** between two variables (whether one increases or decreases as another increases). The calculation will give a value between −1 (perfect negative correlation) and +1 (perfect positive correlation). A SRCC calculation is very often used following the construction of a **scatter graph** which gives a simple visual indication of the strength and direction of any correlation. The calculation gives a numerical value to indicate the strength and direction of correlation and can be tested to see whether the result is **statistically significant**, e.g. in an investigation on how river velocity varies with distance downstream the following results were obtained:

Distance downstream (km)	0.5	2.5	3.7	4.7	5.2	6.7	7.5	10	11.2	13.2
Velocity (m/s)	0.1	0.22	0.17	0.22	0.38	0.18	0.19	0.45	0.61	0.81

Distance downstream	Rank	Velocity	Rank	d	d^2
0.5	1	0.1	1	0	0
2.5	2	0.22	5.5	−3.5	12.25
3.7	3	0.17	2	1	1
4.7	4	0.22	5.5	−1.5	2.25
5.2	5	0.38	7	-2	4
6.7	6	0.18	3	3	9
7.5	7	0.19	4	3	9
10	8	0.45	8	0	0
11.2	9	0.61	9	0	0
13.2	10	0.81	10	0	0

How to carry out the test.

- State the null hypothesis 'there is no relationship between velocity and distance downstream'.
- Draw up a table for the results as shown.
- Give ranks to each site for each variable, starting with the lowest.
- If there is a tie, give the average value of the positions occupied, e.g. in the results opposite two values (0.22) are sharing positions 5 and 6 therefore are awarded 5.5 and the next rank awarded is 7.
- Calculate the differences between the ranks (d) then d^2.
- Add up all the squared differences (Σd^2).
- Calculate the SRCC value (R_s) using the formula.

$$RS = 1 - \frac{6\Sigma d^2}{n^3 - n}$$

$$= 1 - \frac{6 \times 37.5}{n^3 - n}$$

$$= 1 - \frac{225}{990}$$

$$= 1 - 0.227$$

$$= 0.773$$

An SRCC value of 0.773 suggests that there is a reasonably strong positive correlation between distance downstream and mean river velocity. However, the result must now be tested for **significance**, as there is always a possibility that the relationship has occurred by chance. In order to do this the result is compared to a table of **critical values**. If the result equals, or exceeds, the critical value then it can be said to be statistically significant. Two levels of significance are commonly used – the 0.05 (5%) level and the 0.01 (1%) level. If the result is significant at the 0.05 level there is a 5% probability that the relationship has occurred by chance and if significant at the 0.01 level there is just a 1% probability that it is a chance occurrence.

In most geographical investigations the 0.05 level is used so that if the test is significant at that level we can be 95% confident in the relationship that has been established. The critical values depend on the number of pairs of data collected (n) and are found from either tables or graphs. An extract from a critical values table for SRCC is shown below.

You are not expected to learn the formula for SRCC, or the critical values, but should be able to complete the calculation when given the formula and interpret the result in the light of given critical values.

n	0.05 (5%) significance level	0.01 (1%) significance level
10	±0.564	±0.746
12	±0.506	±0.712
14	±0.456	±0.645
16	±0.425	±0.601

In the example above the results of 0.773 is greater than the critical values at both the 0.05 (0.564) and 0.01 (0.746) significance levels so there is only a 1% probability that it is chance occurrence – we can be 99% confident that it is a **statistically significant result and reject the null hypothesis**. We **accept the alternative hypothesis** that there is a relationship between distance downstream and velocity.

Mann–Whitney U Test

The Mann–Whitney U Test is an example of a **comparative test** and is used to determine whether there is a significant difference between two sets of data, e.g. in a rivers investigation the long axis (a-axis) of five randomly selected particles was measured at an upstream site (site A) and a downstream site (site B). To carry out the Mann-Whitney U Test.

- State the null hypothesis 'There is no difference between particle size at the upstream and downstream sites'.
- Arrange the data in two columns, then rank each item of data, starting with the lowest value. Remember to rank across both sets of data.

    Note that tied ranks are dealt with by awarding the mean of the rank positions being occupied (as in the Spearman's Rank Correlation Coefficient calculation).

- Total the ranks for each site (Σr_A and Σr_B), then calculate 'U' values for each site using the formulae;

$$U_A = n_A n_B + \frac{na(na + 1)}{2} - \Sigma R_a$$

Site A (cm)	Rank (r_A)	Site B (cm)	Rank (r_B)
8.9	10	3.5	7
3.4	6	2	3.5
2	3.5	3.2	5
6	8	1	2
7	9	0.2	1
	Σr_A 36.5		Σr_B 18.5

$$= 25 + \frac{30}{2} - 36.5$$

$$= 3.5$$

$$UB = n_A n_B + \frac{nb(nb + 1)}{2} - \Sigma R_b$$

$$= 25 + \frac{30}{2} - 18.5$$

$$= 21.5$$

Size of sample A	Size of sample B				
	5	6	7	8	9
3	0	1	1	2	2
4	1	2	3	4	4
5	2	3	5	6	7
6		5	6	8	10
7			8	10	12
8				13	15

n_A and n_B are the sample sizes (five in each case) and Σr_A and Σr_B are the totals of the ranks for each data set.

When the two U values have been calculated the smaller of the two values must be tested for significance. The critical value for U, for the respective sample sizes and desired significance level (usually 0.05), must be looked up using a table. An extract from a table is shown opposite.

If the smaller value of U is less than the critical value at the desired significance level there is a significant difference between the two samples. In the case of the example, as the smaller value of U (3.5) is greater than the critical value at the 0.05 significance level (2) there is a not a statistically significant difference in particle size between the upstream and downstream sites. We cannot therefore reject the null hypothesis in this case. This does not, of course, mean that the hypothesis that particle size is smaller at downstream sites is incorrect as this is entirely in accord with fluvial theory. The lack of a significant result could be due to bias in the sampling strategy or errors in measurement.

Remember that it is the **smaller** value of U that is tested for significance and it needs to be **less than** the critical value for the result to be significant.

Chi–squared test

The **chi–squared test** is another example of a comparative test and is used when frequency data (number of occurrences in different classes) has been collected. The test establishes whether the **observed pattern** differs from an **expected pattern**, e.g. the orientation of 48 corries was recorded from a map and grouped into four classes relating to their angle of orientation. The results are shown opposite.

Orientation from North	Frequency
0–89°	29
90–179°	6
180–269°	4
270–359°	9
Total	48

In this case the null hypothesis is 'the orientation of corries is random'. If the orientation of corries is random we would expect there to be 48/4 = 12 in each category. This is clearly not the case, but the chi-squared test can be used to establish whether the variation from the expected distribution is statistically significant.

Where O = observed frequencies and E = expected frequencies.

The chi–squared formula is $x^2 = \Sigma \frac{(O - E)^2}{E}$

To calculate the chi-squared value set the data out in a table as below:

	Observed (O)	Expected (E)	O–E	(O–E)²	$\dfrac{(O-E)^2}{E}$
0–89°	29	12	17	289	24.08
90–179°	6	12	–6	36	3
180–269°	4	12	–8	64	5.3
270–359°	9	12	–3	9	0.75

$$x^2 = \Sigma \, \frac{(O - E)^2}{E}$$

$$= 24.08 + 3 + 5.3 + 0.75$$

$$= 33.13$$

As with the other tests this result needs to be tested for significance. First of all calculate what is known as the 'degrees of freedom' for the sample using the formula $(n-1)$ where n is the number of observations, in this case 4, giving (4-1) = 3 degrees of freedom. Use a table to look up critical values at the significance level required.

0.05 (5%) significance level		0.01 (1%) significance level	
Degrees of freedom	Critical value	Degrees of freedom	Critical value
1	3.84	1	6.63
2	5.99	2	9.21
3	7.82	3	11.30
4	9.49	4	13.30
5	11.10	5	15.10

If the calculated value of chi–squared is greater than the critical value then the null hypothesis is rejected and we can state that the observed variance from the expected distribution is statistically significant. From the table we can see that the critical values for the example above are 7.82 at the 0.05 level and 11.30 at the 0.01 level. As the calculated value of 33.13 is greater than 11.30 there is only a 1 per cent probability that the distribution of corrie orientation is random. We can therefore be 99% certain that variation in corrie orientation is statistically significant.

Nearest neighbour analysis

The **nearest neighbour statistic** (R_n) is used to describe the distribution of individual points in a pattern within a given study area, e.g. settlements in a rural area or types of shops/services in an urban area.

The calculation gives a value between 0 and 2.15. A value of 0 represents 'perfect clustering' i.e. all the points are at exactly the same place with no distance between them – a situation that of course will not occur in reality.

Therefore, it is the closeness of the value to 0 that is considered to decide on the degree of clustering. A value of 2.15 represents a pattern of 'perfect regularity' – all distances between all points are exactly the same. Again, this is extremely unlikely to occur in reality, therefore we consider the closeness of the value to 2.15 to indicate the degree of regularity in the spacing of the points.

The statistic is based on measuring the distance between each point in the pattern and its nearest neighbour of the same type. When the measurements have been made the mean distance between each pair of points is then calculated ($\bar{d}$). The nearest neighbour statistic can then be calculated using the formula:

Where N = number of points in the study and A = area of study.

$$R_n = 2\bar{d}\sqrt{\frac{N}{A}}$$

Post office number	Number of nearest neighbour	Distance between points (km)
1	3	1.2
2	1	1.4
3	5	0.8
4	6	1.6
5	2	2.2
6	7	1.4
7	6	1.1
8	4	2.5
9	8	1
		Total distance 13.2

Example – data for the distribution of post offices in a rural area is given below;

Mean distance $= \dfrac{total\ distance}{number\ of\ points}$

$= \dfrac{13.2}{9}$

$= 1.47$

Area $= 45km^2$

$R_n = 2 \times 1.47\sqrt{\frac{9}{45}}$

$= 1.31$

Therefore the distribution is close to random. The statistic is most useful when making comparisons, e.g. when comparing the degree of clustering of different categories of shop within an urban area.

Exam practice answers

2 The challenges of the atmosphere

1. (a) • Named Hadley Cell rising and falling between 0 and 30° north or south.
   • Circular pattern of rising and descending air and indicated for the other cells too.
   • Cells could be named.

(b) Heat source:
   • Influence on the jet stream.
   • Thinning of tropopause towards the poles.
   • Differential heating and cooling of the air.

(c) Reasons for the air rising:
   • Heating of surface.
   • High insolation and uplift.
   • Air masses meeting, etc.
   Type of cloud that usually forms:
   • At 0° cumulus/cumulo-nimbus.
   • Further north stratus.
   • Different heights of clouds.
   Precipitation:
   • Short torrential rainfall.
   • Convectional rainfall/front rainfall.

(d) Any three from (or other suitable answers):
   • Increase in violent and unpredictable weather – flooding and hurricane intensification.
   • When warm water replaces cold there is decimation of fishing grounds.
   • Drought in normally lushly vegetated areas.
   • Agricultural production plummets.

(e) Effects on agriculture, fishing, construction, transport, power, business and retail, leisure and sport, health. Could be positive or negative effects.

2. You must describe and explain here. Importantly, you must ensure that there are locational comments in your answer.
   • Comment on differences in formation.
   • Show you know how, why and where they are formed.
   • Show you know something of the route that they take,
   • Show that there is a degree of seasonality in their formation.
   • Diagrams will aid your answer.
   • Describe the weather associated with each system.
   • Briefly comment on the effect on humans.

3 Ecosystems

1. This extended piece of writing asks simply for you to comment on how man interferes with the nutrient cycle. It is not concerned with how it adapts, neither does it ask for examples. The best answers will comment on how man can influence the sustainability of ecosystems. Clearly, some well-drawn and appropriately placed diagrams will really benefit your answer. You might choose from some of the following:
   • Removal of vegetation by man for crops.
   • Removal of ground cover increases run–off, known as leaching.
   • Changing farming practices, such as rotation, fertilizers and eutrophication.
   • Dumping sewage at sea.
   • Pollution.
   • Over–exploitation.
   The best answers may well range into sustainable development.

4 Hydrological and fluvial challenges

1. Because both play an important part in determining size and frequency of flooding.
   Factors include:
   • *Physical:* Rock type – determines permeability; relief – affects run-off rates; precipitation and climate – severity and frequency important; snow melt – releases large volumes of water at one time; confluence effects – surges of water coinciding in one place.
   • *Human:* Intensity of urbanisation on the floodplain – built surfaces don't absorb water; agricultural land use – affects infiltration; afforestation and deforestation – affect rates of interception; interference and manipulation of the river – affects speed of discharge, levées may prevent flooding.
   • Examples include: In the UK the Ouse in Northamptonshire and Yorkshire; Severn in Shropshire; in the USA the Mississippi.

6 Arid and semi-arid environments

1. (a) Barchan dune: Pebbles or an obstruction might retard saltated sand. Wings/horns are swept forward. Further sand arrives and is shepherded onto the flanks and runs off in streams. Turbulence on the leeward side sweeps the barchan court clear. Sand replaces that which is lost. Winds must blow almost constantly in one direction for barchans to form.
   Seif dune: Has a regular cross-sectional form and orientation. Parallelism is common. There is a correlation between corridor width and dune width, and between dune height and spacing. Shape is the result of complex annual wind regimes. 'Speculation has it that large-scale rotary movement in the air stream' produces the regularly spaced thin sand strips; they are maintained by 'transverse instability', i.e. strong wind is decelerated at ground level and sand is transferred to the 'strips' – Bagnold. Others suggest that seifs may be the result of erosional activity.

(b) Labels to include: unanchored, asymmetrical, windward, leeward, horn, crest, steep/shallow slope, direction of wind, direction of movement, apron and streamer. Explanation as above in (1)(a).

(c) Increasing use of marginal lands means that man has had to inhabit the desert areas of the world. Barchan dunes (transverse dunes) are the most mobile of dunes; they are said to have a 'short memory'. Sand drift in barchan dune areas is rapid, for example Lüderitz, Namibia). Barchans form, grow and migrate, in some cases in hours. Peruvian barchans move on average 1.7 km in their 64–year lifecycle. Barchans don't travel alone!

(d) Dune stabilisation with vegetation. Planting 'sand breaks' – of eucalyptus for instance.

2. Typically this synoptic question links man into a geomorphological process. It asks you to show how irrigation in the desert realm has allowed populations to grow, and settlement and economic development to take place. You might reference the process of population growth to development. In linking the above parts together, you will make the connections to other parts of your specification. You have, of course, to demonstrate an understanding of modern (field channels, canals, pipelines, drip irrigation, boom irrigation, pump irrigation) and traditional (annual flood, the shaduf/Archimedes' screw) irrigation processes. Some reference to actual examples will, of course, be important, for example Alice Springs in Australia's Northern Territory.

7 Cold environments

1. (a) It is likely that freeze thaw is taking place at A (a weathering process). Abrasion (an erosional process) is taking place at B. The rotational movement of material at B forms and maintains the characteristic armchair shape of the cirque.

(b) The feature at X is commonly called the backwall or headwall. It has been steepened by freeze thaw both during and after glaciation. The landform at Y is the rock lip. It forms because of overdeepening of the corrie, by rotational slip. As ice leaves the corrie, by extending flow, there is a lack of glacial energy, the rock lip is left in the landscape.

(c) (i) Seven correctly plotted lines.
   (ii) Most cirques occur between north-east and

Exam practice answers

south–east – there are small numbers of cirques at other orientations.

   (iii) It is colder and therefore less melting (ablation) occurs on slopes facing from north–east through to south–east. These slopes are in shade for the best part of the day. Those in all other areas always face the sun and therefore little snow collects. Wind directions during glacial periods may also have aided snow accumulation.

(d) The weight of ice causes deformation. This deformation causes crystals to slip on planes parallel to the basal plane. This initiates downslope glacial flow. When accumulations of snow fail to keep pace with ablation, the glacial mass responds by retreating up its valley (glacial retreat).

(e) Features include ice wedges, common in the tills of North Norfolk's cliffs. The pingo ramparts (now rounded lakes) of South West Norfolk were former active pingos. The hummocky landscape of North Central Norfolk is the result of thermokarst activity. The patterned ground, stone circles, polygons and stripes are the result of periglacial activity on slopes in Breckland in Norfolk. In higher areas in Britain, screes, blockfields and tor-like features also bear testimony to the effects of periglacial activity. Dry valleys in the southern part of the UK are the final obvious legacy of periglacial activity. The Coldstream area of Scotland also has many of these features. Also asymmetric slopes caused by variations in aspect for example River Exe, Devon.

2. Lowland areas: Soils in glacial and periglacial areas both limit (when soils are thin) and aid agriculture (where thick fertile clay deposits exist). Transport can be aided (following esker and moraine ridges) and be made more difficult (in lake-covered landscapes, for example Finland).

Other uses include: forestry on glacial sands; sands and gravel deposits are used in the building industry; military trainers utilise the Breckland area of Norfolk because of its low agricultural potential.

Highland areas: Soils are thin, slopes are steep – this, combined with a harsh climate, limits agriculture. Such areas are useful for afforestation, hydro-electric power generation, water storage and leisure and tourism.

8 Settlement issues

1. (a) The key here is to learn a simplified land–use map of your settlement morphology case studies with names and some statistical data. If you use a well–known example, such as Mexico City, you will need to be much more precise about the detail, than if you use an uncommon case study such as Johannesburg.

(b) Remember that the easiest models to criticise are the most generalised, for example Hoyt and Burgess, and if you use an area specific model, for example Griffith and Ford for Latin American cities, you might be able to present a more balanced and in–depth argument.

(c) (i) The structure has been given to you, as has the instruction 'with reference to', so you should try to use as much place–specific detail as you can, the more in–depth the better. Environmental problems might be due to the building of poorer housing areas on marginal land such as steep hillsides or marshy areas, with the inevitable landslides and diseases. Social problems are usually to do with a lack of amenities such as electricity and sewerage, but could also be due to a lack of facilities, leading to poor health and sanitation. Economic problems are usually connected with jobs and the lack of formal employment opportunities.

Try to read the question in full first, then you will be able to see the links between questions and use your answer in one part to help you with the next.

   (ii) Problems and solutions are usually linked. Tie in your case studies from the previous section and consider the structure you have been given. The local population can help themselves through a range of self–help schemes, small–scale industrial developments and, most powerful of all, collectivisation. Groups get together to provide amenities, build roads or affordable houses or even to acquire land. From a decision maker's point of view, they have to achieve the same aims of improving the environmental, social and economic situation, but they have greater powers. Usually cooperation between groups is essential, but decision makers can do many things to make housing better by giving over ownership of the land, improving facilities and amenities in the planning stages of development and providing transport links to get to work.

2. (a) Use the picture as a starting point here, but you should certainly brainstorm your ideas as rural and urban are very difficult to define precisely without using a range of different criteria. You should consider ideas to do with density and height of buildings, functions of buildings, land prices, infrastructure and public transport, employment, age structure. Some more alternative ideas such as street lighting are clear on the picture so should be used in contrast with the other ideas. You then have to explain them, and it is important to touch on the ideas behind bid rent, industrial location and migration, and perhaps even make some reference to historical reasons.

(b) Again, an understanding of ideas such as the model of a suburbanised village, or a case study of a village you know that has experienced such change, might give you a framework to answer this question. Most villages are changing due to an influx of new residents and this leads to a change in the social make-up, increased demand for housing and different demands for services and functions.

Remember to answer the question set rather than the one you would like to have been set. It would be easy to answer this question by talking about the reasons for counterurbanisation, but you need to focus on the changing nature of villages.

9 Population and migration issues

1. (a) This is a wide ranging question and requires you to structure your answer:

Housing problems – having to deal with temporary accommodation such as that found in shanty towns, for example bustees of Calcutta, India (legal problems over land, confrontation with authorities and police) or rented accommodation in the city centre that uses up money resources, for example apartments in the centre of Mexico City, Mexico;

Employment – usually informal sector employment, for example prostitutes in Manila, the Philippines (harsh conditions, irregular wage, potential exploitation);

Friends and family – very vulnerable when first arrived, dependant upon welfare net, familial and governmental; crime – the social disarray of the reception areas can often lead to crime,

Alienation and social stress – high crime rates in the Morros of Rio de Janeiro, Brazil;

Disease – vulnerable also to disease and poor sanitation.

(b) Consider a range of case studies that balances reducing the migration stream by improving rural conditions and reducing the migration itself. Include ideas from: development, for example growth poles in Brazil to stimulate development in more rural areas;

- transport, for example develop rural transport systems to reduce isolation and enable goods to be taken to market further away;
- rural/agricultural development, for example irrigation schemes to increase yields;
- alternative/intermediate technology, for example develop better accommodation to improve rural quality of life;
- exploiting local resources, for example local management and exploitation of a resource;
- government policy, for example Indonesian transmigration where people are relocated from densely populated areas to new agricultural areas opened up by the government.

This has to be balanced with the fact that some developments in agriculture and education might actually encourage rural depopulation and not slow down the rate of urbanisation.

10 Development issues

1. (a) Only two marks available, so use a clear and concise definition such as: a quality of life index that uses factors such as life expectancy, adult literacy rate and per capita income as a marker of living standards to determine a comparative index score based upon rankings.

    (b) (i) Degrees of freedom = the number of paired variables – 1 = 9, therefore rs must exceed 0.600 to be 95 per cent sure of the significance of the relationship. The basic relationship is that the higher the GNP pc, the higher the HDI; the HDI is dependent upon the GNP. Stress this relationship using some data from the table.

      (ii) Countries with a higher GNP are able to spend more on health and education and this has a knock-on effect on improving life expectancy, literacy levels and living standards – hence the improved HDI.

    (c) Realistically you could suggest any one of the following options:
    - Dealing with the debt – collective default, debt for equity/nature swaps, repayment, rescheduling.
    - Development – sensible investment of money in healthcare, reduction in military expenditure, accepting technological and monetary aid specifically designed to improve the health of the community.
    - Cut other projects, such as road building.
    - Use schools as platform for teaching about health issues.

    It is likely to be a combination of these two aspects, as once the money becomes available, it may not necessarily be used productively.

11 Worldwide industrial change and issues

1. (a) (i) De-industrialisation takes place when a country's industrial base shrinks.

      (ii) Usually found in peripheral areas where a large proportion of employment in manufacturing is in extremely controlled branch plants.

    (b) E.g. iron and steel industry in Corby.
    Effects:
    - decline in manufacturing employment
    - contribution to GDP drop
    - unemployment
    - dereliction
    - service decline
    - housing stock deteriorates.

    (c) Because, during recession or severe competition large firms implement new policies. Restructuring occurs and branch plants that are labour intensive rather than capital intensive are closed, as decisions are made elsewhere. Branch plants are common in 'peripheral'

Europe. The proper name for this closure is organisational disintegration. All subsections are point marked, but look for sensible development.

12 Agriculture and food supply

1. (a) It is a simple proportional relationship between the two measures. Your answers should make reference to different rates of growth. Use data and actual figures from the graphs.

    (b) Begin by discussing what consumer pressure and environmental concerns may have caused a change:
    - Concerns over beef in the BSE scare.
    - Increasing awareness of the impacts of pesticides and fertilizers on the environment and upon humans.
    - more green awareness after high-profile media events like the Rio summit and GM crop trials.

    Make sure that you stress that such pressures and demands led to the growth in organic production to meet the needs of the new 'green' consumer and the response of supermarkets.

    (c) You do not need to reach a conclusion in this question, but you should consider some of the following points and present a balanced argument:
    Consumer Farmer
    + fewer chemical residues – long time to convert with lower yields afterwards
    + perceived healthier food – greater losses from pests and diseases
    – higher bacterial content – more labour intensive
    – more expensive produce + less damaging to the environment and uses less energy
    + higher returns

2. Clearly you need to take a synoptic view, and include not just agricultural ideas, but also concepts from population. You could begin by outlining the current state of the world's population and where the growth is concentrated. This will enable you to present the split between the less and more economically developed world, both in terms of population growth and ability to deal with such change.
    Then you can address the issues as you perceive them. You need to cover the different aspects of agricultural advancements, such as GM foods and mechanisation, as well as the things that are hindering development, like poor levels of literacy, debt and diseases. You have the ideal opportunity to include some theory here, and could use ideas from Malthus and Boserup as well as under- and overpopulation. Population control is also significant and ultimately could determine a country's ability to feed itself, for example Mauritius and China. Just make sure that your essay is grounded in case studies, and even better in issues, and that you reach a conclusion as the question asks you to.

Index